Use R!

Series Editors:
Robert Gentleman   Kurt Hornik   Giovanni G. Parmigiani

For further volumes:
http://www.springer.com/series/6991

Kenneth Knoblauch • Laurence T. Maloney

Modeling Psychophysical Data in R

Kenneth Knoblauch
Department of Integrative Neurosciences
Stem-cell and Brain Research Institute
INSERM U846
18 avenue du Doyen Lépine
Bron, France

Laurence T. Maloney
Department of Psychology
Center for Neural Science
New York University
6 Washington Place, 2nd Floor
New York, USA

Series Editors:
Robert Gentleman
Program in Computational Biology
Division of Public Health Sciences
Fred Hutchinson Cancer Research Center
1100 Fairview Ave. N, M2-B876
Seattle, Washington 98109-1024
USA

Kurt Hornik
Department für Statistik und Mathematik
Wirtschaftsuniversität Wien Augasse 2-6
A-1090 Wien, Austria

Giovanni G. Parmigiani
The Sidney Kimmel Comprehensive Cancer
Center at Johns Hopkins University
550 North Broadway
Baltimore, MD 21205-2011
USA

ISBN 978-1-4614-4474-9   ISBN 978-1-4614-4475-6 (eBook)
DOI 10.1007/978-1-4614-4475-6
Springer New York Heidelberg Dordrecht London

Library of Congress Control Number: 2012943991

Printed on acid-free paper

Springer is part of Springer Science+Business Media (www.springer.com)

Preface

This book is about modeling psychophysical data with modern statistical methods using the software language R [146]. R is an interactive environment, and the novice user can just enter commands one by one at the command line prompt. It is also a powerful programming language in its own right. R is easily interfaced with other languages, such as C, Fortran, C++, Java, Perl, and Python. Another reason we recommend using R is that it permits access to a growing repository of new and powerful statistical methods, many of which are potentially useful to the psychophysical researcher. In the last chapter, for example, we describe methods that allow the researcher to model individual differences in observers, using mixed-effects models.

Many packages exist to supplement the functions of R. Over 3,000 are currently available on the Comprehensive R Archive Network (or CRAN) at http://www.r-project.org/. Further reasons for using R are that it is free and is supported by a large and enthusiastic online community.

Psychophysics begins with a question, posed to a *psychophysical observer*, human, animal, or neural.[1] The question may be as simple as judging which of two lights is brighter. The experimenter selects the question, constrains the observer's possible responses, and manipulates stimuli. The observer's response is effectively stochastic, and the experimenter's proximate goal is to model the observer's response and the mechanisms underlying that response. The final goal, of course, is to learn how neural systems process sensory information and provide access to the world around us.

The term "psychophysics" is due to Fechner [59], and there are several excellent introductions to psychophysical methods [91, 157–159, 164], the history of psychophysics [17, 115], and psychophysical theory [57, 151]. We note that psychophysics has no connection with psychometric theory. For an excellent introduction to psychometric theory using R we recommend Boeck and Wilson [16].

---

[1] We can treat any device that categorizes stimuli as a psychophysical observer. See [22, 138, 174] for examples of psychophysical models of neural response.

This book is *not* an introduction to psychophysical methods, history, or theory. It is written for those who have a good idea what psychophysics entails and who regularly model psychophysical data or would like to do so—typically established or aspiring psychologists and neuroscientists studying perception. Some of the material is pertinent to researchers in memory using signal detection theory.

The focus of the book is on modern statistical methods readily available in R and how they can be used to analyze psychophysical data. Throughout the book, we illustrate how traditional and recently developed models of psychophysical performance can be viewed as special cases of statistical models of great power and generality, notably the generalized linear model (GLM). In Chap. 1, we use R to explore and model a celebrated set of psychophysical data due to Hecht, Shlaer, and Pirenne [79]. When possible, as in this chapter, we use a running example within the chapter so that readers can appreciate how different methods might be applied to the same data set. Thus, packages and data sets loaded earlier in the chapter as well as variables will be assumed to be defined from the point in the chapter in which they are invoked, but not across the chapters.

The statistical methods employed in most psychophysical laboratories were developed for the most part over the past 50 years and until recently, commonly available statistical packages were of limited use in fitting psychophysical data. That has now changed and much of what the psychophysicist wishes to learn from data and experiment can be learned using off-the-shelf statistical packages developed in the last few decades. Moreover, the statistical methods and packages now available can serve to inspire the psychophysicist to try more sophisticated analyses and consider better experimental designs. If, for example, the experimenter has run an experiment with a factorial design, then he might like to do the nonlinear equivalent of an analysis of variance (ANOVA) and determine whether variation of any of the factors led to changes in observers' performance and whether variations in the factors interact. We illustrate how to carry out ANOVA-like analyses where the data are appropriate for fitting by signal detection models or psychometric functions.

The book should be accessible to upperlevel undergraduate and graduate students and could be used as a textbook for a one-semester course. It presupposes only some knowledge of elementary statistics, some facility with vector and matrix notation, and some familiarity with programming. We do not assume that the reader is familiar with R or is particularly fluent in any other programming language. Almost all of the examples are drawn from studies of human vision only because that is what we know best. The methods presented are readily adaptable to model data in other disciplines ranging from memory to experimental economics, and in testing this book with students we have found that they could readily adapt the code we include as examples to their own data without any extensive programming.

Chapter 2 describes the principal statistical methods we will use: the linear model (LM), the generalized linear model (GLM)[2] the (generalized) additive model

---

[2]The generalized linear model is *not* the same as the general linear model commonly used in cognitive neuroscience. The near-coincidence in terminology is unfortunate. We will not make use of the general linear model at any point.

(GAM), and a brief discussion of methods for fitting nonlinear models in general. The reader familiar with LM in the form of multiple regression or analysis of variance can become familiar with how models are specified and analyzed in R. The remainder of this chapter introduces GLM and GAM by analogy with LM. The treatment of GLM and GAM in this book is self-contained. We provide references to texts on LM, GLM, and GAM using R for readers who want to pursue these topics further. If already familiar with R and its approach to modeling, the reader may choose to skip this chapter on an initial reading and to start instead with one of the later chapters describing particular psychophysical methods. The reader can refer back to Chap. 2 as needed.

We also include two appendices that summarize a few useful terms and results from mathematical statistics (Appendix B) and a brief description of data structures in R as well as basic information about input and output of data in R (Appendix A). We reference particular sections of these appendices in the main text, and they need not be read through if one is already familiar with their content.

The remaining chapters each focus on traditional and not so traditional methods of fitting data to psychophysical models using modern statistical methods. With the exception of Chap. 5 which builds on Chap. 4, the chapters are independent of one another and may be read in any order. We recommend that, wherever you start, you type in and execute the examples in the text as you read them. All of the code is available in scripts in a companion R package **MPDiR** available on CRAN that also contains the data sets not available elsewhere. There are exercises at the end of each chapter that invite the reader to modify the example code or to explore further the R commands used. If you are motivated to go further, you can apply the example code to your own data. Even without extensive experience in programming or programming in R, you should be able to adapt your own data to our examples. All of the data analyses and graphs in this book were generated using R, and we encourage the reader to execute the examples in the book as they occur. In our experience, using R is the best way to learn R.

In Chap. 3 we describe how to use GLM to fit data from signal detection experiments. Signal detection theory (SDT) was introduced in the 1950s to address the problem of observer bias. In a simple Yes–No task, a psychometric function summarizes the probability of a "Yes" response. However, this probability confounds the observer's ability to detect the presence or absence of a signal with any preference the observer may have to respond "Yes." Even when the signal intensity is zero (no signal is presented), the observer may still respond "Yes." In a signal detection experiment, in simplest form, there is only one intensity of signal that is presented or not presented, at random, on each trial. On each trial, the observer is obliged to judge whether a signal was present. The two measures of interest are the rate of responding "Present" when the signal is present (referred to as the Hit rate) and the rate of responding "Present" when the signal is absent, the False Alarm rate. These two separate response measures are combined to allow separate estimates of a measure of the observer's sensitivity to the signal and the observer's bias (tendency to say "Present" independent of the presence or absence of the signal).

In Chap. 4 we illustrate how to fit psychometric functions to data using GLM. The psychometric function describes the relation between the probability of making a specific classification such as "Present" and stimulus intensity. Its earliest formulation dates from at least the beginning of the twentieth century [176]. One long-established use of psychophysical methods is to measure the precision of the human sensory apparatus in detection or discrimination. The researcher's goal nowadays is often to use such data to test models of sensory, perceptual, or cognitive function, and one of the most venerable is the *psychometric function* relating the observer's ability to judge a stimulus to stimulus intensity. There are many variant forms of psychometric function each associated with a particular experimental method. For now we assume that the observer responds "Yes" or "No," judging whether the stimulus was present.

In Chap. 5 we extend the analysis of the psychometric function to more complex situations, for example, when there are multiple choices. We, also, consider here methods for obtaining standard errors and confidence intervals on the parameters estimated for the fitted functions. Finally, we demonstrate some nonparametric approaches to estimating psychometric functions.

In Chap. 6 we describe classification image methods. These methods are a natural extension of signal detection theory and have proven to be of great value in modeling psychophysical data. They are also easy to work with using GLM and GAM.

In the next two chapters, we describe two relatively new methods for representing perception of stimulus intensities well above threshold. That is, these methods are primarily useful in measuring *appearance*. The first method, maximum likelihood difference scaling, is described in Chap. 7, and the second, maximum likelihood conjoint measurement, in Chap. 8. These methods and corresponding psychophysical tasks complement traditional methods. We will illustrate their use.

In many statistical analyses the experimenter wishes to estimate how variable the outcomes of psychophysical measurements are across observers, generalized from results measured for a specific set of observers in an experiment to the population from which they were drawn. In the last chapter, Chap. 9, we focus on how to use newly available R packages to fit mixed-effects models of psychophysical performance. These models take us to the edge of research in statistics and hint at what is to come.

The more advanced programmer will want to know that R is an object-oriented language. Most of the functions that carry out statistical analyses return not just vectors or matrices, but objects summarizing the analyses carried out. Generic functions that print, plot, summarize, or do model comparisons accept these objects and choose the appropriate methods to do what is appropriate. In general, we simply use these functions as needed to illustrate how intuitive they are.

R is a language that is in constant development, and although the core functionality evolves only slowly over time, usually with an eye toward backward compatibility, this is less so for the code in packages. In time, some of the code in this book will be affected by such changes and may not run. We have opted to choose package versions that seem stable but in the event that future changes break some of the code, we will make available supplemental materials on a web page associated with the book that can be accessed at http://extras.springer.com/. The current project home page for the book can be found at http://mpdir.r-forge.r-project.org/.

We thank Angela Brown, Frédéric Devinck, Roland Fleming, Jonathan Pierce, Hao Sun, François Vital-Durand, and Li Zhaoping for generously providing data. We thank Rune H. B. Christensen for extended discussions of ordinal models and Aries Arditi, Patrick Bennett, Reinhold Kleigl, Richard Murray, and, above all, Michael Kubovy for reading and critiquing several chapters. We thank our contacts at Springer, the editors John Kimmel and Marc Strauss, and Hannah Bracken. We also thank all of the R Core and R Help List contributors who responded to many questions along the way. Any remaining errors are, of course, our own.

We especially thank Geneviève Inchauspé and Maria Felicita Dal Martello for infinite patience with our preoccupation with R and this book.

Lyon, France · Kenneth Knoblauch
New York · Laurence T. Maloney
Padova, Italy

Contents

Chapter 1
A First Tour Through R by Example

1.1 Getting Started

In this chapter we illustrate how R can be used to explore and analyze
psychophysical data. We examine a data set from the classic article by Hecht et
al. [79] in order to introduce basic data structures and functions that permit us to
examine and model data in R.

The best way to read this book is sitting at a computer with R running, typing in
the examples as they occur. If you do not already have R installed on your computer,
see Appendix A for instructions on how to install it. If you have properly installed
R and launched it by double clicking on an icon or by running the command "R"
from a terminal command line, you should see some initial information printed
out about the version followed by some initial suggestions of things to explore,
demo(), help(), etc. This is followed by ">" on a line by itself. This is the
command line prompt and indicates where you type in commands. If you enter a
new line (carriage return) before the end of a command, the prompt will become
a "+," indicating a continuation of the previous line. You will continue to receive
continuation lines until the command is syntactically complete, at which point R
will attempt to interpret it.[1] Comments in R are preceded by a "#." Everything on
the line to the right of a comment character is ignored by R. Let's proceed with the
example.

---

[1] Syntactically complete commands may still be incorrect, in which case an error message will be
returned.

K. Knoblauch and L.T. Maloney, *Modeling Psychophysical Data in R*, Use R! 32,
DOI 10.1007/978-1-4614-4475-6_1,
© Springer Science+Business Media New York 2012

1.2   The Experiment of Hecht, Shlaer and Pirenne

In a classic experiment, Hecht et al. [79] estimated the minimum quantity of light that can be reliably detected by a human observer. In their experiment, the observer, placed in a dark room, was exposed at regular intervals to a dim flash of light of variable intensity. After each flash presentation, the observer reported "seen" or "unseen." The data were summarized as the percentage of reports "seen" for each intensity level used in the experiment, for each observer.

Before collecting these data, Hecht et al. performed extensive preliminary analyses in order to select viewing conditions that are described in their article, for which human observers are most sensitive (requiring the fewest number of quanta to detect) and a detailed analysis of their choices is presented in the opening chapter of Cornsweet's text [41].

1.2.1   Accessing the Data Set

The data from their main experiment (Table V in [79]) can be found in the package **MPDiR** that contains data sets used in this book and that are not available in other packages [98]. The package is available from the CRAN web site (http://cran.r-project.org/). If you are connected to the internet, then it can be installed from within R by executing the function

```
> install.packages("MPDiR")
```

Thereafter, to have access to the data sets, the package has to be loaded into memory and attached to the *search path* using the function:

```
> library(MPDiR)
```

The search path is a series of environments (see Sect. A.2.5) in memory in which sets of functions and data objects are stored, for example, those from a package. When a function is called from the command line, R will look for it in the succeeding environments along the search path. The library function attaches the package to the second position on the search path, just behind the work space, called .GlobalEnv. We verify this using the search function

```
> search()
 [1] ".GlobalEnv"        "package:MPDiR"
 [3] "tools:RGUI"        "package:stats"
 [5] "package:graphics"  "package:grDevices"
 [7] "package:utils"     "package:datasets"
 [9] "package:methods"   "Autoloads"
[11] "package:base"
```

which, in this case, returns an 11-element vector of character strings. The specific packages you see will depend on your version of the language R and may not be the same as those shown above.

The data set is called HSP and can now be loaded into the work space by

```
> data(HSP)
```

and we can check that it is there by typing

```
> ls()
[1] "HSP"
```

which returns a vector of the names of the objects in your work space. The data set corresponds to a single object, HSP. The function data is used to load data sets that are part of the default R distribution or that are part of an installed and, as here, loaded package. A second argument indicating a package name can also be used to load data from unloaded but installed packages (see ?data). In Appendix A we discuss how to load data from different sources into the workspace.

R has a rich set of tools for exploring objects. Simply typing the name of an object will print out its contents. This is not practical, however, when the data set corresponding to the object is large. We can view the internal structure of the object HSP using the function str. This extremely useful function allows us to examine any object in R (see ?str for more details).

```
> str(HSP)
'data.frame':    30 obs. of  5 variables:
 $ Q  : num  46.9 73.1 113.8 177.4 276.1 ...
 $ p  : num  0 9.4 33.3 73.5 100 100 0 7.5 40 80 ...
 $ N  : int  35 35 35 35 35 35 40 40 40 40 ...
 $ Obs: Factor w/ 3 levels "SH","SS","MHP": 1 1 1 1 1
          1 1 ..
 $ Run: Factor w/ 2 levels "R1","R2": 1 1 1 1 1 1 1 2 2
          2 2 ..
```

This shows that HSP is an object of class "data.frame." A data frame is a type of list (Appendix A), a data structure that can contain nearly arbitrary components, in this case five of them. Their names are indicated after each of the "$" signs. What distinguishes a data frame from a list is that each component is a vector, and all the vectors must have the same length. The entries in each vector have a class ("numeric," "integer," "logical," "Factor," etc.) indicated next to each component.

In the data set HSP, the first few elements of each component are indicated to the right of the class designation. This lets us quickly verify that data are stored in the correct format. The data set includes two components of class "numeric." Q is flash intensity level measured in average number of quanta. p is the percentage of responses "seen" for that level. N is of class "integer" and indicates the number of flash presentations. Obs and Run are both factor variables encoding categorical information as integer levels. For human consumption, the levels are given labels, which are indicated in the output. The levels of Obs are the initials of the three observers, who were the three authors of the article. For two observers, data are available for a replication of the experiment, indicated by the factor Run.

Even with str, R objects with many components can be impractical to examine. In that case, it may be more convenient simply to list the names of the components, as in:

```
> names(HSP)
[1] "Q"    "p"    "N"    "Obs"  "Run"
```

We can use this function, also, on the left side of an assignment operator, <-, to modify the names of the data frame, which we will take the opportunity to do here to render the meaning of some of the components more obvious.

```
> names(HSP)<- c("Quanta", "PerCent", "N", "Obs", "Run")
```

Here, we have used the function c() to create a five element vector of class 'character' and stored it as the "names" attribute of the columns of the data frame. The change in names occurs only on the copy of HSP in our work space, *not* on the original version stored in the package. Note that the assignment operator, "<-," is entered with 2 keystrokes.[2]

The first ten lines of the data frame can be displayed by

```
> head(HSP, n = 10)
   Quanta PerCent  N Obs Run
1    46.9     0.0 35  SH  R1
2    73.1     9.4 35  SH  R1
3   113.8    33.3 35  SH  R1
4   177.4    73.5 35  SH  R1
5   276.1   100.0 35  SH  R1
6   421.7   100.0 35  SH  R1
7    37.1     0.0 40  SH  R2
8    58.5     7.5 40  SH  R2
9    92.9    40.0 40  SH  R2
10  148.6    80.0 40  SH  R2
```

Notice the modified column names. Note, also, that inside a function argument, we use "=" to bind the value to the formal argument n, which specifies the number of lines of output to print. Use of "<-" here is not advised as it would produce the side effect of creating a variable n in the work space (see Problem 1.4). There is also a function tail that does what you might expect.

Another useful view of the data is obtained with the summary function.

---

[2]The more intuitive "=" symbol is also valid for assignment at the command line, but the two symbols are not interchangeable in all situations, so we recommend using "<-" on the command line.

```
> summary(HSP)
     Quanta         PerCent              N           Obs
 Min.    : 24   Min.    :  0.0   Min.    :35   SH :12
 1st Qu.: 58    1st Qu.:  6.4    1st Qu.:40    SS :12
 Median : 93    Median : 42.0    Median :50    MHP: 6
 Mean    :138   Mean    : 48.4   Mean    :45
 3rd Qu.:210    3rd Qu.: 94.0    3rd Qu.:50
 Max.    :422   Max.    :100.0   Max.    :50
 Run
 R1:18
 R2:12
```

Summary statistics are displayed for numerical and integer components. The levels of factor components are tabulated. Thus, the imbalance in the data set is clearly indicated by the unequal numbers of observations per Obs and per Run. Such tabulations can also be generated with the table function.

```
> with(HSP, table(Run, Obs))
    Obs
Run  SH SS MHP
  R1  6  6   6
  R2  6  6   0
```

Here, we have also introduced the handy function with that creates a temporary environment within which the component names of its first argument are visible.

1.2.2   Viewing the Data Set

A graphical representation is often effective for gaining a sense of the data. When a data frame is used as the argument of the function plot, R creates a square array of scatterplots, one for each pair of columns in the data frame.

```
> plot(HSP)
```

The resulting plot for the HSP data set, shown in Fig. 1.1, is called a *pairs plot*. The column names are conveniently displayed along the diagonal. The scatterplots above the diagonal are based on the same pair of variables as those below but with the axes exchanged. The scatterplot at position $(2,1)$ in the square array displays the variables of greatest experimental interest, the proportion of seen flashes as a function of average number of quanta in the flash, the frequency of seeing data. Although the values from all observers and runs are confounded in this plot, the basic trend in the data is evident: the frequency of seeing is at 0% for the lowest levels of the flash and increases to asymptote at 100% as the flash intensity increases.

Three other scatterplots display useful features of the data. The graph at position $(3,4)$ shows that the number of presentations of each flash, N, is the same for two

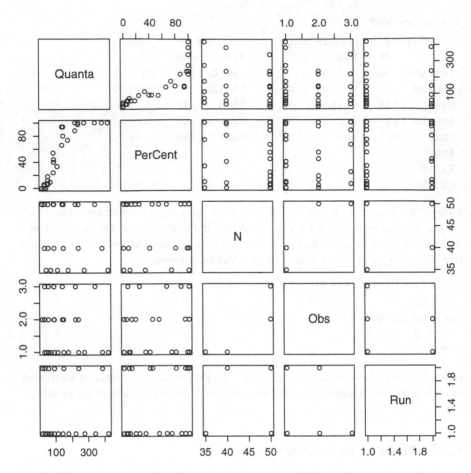

Fig. 1.1 Example of a pairs plot obtained by using a data frame as the argument to the `plot` function

observers, but differs for the third one. Examining the scatterplot just to its right, at position $(3,5)$, shows that this observer completed different numbers of trials in two runs. Finally, the graph just below, position $(4,5)$, reveals that two of the observers completed two runs and a third only one. Pairs plots are also useful for detecting errors in recording data or coding factors.

This example demonstrates the object-oriented nature of R. The `plot` function is an example of a *generic function*. Associated with a generic function is a set of method functions that are tailored to operate on specific classes of objects. In the case of a data frame, the generic `plot` function dispatches the execution to a plot method for data frames, called `plot.data.frame`, that produces the pairs plot. Other classes of objects will result in dispatch to plot methods specific to their class and result in plots that are meaningful for that class of object. We will see numerous examples of using the `plot` function on different kinds of R objects in subsequent chapters, including the case of a simple scatterplot. To see the methods that are available for `plot`, try

```
> methods(plot)
```

The `str` and `summary` functions used above are other examples of generics, and we will exploit many more throughout the book.

1.2.3 One More Plot Before Modeling

As indicated above, the frequency of seeing data in the pairs plot confounds the variables Obs and Run. It could be useful to examine the individual plots for irregularities across these conditions. The **lattice** [156] package has functions that quickly generate a series of plots of variables across the levels of a factor (or factors). It is among the nearly 30 packages that are part of the base distribution of R; you do not normally have to install it. To load it, we use the `library` function. The following code fragment creates Fig. 1.2:

```
> library(lattice)
> xyplot(PerCent/100 ~ Quanta | Obs + Run, data = HSP,
+    xlab = "Quanta/flash", ylab = "Proportion s̈een",
+    scales = list(x = list(log = TRUE, limits
                  = c(10, 500),
+      at = c(10, 20, 50, 100, 200),
+      labels =  c(10, 20, 50, 100, 200))),
+    skip = c(rep(F, 5), T), layout = c(3, 2, 1),
+    as.table = TRUE)
```

The **lattice** function `xyplot` enables the efficient generation of scatterplots across the factor levels. The key feature in the **lattice** plotting commands is the specification of the variables as a formula. In this case, we specify that PerCent divided by 100 will be plotted as a function of Quanta for each combination of the levels of the factors Obs and Run. The first argument to `xyplot` is an example of a formula object. We will look at more examples of formula objects in the next chapter. For now, the tilde, '~', can be interpreted here as 'is a function of' and the vertical bar, '|', separates the functional relation from the factors on which it is conditioned. The `data` argument specifies the data frame within which the terms in the formula can be found, i.e., each of the terms in the formula corresponds to the name of a column of the data frame. Lattice plotting functions produce a default format that has been chosen to yield generally pleasing results in terms of information display. It can be easily customized. The rest of the arguments just control details of the formatting of the graphs and their layout. The `scales` argument is used to set the tick positions and labels in the figure, the `skip` argument prevents an empty graph from being displayed for observer MHP who only performed one run, and the `layout` and `as.table` arguments control the arrangement of the graphs. These additional arguments are not absolutely necessary. We include them

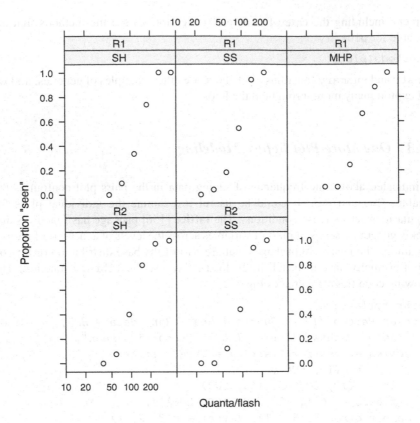

Fig. 1.2 Frequency of seeing data for each observer and run as a function of the average number of quanta/flash. The data are displayed as a lattice

to demonstrate that one can control fine details with the command, but you should try to redo the graph without them to see what happens.

The factor levels are indicated in strips above each graph, allowing easy identification of the condition. Notice that the abscissa has been specified to be logarithmically spaced. The crude scatterplot of PerCent vs Quanta in Fig. 1.1 was plotted on a linear scale. The main tendencies noted in the pairs plot are still present in the individual graphs, however.

1.3   Modeling the Data

Our aim is to fit the frequency of seeing data displayed in Fig. 1.2 with a psychometric function. The psychometric function, for our purposes, is a smooth underlying curve describing the probability of detecting a flash as a function of flash intensity. The curve is usually part of a parameterized family of functions and we seek the parameter values that specify a curve that best matches the data. As we will see in Chap. 4, there are several possibilities as to the choice of functional form to fit the data. For our example, we will use a Gaussian cumulative distribution function (see Sect. B.2.5).[3] This corresponds to a probit analysis [60] and has the advantage that it can be performed simply in R as a generalized linear model (GLM) [127] with the glm function, specifying a binomial family. We postpone discussing the GLM until Chap. 2 and Chaps. 4 and 5 will describe how we use the GLM to fit psychometric functions in detail.

Our goal is to fit a model that best predicts the probability that the observer will respond "seen" (or from here on, "Yes" for "I saw the stimulus on that trial") when a stimulus of intensity Q (log quanta) is presented

$$P[\text{Yes}] = \Phi\left(\frac{Q - Q_{.5}}{\sigma}\right) \tag{1.1}$$

The function Φ maps the physical quantity in parentheses to the range $(0, 1)$. It is typically a cumulative distribution function for a random variable (Sect. B.2.3), here the cumulative distribution function of a Gaussian (normal) random variable with mean 0 and standard deviation 1. We allow ourselves two free parameters to fit data. The first is $Q_{.5}$, the intensity that results in the observer responding Yes with probability 0.5. This measure is a convenient characterization of the observer's sensitivity. The second parameter σ controls how steep the fitted psychometric function will be. The slope of the Gaussian psychometric function at $Q_{.5}$ is inversely proportional to σ. We rewrite the function above as

$$\Phi^{-1}(\text{E}[R]) = \frac{Q - Q_{.5}}{\sigma} = \beta_0 + \beta_1 Q \tag{1.2}$$

where R denotes the observer's response on a trial coded so that it takes on the value 1 if the observer responds Yes and 0 if the observer responds No and Q is the intensity of the stimulus.

The expected value of R is, of course, the probability that the observer responds Yes. We recode the variable in this way because the GLM is used to predict not the observer's exact response (which is 0 or 1) but its expected value across many trials. The fixed function Φ is replaced by its inverse on the left-hand side of the equation. The inverse of a cumulative distribution function is called a *quantile function* (see Sect. B.2.3). Moving Φ from one side of the equation to the other is simply cosmetic. It makes the resulting equation look like a textbook example of a GLM. We then replaced the parameters $Q_{.5}$ and σ by β_0 and β_1 where it is easy to see that $\beta_0 = -Q_{.5}/\sigma$ and $\beta_1 = \sigma^{-1}$. Again, we are simply increasing the resemblance of the resulting equation to a textbook example of a GLM. We will use the GLM to estimate β_0 and β_1 and then solve for estimates of $Q_{.5}$ and σ. In the resulting equation, the right-hand side is called the *linear predictor* and looks

---

[3]For theoretical reasons, Hecht et al. chose a different function, a cumulative Poisson density. We will return to consider fitting the data with their model in Sect. 4.2.1.

like a linear regression equation, the linear model (LM) of which the GLM is a generalization.[4] We discuss the GLM in detail in the following chapter. For now it is enough to recognize that (1.2) is an example of a GLM and that we can estimate parameters using the R function `glm`. Next we set up the GLM by first specifying a formula object.

In the current problem, the data are specified as the percent correct for N trials at each intensity. It would be equivalent for our analyses to recode them as N 1's and 0's, indicating the trial-by-trial responses, with the number of 1's equal to the number of successes. We recode the percent correct as a two-column matrix indicating the numbers of successes and failures. The resulting counts of successes are binomial random variables (see Sect. B.3.2) with means equal to $N_i\mathsf{E}[R_i]$ for each stimulus intensity, i. We calculate these values from the data and add two new columns to the data frame:

```
> HSP$NumYes <- round(HSP$N * HSP$PerCent/100)
> HSP$NumNo <- HSP$N - HSP$NumYes
```

To add the new columns, we just specify them with the list operator $.

We could reduce the number of times that we write HSP by using `within` which works like `with` but makes changes within the context of a specified data frame and returns the entire, modified data frame. Then the above code fragment becomes

```
> HSP <- within(HSP, NumYes <- round(N * PerCent/100)
+       NumNo <- N - NumYes )
```

Notice that we still need to to assign the result back to the variable in memory to make the change.

We rounded the data to guarantee that the resulting counts of Yes and No responses are integers.[5]

1.3.1   Modeling a Subset of the Data

There are data for five frequency of seeing curves in HSP. To begin, we will only fit one of them. We extract a subset of the data, the first run of observer SH, using the `subset` function

```
> ( SHR1 <- subset(HSP, Obs == "SH" & Run == "R1") )
```

---

[4]The reader may be wondering whether fitting estimates of parameters in one form such as β_0, β_1 and then transforming the estimates to estimates of the parameters we want ($Q_{.5}$ and σ) are legitimate. It is. See Sect. B.3.3.

[5]Interestingly, there is an error in Hecht et al.'s table, pointed out to us by by Michael Kubovy. The entries in the resulting counts of Yes and No responses for one of the conditions are not close to integers. It is possible that the number of trials for this observer was not the same at all intensity levels.

	Quanta	PerCent	N	Obs	Run	NumYes	NumNo
1	46.9	0.0	35	SH	R1	0	35
2	73.1	9.4	35	SH	R1	3	32
3	113.8	33.3	35	SH	R1	12	23
4	177.4	73.5	35	SH	R1	26	9
5	276.1	100.0	35	SH	R1	35	0
6	421.7	100.0	35	SH	R1	35	0

Enclosing the line in parentheses causes the result to be printed out.

The second argument of subset creates a logical vector that determines the choice of rows to be those for which the observer is SH and the run is R1. An alternative method for obtaining a subset of the data frame uses the fact that data frames can be indexed like matrices using square brackets (see Appendix A). We explicitly specify which row entries and which column entries (all of them, signified by leaving the second coordinate blank) to select using a logical vector

```
> with(HSP, HSP[Obs == "SH" & Run == "R1", ])
```

A third method of subsetting is possible, using the subset argument that many modeling functions, such as glm, include that would exploit the same logical expression. Which method to use here is largely a matter of convenience.

The call to glm requires three arguments: a formula object, a family object, and the data frame to use in interpreting the terms of the formula. Hecht et al. used the logarithm of the covariate Quanta and so do we

```
> SHR1.glm <- glm(formula = cbind(NumYes, NumNo) ~
    log(Quanta),
+    family = binomial(probit), data = SHR1)
```

The first argument is the formula object, a specification of how we are modeling the observer's responses in a short-hand language used by most R modeling functions. The formula object expresses the relation between the response variable on the left-hand side of ~ and the linear predictor on the right-hand side. We saw an example of such a formula object as the first argument to xyplot above. Here, the tilde can be read as "is modeled as."

The response term in the function call combines the two columns, NumYes and NumNo, into a 6×2 matrix using the function cbind. The term log(Quanta) indicates the covariate with the intercept implicit in the model formulation. With the family argument, we first specify that our responses are distributed as a binomial variable. The function binomial takes one argument that indicates the form of the psychometric function. The argument is called the link function. The default link function for the binomial family is the log odds ratio or logit transformation, so we must specify the probit (Gaussian) transform explicitly. The data argument indicates the data frame within which the column vectors corresponding to the terms in the formula are defined.

When arguments are named, as here, they may be specified in any order. Because the arguments are in the default order specified for the glm function (verify by typing `help(glm)` or `?glm`), we could also leave out the names. Including them makes the code easier to read. The result returned by `glm` is a model object of class "glm" assigned to a variable named `SHR1.glm` that we will begin to examine now.

1.3.1.1 Examining the Model Object

Typing in the name of the model object (or, for that matter, any object in R) invokes its `print` method, equivalent to typing in `print(SHR1.glm)`:

```
>  SHR1.glm

Call: glm(formula = cbind(NumYes, NumNo)~log(Quanta),
               family = binomial(probit), data = SHR1)

Coefficients:
(Intercept)   log(Quanta)
    -13.02          2.67

Degrees of Freedom: 5 Total (i.e. Null);   4 Residual
Null Deviance:        185
Residual Deviance: 2.78      AIC: 17.4
```

The result is a summary of basic information about the fit. The first line displays the function call that created the model object. The coefficients are the intercept and slope for the covariate from (1.2), i.e., β_0 and β_1, respectively. Information on the number of degrees of freedom and measures related to the goodness of fit are also given which we will discuss further below.

R provides powerful methods for comparing different models of the same data. Deviance is a measure related to the log likelihood of a model (see Sect. B.4.2) and is used to assess goodness of fit. We use it to compare complex models to simpler versions of the same model. The printed results include the *residual deviance* that is for the fitted model with intercept and slope terms, β_0 and β_1, respectively, and the *null deviance*, for a simpler model with just the intercept term β_0, a special case of the first model with β_1 forced to be 0. The comparisons of the deviances of the models with and without each term lead to nested hypothesis tests (see Sect. B.5).

Finally, the AIC, or Akaike Information Criterion, is a second measure of goodness of fit useful for comparing several models that attempt to take into account model complexity (indicated by the number of parameters) as well as the closeness of the data to the estimated model [6] (see Sect. B.5.2).

The specific test used by GLM is based on a χ^2 distribution that someone familiar with nested hypothesis testing might expect (see Sect. B.5 for an explanation). Conveniently, this test is provided directly by the `anova` method that generates a table in the conventional form of an analysis of variance (or analysis of deviance, here) table.[6]

```
> anova(SHR1.glm, test = "Chisq")
Analysis of Deviance Table

Model: binomial, link: probit

Response: cbind(NumYes, NumNo)

Terms added sequentially (first to last)


            Df Deviance Resid. Df Resid. Dev Pr(>Chi)
NULL                          5       185.1
log(Quanta)  1      182       4         2.8   <2e-16
                                                 ***
---
Signif. codes:  0 `***' 0.001 `**' 0.01 `*' 0.05 `.'
                0.1 ` ' 1
```

The summary method provides a more detailed set of results:

```
> summary(SHR1.glm)

Call:
glm(formula = cbind(NumYes, NumNo) ~ log(Quanta),
    family = binomial(probit), data = SHR1)

Deviance Residuals:
     1       2       3       4       5       6
-0.464   0.605  -0.145  -0.694   1.279   0.250

Coefficients:
            Estimate Std. Error z value Pr(>|z|)
(Intercept)  -13.017      1.618   -8.04  8.8e-16 ***
log(Quanta)    2.671      0.331    8.07  6.8e-16 ***
---
Signif. codes:  0 '***' 0.001 '**' 0.01 '*' 0.05 '.'
                0.1 ' ' 1

(Dispersion parameter for binomial family taken
   to be 1)
```

---

[6]The anova method might be better named something like model.compare but, alas, anova it is.

```
     Null deviance: 185.053  on 5  degrees of freedom
Residual deviance:   2.781  on 4  degrees of freedom
AIC: 17.36

Number of Fisher Scoring iterations: 6
```

and includes the deviance residuals as well as additional statistics for the estimated coefficients, β_0 and β_1 described above.

The coefficients for the intercept and slope can be extracted as a vector from the model object with the coef method. Confidence intervals for the fitted coefficients are obtained with the confint method.

```
> coef(SHR1.glm)
(Intercept) log(Quanta)
    -13.02        2.67
> confint(SHR1.glm)
            2.5 % 97.5 %
(Intercept) -16.55  -10.1
log(Quanta)   2.08    3.4
```

The estimated responses for the data are obtained with the fitted method taking the model object as an argument, but typically one wants to generate enough values to produce a smooth curve for a plot. These can be obtained with the predict method. To use this, we supply the model object and a data frame with the same column names used in performing the fit. Additionally, we must specify that the predicted values are on the scale of the response instead of on that of the linear predictor, which is the default. At no extra cost, we obtain standard errors for the fit by setting the argument se.fit = TRUE.

We illustrate this using graphic functions from the base **graphics** package [146]. Note how with the plot function from base graphics, we create the graphic with successive function calls that add the embellishments.

```
> plot(PerCent/100 ~ Quanta, SHR1,
+   xlab = "Quanta/Flash", ylab = "Proportion šeen",
+   main = "Obs: SH, Run:  1", log = "x",
+   xlim = c(20, 440))   # set up plot
> xseq <- seq(20, 450, len = 100)
> SHR1.pred <- predict(SHR1.glm, newdata =
+    data.frame(Quanta = xseq), type = "response",
+    se.fit = TRUE)    # obtain predicted values
> polygon(c(xseq, rev(xseq)), c(SHR1.pred$fit
   + SHR1.pred$se.fit,
+   rev(SHR1.pred$fit - SHR1.pred$se.fit)),
+   border = "white", col = "grey")  # plot SE envelope
> lines(xseq, SHR1.pred$fit, lwd = 2)# add fitted curve
```

Fig. 1.3 Psychometric
function fit by probit analysis
to frequency of seeing data
(*solid curve*) with standard
error limits (*grey envelope*)
for the first run of observer
SH. The *dotted lines* indicate
a threshold criterion

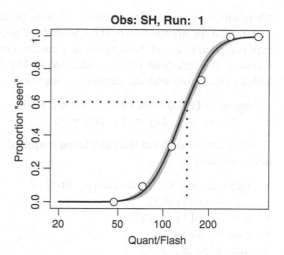

```
> points(PerCent/100 ~ Quanta, SHR1, cex = 1.5,
    pch = 21,
+   bg = "white")    #add points
```

Notice that we did not have to worry about the fact that the covariate was
transformed logarithmically in obtaining the predicted values displayed in Fig. 1.3
as the solid curve. The dotted lines obtained with the segments function are
described below.

1.3.2   Estimating Thresholds and jnd's

Next we reparameterize the Gaussian function fit to the data in terms of the
parameters $Q_{.5}$ and σ that we began with.

$$Q_{.5} = -\frac{\beta_0}{\beta_1} \tag{1.3}$$

$$\sigma = \frac{1}{\beta_1} \tag{1.4}$$

We extract the model coefficients from the glm object with the coef method and
obtain the inverse Gaussian from the normal quantile function, qnorm.

Hecht et al. reported observer's performance level not as the log(Quanta) value at
which the observer would say Yes half of the time, $Q_{.5}$, but at the log(Quanta) value
at which the observer would be expected to say Yes on 60% of the trials. We can use
the Gaussian quantile function qnorm to compute this value for this observer:

```
> (thresh <- exp(qnorm(p = 0.6,
+   mean = -coef(SHR1.glm)[1]/coef(SHR1.glm)[2],
+   sd = 1/coef(SHR1.glm)[2])))
[1] 144
```

We use the `coef` method that returns the fitted parameters from the model object as a vector. To specify the threshold on the scale of quanta, the estimated value must be exponentiated to invert the logarithmic transformation of the covariate in the model formula. The threshold value is indicated in Fig. 1.3 by the dotted lines and was added to the figure with the `segments` function.

```
> segments(c(20, thresh), c(0.6, 0.6), c(thresh, thresh),
+   c(0.6, 0), lty = 3, lwd = 3)
```

If we need to perform this calculation frequently, it would be practical to make it into a function

```
>   thresh.est <- function(p, obj) {
+   cc <- coef(obj)
+   m <- -cc[1]/cc[2]
+   std <- 1/cc[2]
+   qnorm(p, m, std)
+   }
```

where p is the probability and `obj` is the glm object.[7] An R function returns the value of its last executed line.

Then, we simply execute:

```
> thresh.est(0.6, SHR1.glm)
[1] 4.97
```

to get the estimate in log(Quanta) or

```
> exp(thresh.est(0.6, SHR1.glm))
[1] 144
```

Although not of interest for these data, this function can be used as is to calculate a jnd, which we will define here as the intensity difference[8] between 50 and 75% proportion seen. We use a vector for the argument p and apply the function `diff` to the result.

```
> diff(thresh.est(c(0.5, 0.75), SHR1.glm))
[1] 0.253
```

1.3.3   Modeling All of the Data

The `HSP` data set contains the results for five experiments, representing three observers, two of whom were tested twice. Using a model formula, it is easy to

---

[7]See the function `dose.p` in the package **MASS** for a more general version [178].

[8]Actually, here it corresponds to the log of the ratio because of the log transform of `Quanta`.

model all of the data at once. For the sake of demonstration, we will consider the five experiments as independent replications with a single source of variance. To facilitate this point of view, we define a new factor with five levels from the interaction of Obs and Run:

```
> HSP$id <- with(HSP, interaction(Obs, Run,
+    drop = TRUE))
```

This code adds a new column to the data frame in memory that codes each combination of observer and run as a separate level of a categorical or factor variable. The argument drop = TRUE avoids the creation of a level for the missing run of observer MHP.

We will compare two models. The simplest model assumes that a single psychometric function fits all five runs, i.e., a single intercept and slope describe all five runs. The call is identical to the one that we used earlier except that we specify the data frame HSP instead.

```
> HSP0.glm <- glm(cbind(NumYes, NumNo) ~ log(Quanta),
+    binomial(probit), HSP)
```

As an alternative, we consider a model in which each run has a separate intercept and slope. The update function simplifies fitting a new model. It takes a model object and a modification of its specification as arguments and returns the new fit.

```
> HSP1.glm <- update(HSP0.glm,  . ~ id/log(Quanta) - 1)
```

The '.' stands for what was in the model previously. The '/' indicates that a separate model should be considered for each level of the factor id. This is really a short hand for the equivalent model: id + id:log(Quanta), where the ':' indicates an interaction between the two terms that it separates. Since there will be one intercept for each curve, we remove the overall intercept with the term '−1'.

We compare the models by calling anova with more than one argument.

```
> anova(HSP0.glm, HSP1.glm, test = "Chisq")
Analysis of Deviance Table

Model 1: cbind(NumYes, NumNo) ~ log(Quanta)
Model 2: cbind(NumYes, NumNo) ~ id +
           id:log(Quanta) - 1
  Resid. Df Resid. Dev Df Deviance Pr(>Chi)
1        28       81.3
2        20       16.2  8     65.1  4.5e-11 ***
---
Signif. codes:  0 `***' 0.001 `**' 0.01 `*' 0.05 `.'
                0.1 ` ' 1
```

The analysis of deviance table rejects the hypothesis that a single psychometric function fits the data in favor of the alternative.

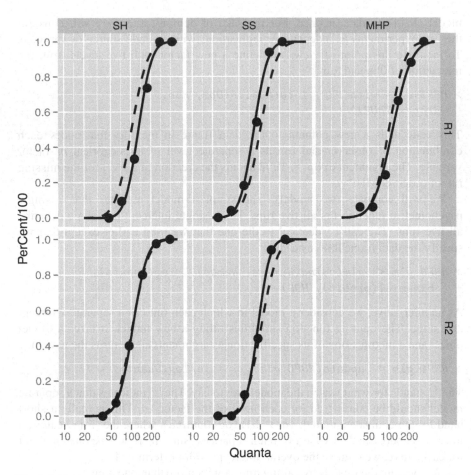

Fig. 1.4 Frequency of seeing data with a single psychometric function (*dashed curves*) and with individual curves per data set (*solid curves*), using the **ggplot2** package

The fitted curves under the two models are compared in Fig. 1.4 using the **ggplot2** package [190] which is based on *The Grammar of Graphics* [192]. It provides an alternate approach and look to visualizing data sets than **lattice**. Note that the factor levels by default are indicated on both vertical and horizontal margins. From this graph, we readily appreciate that a single psychometric function fits the data less well.

```
> nd <- data.frame(
+   Quanta = rep(seq(20, 500, len = 100), 5),
+   id = rep(levels(HSP$id), each = 100)
+   )
> levels(nd$id) <- levels(HSP$id)
> nd$pred0 <- predict(HSP0.glm, newdata = nd, type =
    "response")
```

```
> nd$pred1 <- predict(HSP1.glm, newdata = nd, type =
  "response")
> mm <- matrix(unlist(strsplit(as.character(nd$id),
+   "              ."))), ncol = 2, byrow = TRUE)
> nd$Obs <- factor(mm[, 1], levels = c("SH", "SS",
  "MHP"))
> nd$Run <- factor(mm[, 2])
> library(ggplot2)
> qplot(Quanta, PerCent/100, data = HSP, facets = Run
  ~ Obs) +
+   geom_point(size = 4) +
+   geom_line(data = nd, aes(x = Quanta, y = pred1),
    size = 1) +
+   geom_line(data = nd, aes(x = Quanta, y = pred0),
    size = 1,
+   linetype = "dashed") +
+   scale_x_log10(limits = c(10, 500),
+     breaks = c(10, 20, 50, 100, 200),
+     labels =  c(10, 20, 50, 100, 200))
```

1.3.4 Odds and Ends

Before ending this chapter, we include a few additional household functions that are essential. First, the function rm will delete objects from the work space. To remove all visible objects in the workspace use:

```
> rm(list = ls())
```

To end an R session, use the function q.[9] By default, this will open a dialog requesting whether you really want to quit or not and whether you want to save the variables from your work space. If these are saved, then the next time that you open R from that directory, they will be reloaded into memory.

In this chapter, we demonstrated an approach to modeling data in R through an extended example. Basic functions were introduced for examining objects and modeling data. In subsequent chapters, we will go deeper into the theory underlying several psychophysical paradigms. We will show that a common modeling framework links them and that R provides a remarkably rich, powerful, and extensible environment for implementing these analyses.

---

[9]In the GUI interface for Mac OS X, a platform dependency rears its ugly head. It is recommended to quit using the Quit button (in the upper right corner of the console window, resembling a switch for turning the room lights out) as otherwise, the history of past commands will not be saved.

1.4   Exercises

1.1. Explore the help facilities of R. At the command line, type `help(help)` and read over the documentation. Try the same with other help-related commands cited in the text, e.g., `help(apropos)`. Try the command `help.start()`.

1.2. After having executed the code in this chapter, examine the search path. How has it changed? Why?

1.3. Check if the **MPDiR** package is in the search path. Examine its contents using the pos argument of `ls`.

1.4. We indicated that the operators "`<-`" and "`=`" are not identical. To appreciate the difference, consider the following exercise. Make sure that the HSP data frame is loaded in your work space and then assign a variable with the name n the value of 10, i.e., a 1 element integer vector. Now, examine the value of n after executing each of the following function calls:

```
> head(HSP, n = n)
> head(HSP, n = 15)
> head(HSP, n <- 15)
```

1.5. The function `splom` in the **lattice** package generates a matrix of scatterplots, similar to the pairs plot of the data frame in Sect. 1.2.2. Plot the HSP data set with this function and compare it with Fig. 1.1 to appreciate the difference in presentation style.

1.6. Assign the results of `summary(SHR1.glm)` to a variable. What is its class? How many components does it have? How would you access the p-values of the fitted coefficients?

1.7. Refit the data of `SHR1` using the default logit link. Plot the predicted values on the same graph with the predicted values of the probit fit. How do they differ? How do the thresholds differ?

1.8. Examine the summary of `HSP1.glm`. What do the different terms represent in the model? Estimate the threshold at proportion seen = 0.6 for each of the five experimental runs.

1.9. The data set `Vernier` in the package **MPDiR** contains results from an experiment on the detection of misalignment (phase difference) between two adjacent, horizontal gratings that were drifting either upward or downward [167]. Extract the subset of the data for which the waveform is "Sine" and the temporal frequency is 2 cycles/degree and the direction is upward. Fit a psychometric function to this subset of the data using `glm`. Then, plot the data points and the fitted curve on the same graph. Using the fitted object, calculate the discrimination threshold as the phase shift difference between 0.5 and 0.75 detection upward.

Chapter 2
Modeling in R

2.1 Introduction

One of the many strengths of R is in the diversity and convenience of its modeling functions. In this chapter, we describe several standard statistical models and show how to fit them to data using R.

The key to modeling in R is the *formula object*, which provides a shorthand method to describe the exact model to be fit to the data. Modeling functions in R typically require a formula object as an argument. The modeling functions return a *model object* that contains all the information about the fit. Generic R functions such as `print`, `summary`, `plot`, `anova`, etc. will have methods defined for specific object classes to return information that is appropriate for that kind of object.

We first present the *linear model* (LM) and corresponding modeling functions. We begin with these precisely because the reader is likely familiar with the linear model in some form (multiple regression, analysis of variance, etc.) and it is easier to get used to working with objects and R in a familiar context. Moreover, all the other models considered here are extensions of the linear model. Second, we describe *linear mixed-effects models*. These permit modeling random effects to account appropriately for variation due to factors such as observer differences. Then, we consider some tools for fitting *nonlinear models* in R. Finally, we describe *generalized linear models* (GLM). These models allow us to fit many different kinds of psychophysical data. They will play a central role in succeeding chapters.

There are several excellent books that describe the theory and fitting of the models discussed here in R, notably [58, 178, 198].

2.2 The Linear Model

In the basic linear model, a response vector (dependent variable) is modeled as a weighted linear combination of explanatory variable vectors (independent variables)

K. Knoblauch and L.T. Maloney, *Modeling Psychophysical Data in R*, Use R! 32,
DOI 10.1007/978-1-4614-4475-6__2,
© Springer Science+Business Media New York 2012

and an additive error vector, $\varepsilon = (\varepsilon_1, \ldots, \varepsilon_n)$, where each ε_i is an independent, identically distributed (iid) Gaussian random variable with mean 0 and variance σ^2 (hereafter abbreviated by $\varepsilon \sim N(0, \sigma^2)$).

For example, in a multiple regression model with dependent variable $Y = (Y_1, \ldots, Y_n)$ and independent variables $X_1 = (X_{1,1}, \ldots, X_{1,n})$ and $X_2 = (X_{2,1}, \ldots, X_{2,n})$ the linear model is

$$Y_i = \beta_0 + \beta_1 X_{1,i} + \beta_2 X_{2,i} + \varepsilon_i, \quad i = 1, n \tag{2.1}$$

where n is the number of observations of Y, the β's are coefficients to be estimated when fitting the model and $\varepsilon_i \sim N(0, \sigma^2)$. We use subscripts, $1, 2$, here to index the independent variables, but we will see below that they correspond to columns in a matrix.

The model is simply summarized in vector form as

$$Y = \beta_0 + \beta_1 X_1 + \beta_2 X_2 + \varepsilon \tag{2.2}$$

and even more simply in matrix form as

$$Y = \boldsymbol{X}\beta + \varepsilon \tag{2.3}$$

where $\boldsymbol{X}$, the *model* or *design matrix*, is composed of a column of ones followed by columns with the values of X_1 and X_2, and $\beta = (\beta_0, \beta_1, \beta_2)$. Fitting the model involves estimating $\beta = (\beta_0, \beta_1, \beta_2)$ and the variance σ^2 of the Gaussian errors. Once we have the estimate $\hat{\beta} = (\hat{\beta}_0, \hat{\beta}_1, \hat{\beta}_2)$, we can also estimate the *residual vector*

$$e = Y - \boldsymbol{X}\hat{\beta} \tag{2.4}$$

Rearranging, and comparing to (2.3), we see that the residual vector e is an estimate of the error vector ε

$$Y = \boldsymbol{X}\hat{\beta} + e \tag{2.5}$$

We use maximum likelihood estimation to estimate β. In the case of the linear model this choice of $\hat{\beta}$ also minimizes the sum of the squared residuals. This solution is often referred to as the *least squares solution*.

In the basic linear model, the independent variables contain numbers that we weight and add to form estimates of the dependent variable. In the case that the explanatory variables are continuous numerical values, we call them *covariates* and we are engaged in *multiple regression*.

Explanatory variables can also have entries that are categorical labels such as "Male," "Female." When the explanatory variables are categorical, they are called *factors* and the linear model is referred to as *analysis of variance* (ANOVA). When the model contains both factor and continuous explanatory variables, it is called *analysis of covariance* (ANCOVA).

To use the modeling functions for the linear model in R, we first need to specify the model with dependent and independent variables. R uses a formula language developed by Wilkinson and Rogers [191]. We first create a data frame that contains the vectors of the response and explanatory values as columns.

If we have such a data frame containing columns named Y, X1, X2, each of class "numeric," then the formula object corresponding to (2.3) would be

```
> Y ~ X1 + X2 + 1
```

The tilde, "~," is an operator that separates the response to be modeled from the explanatory variables. It can be read as "is modeled by." The term corresponding to the intercept β_0 is represented in the formula object by + 1. We can simplify the formula object slightly as + 1 is the default case:

```
> Y ~ X1 + X2
```

Either formula indicates to a modeling function that a 3 column model matrix should be constructed with a first column of 1's and the second and third from the column vectors X1 and X2. In general, formula objects for linear models have an intercept unless we explicitly suppress it by writing

```
> Y ~ X1 + X2 - 1
```

or

```
> Y ~ X1 + X2 + 0
```

From here on, we omit the + 1.

The formula object doesn't explicitly mention the coefficients β or the error term. The formula object simply specifies the relations among the dependent and independent variables.

When the explanatory variables are factors, the notation of the linear model is typically different. For example, with factors A and B, where A has levels "Male" and "Female" and B has levels "Old" and "Young," each observation might be modeled by

$$Y_{ijk} = \mu + \alpha_i + \beta_j + \alpha\beta_{ij} + e_{ijk} \tag{2.6}$$

where μ is the grand mean of the Y_{ijk}, α_i is the effect of level i of factor A, β_j is the effect of level j of factor B, $\alpha\beta_{ij}$ is the interaction of levels i and j of the two factors and, as before, $\varepsilon_{ijk} \sim N(0, \sigma^2)$. In other words, we think of each response as the grand mean "adjusted" by adding α_1 if the subject is male or adding α_2 if female, "adjusted" further by β_1 if the subject is "Old," etc. The term $\alpha\beta_{ij}$ captures possible interactions between the factors. Any text on analysis of variance will develop this sort of model at length.

ANOVA is just a form of the linear model and the notation of ANOVA in R makes this clear. The model in (2.6) can be specified in a formula object as

```
> Y ~ A + B + A:B
```

where the term with : indicates the interaction between all of the levels of A and B. A shorthand notation that is equivalent employs the * symbol

```
> Y ~ A * B
```

which means A, B, and the interaction term from above. There are simple ways to include only some interactions and not others, and that allow the user to explore alternative models for a given data set (see Sect. 11.1 of *An Introduction to R*, in the documentation that comes with R) [146].

The only thing special about ANOVA is the use of independent variables that are factors. This difference, however, leads to differences in how the model matrix **X** is constructed. In the model matrix for an ANOVA, each categorical (or factor) variable is expanded into columns of indicator variables with one column allocated to each level. For example, a model with one factor with 3 levels and 2 replications per level is written out as

$$
Y = \begin{pmatrix} 1\;1\;0\;0 \\ 1\;1\;0\;0 \\ 1\;0\;1\;0 \\ 1\;0\;1\;0 \\ 1\;0\;0\;1 \\ 1\;0\;0\;1 \end{pmatrix} \begin{pmatrix} \mu \\ \alpha_1 \\ \alpha_2 \\ \alpha_3 \end{pmatrix} \tag{2.7}
$$

The first column of 1's corresponds to the grand mean, μ, and each succeeding column to a level of the factor variable. The difficulty here is that the columns are linearly dependent; the sum of columns 2–4 equals column 1. If we attempted to fit a model with this model matrix, we would not succeed as there is no unique solution to the equation. This issue does not arise in typical multiple regression models unless there is a linear dependence among some subset of the explanatory variables including the constant term.

To deal with the inherent dependence in the model matrix of an ANOVA, constraints are placed on the model matrix. Many different choices for the constraints are possible, and we illustrate some alternatives, by way of which we also introduce the useful function `model.matrix` that computes a model matrix for a formula object.

Continuing with the example from (2.7), we use the function `factor` to create a variable of class "factor" with 3 levels and 2 replications per level that we assign to a variable A.

```
> (A <- rep(factor(1:3), each = 2))
[1] 1 1 2 2 3 3
Levels: 1 2 3
```

The function `rep` is useful for generating repeating patterns.

With the responses denoted as Y, the formula object for the linear model in (2.7) is

```
> Y ~ A
```

To view the default model matrix that is generated for this model, we use the function model.matrix with a one-sided formula object.

```
> model.matrix( ~  A)
  (Intercept) A2 A3
1            1  0  0
2            1  0  0
3            1  1  0
4            1  1  0
5            1  0  1
6            1  0  1
attr(,"assign")
[1] 0 1 1
attr(,"contrasts")
attr(,"contrasts")$A
[1] "contr.treatment"
```

The resulting matrix contains a column of all 1's corresponding to the intercept term and columns corresponding to just two of the levels, A2 and A3. This model matrix has linearly independent columns and R has achieved this outcome by dropping the first level, A1.

The choice of model matrix changes the parameterization of the model (the price of being able to identify uniquely the parameters, a model property referred to as *identifiability*). The default choice in R is called *treatment contrasts*. The first column of 1's will generate an estimate of the mean of the first level of the factor. The succeeding columns generate estimates of the difference of means between each level and the first level. If the first level corresponds to a control condition, then the subsequent levels could be used to contrast each with the control.

We could alternatively remove the intercept term to yield the following model matrix:

```
> model.matrix( ~  A - 1)
  A1 A2 A3
1  1  0  0
2  1  0  0
3  0  1  0
4  0  1  0
5  0  0  1
6  0  0  1
attr(,"assign")
[1] 1 1 1
attr(,"contrasts")
attr(,"contrasts")$A
[1] "contr.treatment"
```

The new model matrix still has three columns but now each column corresponds
to one level of the factor. Recall that the syntax (-1) tells R to omit the intercept
term in a linear model, here the mean of the first level. In this parameterization, the
estimates will correspond to the means of each of the factor levels. This is referred
to as the *cell means model*.

Several other options are available in R (see, ?contr.treatment). For
example, the contr.sum function can be used to generate a model matrix in
which the sum of contrasts across levels equals 0. The contrasts argument of
model.matrix requires a list whose components are named by the variables in
the formula and are assigned character strings indicating the name of a function to
calculate the contrasts for that variable.

```
> model.matrix( ~ A, contrasts = list(A = "contr.sum"))
  (Intercept) A1 A2
1           1  1  0
2           1  1  0
3           1  0  1
4           1  0  1
5           1 -1 -1
6           1 -1 -1
attr(,"assign")
[1] 0 1 1
attr(,"contrasts")
attr(,"contrasts")$A
[1] "contr.sum"
```

Studying the built-in functions will be of use if the reader needs to tailor the contrast
functions for specialized situations.

2.2.1 Development of First- and Second-Order Motion

As a linear model example, we consider the data of Thibault et al. [171] who
examined the minimum modulation contrast to detect two kinds of motion as a
function of age, based on the spatiotemporal modulation of luminance (first order)
or of contrast (second order). The data set Motion, obtained from the **MPDiR**
package by,

```
> data(Motion)
```

contains five components, Subject, a 70-level factor identifying individual observers, a covariate LnAge, the (natural) logarithm of the age in months, Mtype,
a 2-level factor with levels "FO" and "SO" coding the type of motion stimulus,
Sex, a 2-level factor indicating the sex of the observer, and LnThresh, the (natural)
logarithm of the minimum contrast modulation of the stimulus detected by the
observer. We check that all is as expected by

```
> str(Motion)
'data.frame':    112 obs. of  5 variables:
 $ Subject : Factor w/ 70 levels "S01","S02","S03",..:
                1 2..
 $ LnAge   : num  2.2 2.2 2.2 2.2 2.2 ...
 $ Mtype   : Factor w/ 2 levels "FO","SO": 1 1 1 1 1 2
                1 2..
 $ Sex     : Factor w/ 2 levels "f","m": 1 1 2 2 1 1 2
                2 1..
 $ LnThresh: num  3.22 2.19 2.19 2.88 2.1 ...
```

The data were collected using the preferential looking technique [170]. Only the estimated thresholds are reported, so we confine attention to them.

We can ask several questions about these data. Does threshold depend upon age? Are there differences in threshold as a function of sex or type of motion? Does a change in threshold with age depend on either of these factors? A first appreciation of the data is obtained by plotting the age dependence of threshold for each level of the two factors in separate panels, using the following code fragment:

```
> library(lattice)
> xyplot(LnThresh ~ LnAge | Mtype + Sex, data = Motion,
+     xlab = "Log Age (months)",
+     ylab = "Log Threshold (contrast)",
+     panel = function(x, y) {
+         panel.lmline(x, y)
+         panel.loess(x, y, lty = 2, lwd = 2,
+          col = "black")
+         panel.xyplot(x, y, col = "black", pch = 16,
+          cex = 1.5)
+        })
```

The first argument is a formula object that specifies what we want to plot. The term LnAge | Mtype + Sex specifies that we want separate plots for each combination of Mtype and Sex. The panel argument allows us to customize what will be plotted in each graph with a panel function. If it is left out, the default panel function generates scatterplots using the part of the formula to the left of the vertical bar for each combination of the factors indicated to the right of the bar. In the panel function above, we first add a least squares regression line to each plot using panel.lmline and then add a local (loess) regression curve, using panel.loess. The loess curve follows a smoothed average trend in the data. Adding a smooth curve like the loess to the data provides a quick visual assessment of whether a line is a reasonable description of the data. Finally, we add the points with the panel.xyplot function. Each of the plot functions can take additional arguments to control the appearance of what it plots.

Examination of the plots suggests that a linear model would be appropriate for the first-order motion data, but the loess curves hint that there may be a ceiling

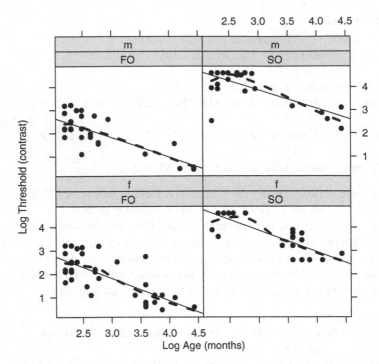

Fig. 2.1 Contrast modulation thresholds as a function of age for male (m) and female (f) observers and for first-order (FO) and second-order (SO) motion stimuli. The *solid lines* are from linear regression and the *dashed line* using local regression

effect present at the youngest ages for the detection of second-order motion. We will ignore this possibility here but note that tools are available in R that would allow us to fit censored data (see, e.g., the function `tobit` in the **AER** package [92]) (Fig. 2.1).

The function `lm` is the principle tool for analyzing linear models. It can handle models in which the explanatory variables are all continuous (*covariates*) or categorical (*factors*). In most applications, `lm` has at least two arguments, a formula object specifying the model we wish to fit and a data frame which provides the environment within which the names in the formula are interpreted. Each variable in the formula object corresponds to a column in the data frame with the same name.

The first model we consider is

$$Y = \beta_0 + \beta_1 \text{LnAge} + \beta_2 \text{Mtype} + \beta_3 \text{Sex} \tag{2.8}$$

It includes only additive effects of (log transformed) Age, Sex and Motion type. The model contains both covariates (`LnAge`) and factors (`Sex` and `Mtype`) and, thus, is an example of an ANCOVA.

It is often recommended to begin with the most complex model, including all explanatory variables and their interactions of all orders. Successive comparisons with simpler models will then allow us to determine the minimum set of terms that adequately explains the data. For didactic purposes, however, we start with a simpler model. The model in (2.8) is fit as follows:

```
> Motion.lm0 <- lm(LnThresh ~ Mtype + Sex + LnAge,
+   data = Motion)
```

The first argument is the formula object, the second argument, the data frame containing all of the variables.

In applications of the linear model, it is typically recommended to transform all covariates to be centered around their respective means. We can do so by using the scale function, replacing the covariate LnAge by scale(LnAge, center = TRUE, scale = FALSE). This subtracts the mean from each value. If we set scale = TRUE, then the values will also be scaled by the root mean square of the vector. When center = TRUE, this is simply the standard deviation. After centering the means, the estimated intercept is then centered with respect to the sampled data, rather than to an extrapolated point and is more easily interpretable. Here, the curves are nearly parallel, and this procedure has little effect on the interpretation (see Exercise 2.3).

The results of the fit are stored in the object Motion.lm0 which is of class "lm" and can be probed with methods defined for objects of this class. For example, the summary method displays the table of estimated coefficients, their standard errors, t- and p-values among other information. To extract just the table, we use the coefficient method which is aliased to the abbreviation coef.

```
> coef(summary(Motion.lm0))
             Estimate Std. Error t value Pr(>|t|)
(Intercept)   4.4698     0.2373   18.84 6.54e-36
MtypeSO       2.0404     0.1042   19.57 2.62e-37
Sexm         -0.0137     0.1047   -0.13 8.96e-01
LnAge        -0.8822     0.0749  -11.78 4.34e-21
```

The values for each model coefficient are the estimated differences of successive levels from the first level of the factor Mtype (i.e, treatment contrasts). Thus, MtypeSO is the effect of second-order motion with respect to first and Sexm is the difference in effect between males and females. There appears to be no main effect of Sex though we cannot yet exclude an effect of Sex on the slopes of the lines. That is, there is no hint that thresholds for male subjects are consistently higher or lower by a fixed amount.

We test for a possible effect of Sex on slope by adding all second-order interactions and performing a likelihood ratio test with the anova method. We could do this by rerunning lm with a new formula object specifying all the interaction terms but R provides an efficient way to update and rerun a linear model fit after modifying the formula. We want to add all second-order interactions. We could explicitly add interaction terms of the form LnAge:Mtype, LnAge:Sex, etc. to the model, but the formula language provides a convenient shortcut. Raising the model to the nth power generates all interaction terms up to order n.

The new model is fit by

```
> Motion.lm1 <- update(Motion.lm0, . ~ .^2)
```

The "." in this context indicates that the previous value should be substituted. The formula says to include the same dependent variable as used in Motion.lm0 and to the explanatory variables add all of their second-order interactions.

After recomputing the model fit with second-order interactions added, we use the anova method to compare the two models, with and without the second-order interactions.

```
> anova(Motion.lm0, Motion.lm1)


Analysis of Variance Table

Model 1: LnThresh ~ Mtype + Sex + LnAge
Model 2: LnThresh ~ Mtype + Sex + LnAge + Mtype:Sex +
    Mtype:LnAge + Sex:LnAge
  Res.Df  RSS Df Sum of Sq     F Pr(>F)
1    108 31.5
2    105 31.2  3     0.207 0.23   0.87
```

The reader can also try anova(Motion.lm1) to obtain an anova table for the new fit with second-order interactions. The table shows the reduction in the residual sum of squares as each term from the formula is added in. When the data are unbalanced, this depends on the order of the terms in the formula. The results of the nested hypothesis tests performed by anova on a set of models do not depend on the order of the terms in the formulae, however.

Formally, we are testing whether we can reject the first model in favor of the second. We want to compare the models and see if any of the second-order interactions are significant. The comparisons of the models with and without each interaction lead to likelihood ratio tests (see Sect. B.5). It is based on an F-statistic rather than the χ^2 that someone familiar with nested hypothesis testing might expect. Intuitively, we can draw an analogy to the comparison of two means where the t-statistic is used instead of the z when the sample standard deviation is estimated from data. Examining the output of the comparison, we see that the models do not differ significantly: that is, there are no significant interactions.

Next we test whether there is any effect of Sex by updating the model and adding the term - Sex. The minus sign in the added term specifies that we are removing the term Sex from the model:

```
> Motion.lm2 <- update(Motion.lm0, . ~ . - Sex)
> anova(Motion.lm2, Motion.lm0)
Analysis of Variance Table
```

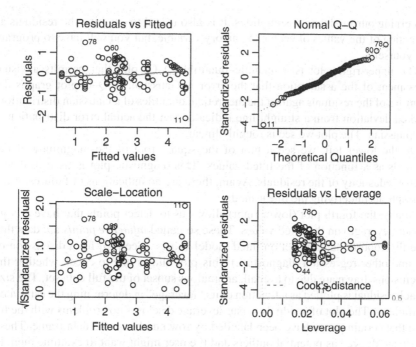

Fig. 2.2 Diagnostic plots for the linear model stored in object `Motion.lm2`

```
Model 1: LnThresh ~ Mtype + LnAge
Model 2: LnThresh ~ Mtype + Sex + LnAge
  Res.Df  RSS Df Sum of Sq    F Pr(>F)
1    109 31.5
2    108 31.5  1   0.00495 0.02    0.9
```

and the call to `anova` tests for a significant effect of including `Sex`. We find no evidence for an effect of `Sex`.

R provides several tools for assessing whether the linear model is appropriate for a particular application. The method `plot` for "lm" objects generates by default a set of four graphs that are useful in diagnosing whether the model assumptions underlying the fit are reasonable.

```
> plot(motion.lm2)
```

The two plots in the first column of Fig. 2.2 display the pattern of residuals with respect to the fitted values. These are useful for detecting inhomogeneity in the distribution of residuals. If the model is appropriate, we expect residuals that are unpatterned with no trends in magnitude.

Patterns or trends in magnitude would indicate that the model is inappropriate and that we might need to consider additional explanatory variables, transform the response nonlinearly, or consider a nonlinear model. All but the last option can be performed with `lm`. The results of the first diagnostic plot raise no warning flags concerning our modeling assumptions. It is also useful to examine the residuals as a function of the values of each explanatory variable, but you will have to program that yourself!

The upper-right plot is a quantile–quantile (Q–Q) plot that permits a visual assessment of the assumption that the error is Gaussian. A Q–Q plot graphs the quantiles of the residuals against the theoretical quantiles of a Gaussian distribution. Marked deviation from a straight line indicates that the actual error distribution is not Gaussian. The plot we see is roughly linear.

On the lower left, we see a plot of the square root of the magnitude of the residuals as a function of the fitted values. This diagnostic plot is used to detect heteroscedasticity of the residuals. Again, there are no indications of failures of the assumptions underlying the linear model.

Finally, the fourth plot (lower right) allows us to detect points that have a large impact (leverage) on the fitted values. These so-called *influence points* are data that have disproportionate impact on fitted model values. (See [40] for a discussion of this and other regression diagnostics.) This plot allows one to check whether the conclusions drawn are due to a small, anomalous subset of the full data set. The size of each residual is plotted as a function of its "leverage" in determining the estimated coefficients. The plot of residuals versus leverage discloses no problems with the fit. Note that certain points have been labelled by row number in the data frame. These have been flagged as potential outliers and the user might want to examine them in more detail.

Based on our analyses and model comparisons, we arrive at a model in which the logarithm of contrast threshold is a linear function of the logarithm of age that does not depend on the sex of the observer. The linear functions for all conditions have the same slope but observers are about 8 times more sensitive to the first order motion stimulus than to the second. We leave as an exercise plotting the data with the fitted lines.

2.3   Linear Mixed-Effects Models

Thus far in this chapter, we have considered only fixed-effect linear models. These are linear models in which the explanatory variables are *not* themselves random and the only random source of variation in the model is the residual variation that is not accounted for by the explanatory variables. In certain circumstances, however, some of the explanatory variables may represent random samples from a larger population and we would like to draw conclusions not just about the data set that we have but also the population from which it was drawn.

A typical example would be data for a group of subjects in an experiment who are drawn from a population of all potential subjects who might have participated in the experiment. Other examples include a set of natural images used as stimuli considered as a sample from the population of all natural images or a set of tone sequences sampled from the population of all melodies. If we model the effects due to these variables (people or stimuli) as fixed effects, then the conclusions that we draw pertain to the specific characteristics of the sample used in the study and cannot be attributed to the characteristics of the population as a whole.

If we wish to generalize the conclusions drawn from the sample to the population, such variables must be treated as *random effects*. If all the explanatory variables are treated as random effects, then our model is a *random effects model*. If only some of them are, then the resulting model is a *mixed-effects model*.

In fitting a random-effects or mixed-effects model, we estimate the variability of the population from which the sample is drawn rather than just the effects attributed to the individual members of the sample. Such models then include multiple sources of variation, those due to the random explanatory variables as well as the residual variation unexplained by the fixed and the random effects. A remarkably clear introduction to mixed-effects models and how to fit them can be found in Pinheiro and Bates [141].

In a mixed-effects model, the response vector is taken conditionally on the random effects and is modeled as the sum of a fixed effects term X and a random effects term, Z.

$$(Y|b) = X\beta + Zb + \varepsilon, \tag{2.9}$$

where $(Y|b)$ is the response vector conditional on b, the random effects vector, $X\beta$ is the fixed effects term, the product of a design matrix and a vector of fixed effects coefficients, Zb is the random effects term, the product of a design matrix for the random effects and the vector of random-effects coefficients such that $b \sim N(0, \sigma_b^2 I)$ and $\varepsilon \sim N(0, \sigma^2)$ is the residual variation unaccounted for by the rest of the model. In fitting a mixed-effects model the fixed effects coefficients, β and the variance of the random effects, σ_b^2 are both estimated. Mixed-effects models introduce an important simplification permitted by random effects. Suppose, for example, 1,000 observers participated in a study. If a fixed intercept was associated with each subject, then the model would require 1,000 parameters to be estimated. If instead, the observers were considered a random effect, then only a single extra parameter, the variance due to observers, would be added to the model.

2.3.1   The ModelFest Data Set

The ModelFest data set originates from a study conducted by a consortium of researchers to produce a data base under fixed conditions from several laboratories that would serve as a reference data set for models of human spatial vision [26,183]. The data set consists of contrast thresholds for 43 stimuli from each of 16 observers with each condition repeated 4 times, yielding a balanced data set. The stimuli can be viewed at http://vision.arc.nasa.gov/modelfest/stimuli.html and consist of a variety of luminance patterns of varying complexity. The data can be accessed in several fashions from links at the web site http://vision.arc.nasa.gov/modelfest/data.html.[1]

---

[1] An alternate source for the data is at http://journalofvision.org/5/9/6/modelfestbaselinedata.csv.

We read the data into R directly from the URL in "comma separated value" or csv format, using the function `read.csv`. This is a wrapper function for `read.table` that presets the arguments necessary for reading files in csv format. See Appendix A.

The structure of each record consists of a character string identifying the observer followed by the four thresholds of the first stimulus, then the second, etc. This produces a data frame with one row for each observer, i.e., a wide format, but we would prefer the long format, in which there is one observation per line with the stimulus and observer variables provided in separate columns. The code below retrieves the data from the web site and effects the required modifications.

```
> site <- file.path(url{"http://vision.arc.
    nasa.gov/modelfest/",
+   "data/modelfestbaselinedata.csv"})
> ModelFest <- read.csv(site, header = FALSE)
> Obs <- ModelFest$V1
> ModelFest.df <- data.frame(LContSens =
    c(t(ModelFest[, -1])),
+   Obs = rep(Obs, each = ncol(ModelFest) - 1),
+   Stim = rep(paste("Stim", 1:43, sep = ""), each = 4)
+   )
> ModelFest.df$Stim <- with(ModelFest.df, factor(Stim,
+       levels = unique(Stim)))
```

The `file.path` function is used here simply as a convenience to construct a path that is valid under any operating system and to permit us to break up the URL address so that it stays within the margins of our page. The thresholds are given as $-\log_{10}(\text{contrast})$ or log contrast sensitivity, which explains the variable name that we chose, `LContSens`. The last line of the code fragment above reorders the factor levels so that they are in numerical rather than lexical order.

For simplification, we restrict our analyses to the data for the first ten stimuli. We use the `paste` function to construct the labels of the first ten levels of the factor `Stim` and then the `subset` function to extract the rows of the data frame corresponding to those levels. Then, we redefine the factor so as to drop the unused levels from its definition.

```
> mfGab <- subset(ModelFest.df,
+   Stim %in% paste("Stim", 1:10, sep = ""))
> mfGab$Stim <- mfGab$Stim[, drop = TRUE]
> SpatFreq <- c(1.12, 2^seq(1, 4.5, 0.5), 30)
> mfGab$SpatFreq <- rep(SpatFreq, each = 4)

> with(mfGab, interaction.plot(SpatFreq, Obs, LContSens, mean,
+ type = "b", xlab = "Spatial Frequency (c/deg)",
+ ylab = "Log Contrast Sensitivity", cex.lab = 1.5))
```

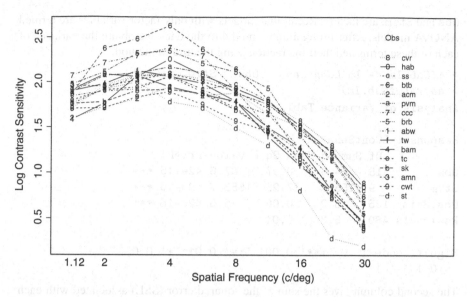

Fig. 2.3 Log contrast sensitivity as a function of spatial frequency for the Gabor patterns of the `ModelFest` data set for each of the 16 observers

The stimuli are Gabor patterns (Gaussian-damped, sine-wave gratings) of fixed size and increasing spatial frequency and their associated responses define a contrast sensitivity function over the spatial frequency of the carrier frequency. We add a column to the data frame including these spatial frequencies. The spatial frequency variable will be convenient for plotting the data and also if we decide to use spatial frequency as a covariate (see Problem 2.10). The spatial frequencies of the Gabor carrier frequencies are separated by half-octave steps except for the first and last ones.

The mean log contrast sensitivity is plotted for each observer as a function of spatial frequency in Fig. 2.3, using the function `interaction.plot`. Each observer's data are conveniently plotted with a different symbol and line type, indicated in the legend at the right. The dependence on spatial frequency is qualitatively similar for each observer, but the average heights of the points vary between observers. Some observers are overall more sensitive than others. The average SD by stimulus and observer is 0.1 log contrast (maximum of 0.23) while SD across observers for each stimulus is nearly twice as large. This increased variability across observers provides additional justification for considering the factor `Obs` as a random effect.

As mentioned above, the data set is balanced with each observer having the same number of replications at each spatial frequency. Balanced data sets possess the nice property that the contributions of the explanatory variables to the total variation in the response are orthogonal. This means that the response can be partitioned into independent and orthogonal sources of variation. If we ignore `SpatFreq` in the `mfGab` data frame, for the moment, then the remaining two explanatory variables,

Obs and Stim, are factors. Recall that models with only factor variables are termed ANOVA models. After fitting a linear model to the data, we obtain the variance of each of these terms and their interaction using the anova method.

```
> mfGab.lm <- lm(LContSens ~ Obs * Stim, mfGab)
> anova(mfGab.lm)
Analysis of Variance Table

Response: LContSens
            Df Sum Sq Mean Sq F value Pr(>F)
Obs         15   14.6    0.97    87.8 <2e-16 ***
Stim         9  155.2   17.25  1553.7 <2e-16 ***
Obs:Stim   135    8.2    0.06     5.5 <2e-16 ***
Residuals  480    5.3    0.01
---
Signif. codes:  0 `***' 0.001 `**' 0.01 `*' 0.05 `.'
    0.1 ` ' 1
```

The second column gives the sum of the squared error (SSE) associated with each term. It is simple to verify that this sums to the total variation in log contrast sensitivity.

```
> sum(anova(mfGab.lm)[, 2])
[1] 183
> sd(mfGab$LContSens)^2 * (nrow(mfGab) - 1)
[1] 183
```

The third column indicates the mean squared errors (MSE), obtained by dividing each term of column 2 by its degrees of freedom in column 1. As can also be verified, however, the F-values were all obtained by dividing each MSE by the MSE of the last row, Residuals which corresponds to a test treating both Stim and Obs as fixed effects.

In order to model Obs as a random effect, we would normally calculate an F-value by using the MSE for the interaction term, Obs:Stim in the denominator [88, 195]. Doing so is equivalent to calculating the error with respect to each level of Stim within each level of Obs, stratifying or nesting the error calculation with respect to observers, as one does in a paired t-test to increase the sensitivity of the test. For balanced data sets, as in this case, the aov function provides a convenient interface for specifying error strata and obtaining the appropriate F-test for such circumstances. The format is similar to lm but permits inclusion of a term Error for specifying the nesting factor.[2]

```
> mfGab.aov <- aov(LContSens ~ Stim + Error(Obs/Stim),
    mfGab)
> summary(mfGab.aov)
```

---

[2]In fact, it works by calling lm, but isolating the correct error terms for the F calculation.

```
Error: Obs
              Df Sum Sq Mean Sq F value Pr(>F)
Residuals 15    14.6    0.975

Error: Obs:Stim
              Df Sum Sq Mean Sq F value Pr(>F)
Stim           9  155.2   17.25      283 <2e-16 ***
Residuals 135    8.2    0.06
---
Signif. codes:  0 `***' 0.001 `**' 0.01 `*' 0.05 `.'
                0.1 ` ' 1

Error: Within
              Df Sum Sq Mean Sq F value Pr(>F)
Residuals 480   5.33  0.0111
```

Note how we specify the Error term as Stim nested within Obs. Some additional insight may come by noting that a term Error(Obs + Obs:Stim) is an equivalent formulation and yields identical results.

The summary method outputs a set of anova tables using the same df, SSE and MSE values as obtained using lm but reorganized with respect to different error strata, Obs, Obs:Stim and Within. The second table contains the appropriate F-test calculated using the interaction term. More detailed explanation of the Error term and further examples of its usage can be found in Chambers and Hastie [28] and in Li and Baron [112] and in the notes by Revelle at http://personality-project. org/r/r.guide.html.

Data sets are often unbalanced, however, in which case the simple methods presented above are not easily applied. The SSE terms will no longer be orthogonal so the decomposition may, for example, depend on the order or presence/absence of other terms in the model. It is still possible to analyze such data, but one must take recourse in methods that calculate maximum likelihoods or variants thereof. R contains several packages that permit such modeling.

Two of the principal packages devoted to analyzing mixed-effects models are the recommended package **nlme** [142] described in detail in the book by its authors [141] and the newer **lme4** package [8, 9, 11]. While the older **nlme** provides a more complete set of methods and options for linear and nonlinear mixed-effects models, the newer **lme4** contains a function for analyzing generalized linear mixed-effects models whose use we will explore in Chap. 9. At the time of writing this book, however, this package is under rapid development with changes and extensions of some of its current functionality. For this reason, we limit our discussion here to the lmer function from the **lme4.0** package [10] (currently available from Rforge (https://r-forge.r-project.org/projects/lme4/) but eventually to be released on CRAN), which is a stable version of the current package.

The lmer function works similarly to the other modeling functions in R, requiring minimally a formula object. The terms in the formula are obtained from a data frame, if present, specified as the second argument, and otherwise are searched for in the environment in which lmer was called. The formula object must include at least one fixed effect and one random effect term. Note that this required fixed effect could be as simple as the implicit intercept term in the formula specification. The fixed effects terms are defined as shown previously. The random effect terms are specified by a two part expression separated by a vertical bar (or pipe symbol), "|," and enclosed in parentheses.

```
(1 | Obs)
```

The left-hand term indicates the random component and the right the grouping or nesting factor.

For the contrast sensitivity data, a random intercept associated with each observer would indicate a model in which the shape of the contrast sensitivity function across frequency was independent of observers but the height of the data depended on the observer. In the following code fragment, we load the **lme4.0** package in memory and fit the model, after which we explicitly evoke the print method for the object in order to suppress the print-out of the large correlation matrix for the fixed effects.

```
> library(lme4.0)
> mfGab.lmer <- lmer(LContSens ~ Stim + (1 | Obs),
    mfGab)
> print(mfGab.lmer, correlation = FALSE)
Linear mixed model fit by REML
Formula: LContSens ~ Stim + (1 | Obs)
   Data: mfGab
  AIC  BIC logLik deviance REMLdev
 -492 -439    258     -574    -516
Random effects:
 Groups   Name        Variance Std.Dev.
 Obs      (Intercept) 0.0238   0.154
 Residual             0.0221   0.149
Number of obs: 640, groups: Obs, 16
Fixed effects:
            Estimate Std. Error t value
(Intercept)   1.8210     0.0428    42.5
StimStim2     0.1394     0.0263     5.3
StimStim3     0.2422     0.0263     9.2
StimStim4     0.2855     0.0263    10.9
StimStim5     0.1710     0.0263     6.5
StimStim6     0.0227     0.0263     0.9
StimStim7    -0.2001     0.0263    -7.6
StimStim8    -0.5232     0.0263   -19.9
StimStim9    -0.8615     0.0263   -32.8
StimStim10   -1.2535     0.0263   -47.7
```

The model is fit with 11 parameters, the 10 fixed effects associated with the factor Stim and the single random effect associated with the factor Obs. This contrasts with the model fit to these data using lm which requires the estimation of 160 coefficients. Each random component of the model is indicated by its variance and its square-root, or standard deviation. It is not the standard deviation of the variance.

Being in the same units as the response variable, the standard deviation is usually more easily related to the variation visible graphically when the data are plotted. For example, the contrast sensitivities at a given spatial frequency in Fig. 2.3 vary over a bit more than 0.5 log contrast sensitivity which is close to ± 2 SDs the random term Obs. The residual variance, however, is about double that of the models fit with lm and aov. This is because the model that we fit with lmer contains only a random effect for Obs whereas above we grouped the data by Stim in Obs. We can fit that model, too, with lmer using the update method as follows.

```
> mfGab.lmer2 <-update(mfGab.lmer, . ~ . +
     (1 | Obs:Stim))
> print(mfGab.lmer2, correlation = FALSE)
Linear mixed model fit by REML
Formula: LContSens ~ Stim + (1 | Obs) + (1 | Obs:Stim)
   Data: mfGab
  AIC  BIC logLik deviance REMLdev
 -683 -625    354     -757    -709
Random effects:
 Groups    Name        Variance Std.Dev.
 Obs:Stim  (Intercept) 0.0125   0.112
 Obs       (Intercept) 0.0228   0.151
 Residual              0.0111   0.105
Number of obs: 640, groups: Obs:Stim, 160; Obs, 16

Fixed effects:
            Estimate Std. Error t value
(Intercept)   1.8210     0.0488    37.3
StimStim2     0.1394     0.0437     3.2
StimStim3     0.2422     0.0437     5.5
StimStim4     0.2855     0.0437     6.5
StimStim5     0.1710     0.0437     3.9
StimStim6     0.0227     0.0437     0.5
StimStim7    -0.2001     0.0437    -4.6
StimStim8    -0.5232     0.0437   -12.0
StimStim9    -0.8615     0.0437   -19.7
StimStim10   -1.2535     0.0437   -28.7
```

Now the residual variance does match that obtained with aov and lm. The variances of the estimated random components differ, but evaluation of the model with the anova method demonstrates that the calculated F values are the same.

```
> anova(mfGab.lmer2)
Analysis of Variance Table
      Df Sum Sq Mean Sq F value
Stim   9  28.2    3.14     283
```

The additional random effect, Obs:Stim provides for a random variation of the shape of the contrast sensitivity in addition to the height. The reduction in the AIC value as well as the results of a likelihood ratio test obtained using the anova method indicates that the data are indeed better described with this additional random component.

```
> anova(mfGab.lmer, mfGab.lmer2)
Data: mfGab
Models:
mfGab.lmer: LContSens ~ Stim + (1 | Obs)
mfGab.lmer2: LContSens ~ Stim + (1 | Obs) +
   (1 | Obs:Stim)
           Df  AIC  BIC logLik Chisq Chi Df Pr(>Chisq)
mfGab.lmer  12 -550 -497    287
mfGab.lmer2 13 -731 -673    379   183      1     <2e-16
                                                    ***
---
Signif. codes:  0 `***' 0.001 `**' 0.01 `*' 0.05 `.'
                0.1 ` ' 1
```

The AIC, BIC and log likelihood values displayed in the output of the anova method differ from those reported by the print and summary methods. The default method for estimating the parameters of the model is termed REML which stands for restricted or residual maximum likelihood. It is described by Pinheiro and Bates [141] as corresponding "to assuming a locally uniform prior distribution for the fixed effects β and integrating them out of the likelihood." The REML method is often preferred because it leads to an unbiased estimate of the variance [11]. Unlike the maximum likelihood criterion, however, REML is not invariant to monotone transformations of the fixed effects, so that two REML fits with different fixed effects structures cannot be compared as nested models using likelihood ratio tests. When evaluating fixed effects, one is advised to refit the models with the argument REML = FALSE set to obtain fits by MLE before comparing them using anova. Even doing this, however, the test is only approximate, and the p-values tend to be anti-conservative (too low), so that more stringent cut-offs should be employed for evaluating significance. The anova method appears to adjust the likelihoods to the maximum likelihood values before performing the test.

An astute reader will have noticed that the printed summary of the results of lmer does not include p-values for the fixed effect coefficients. The reasons for not including these in the output are outlined by one its authors, Douglas Bates, at https://stat.ethz.ch/pipermail/r-help/2006-May/094765.html. In brief, for general applications of these models, i.e., with data that can be unbalanced, nested at multiple levels and/or with crossed random effects, there is a difficulty in knowing how to calculate the degrees of freedom for the t-statistics that are shown (or the denominator degrees of freedom for the F-statistic, which would be the square of the t-statistic). Without these, the p-values cannot be computed. Other methods do exist, however, for assessing the significance of terms or calculating confidence intervals, using simulation. For example, the function mcmcsamp uses Markov Chain Monte Carlo methods to sample from the posterior distribution of the parameters of a fitted model. This produces an object that can be passed to the function HPDinterval to return intervals based on the posterior density estimates. These are referred to as Highest Posterior Density (or, sometimes, credible) intervals to distinguish them from classical confidence intervals. As a simple example, we evaluate the parameter distributions from the mfGab.lmer2 model based on 1,000 simulated samples from the posterior distribution and print out their 95% credible intervals.

```
> mfGab.mcmc <- mcmcsamp(mfGab.lmer2, n = 1000)
> HPDinterval(mfGab.mcmc)
$fixef
                 lower    upper
(Intercept)     1.7572   1.8844
StimStim2       0.0782   0.1995
StimStim3       0.1839   0.3095
StimStim4       0.2219   0.3496
StimStim5       0.1071   0.2311
StimStim6      -0.0408   0.0834
StimStim7      -0.2619  -0.1434
StimStim8      -0.5888  -0.4624
StimStim9      -0.9202  -0.7991
StimStim10     -1.3155  -1.1908
attr(,"Probability")
[1] 0.95

$ST
      lower upper
[1,] 0.441 0.634
[2,] 0.574 0.960
attr(,"Probability")
[1] 0.95

$sigma
      lower upper
[1,] 0.117 0.134
attr(,"Probability")
[1] 0.95
```

HPDinterval returns a list of three components providing the intervals for the fixed-effects coefficients, the random effects and the residual variance.

In a mixed-effects model, there are multiple levels of variation, corresponding to the fixed effects and however many levels of random effects are included in the model. For example, in the second model, there are two levels of random effects: at the level of the observer and at the level of the stimulus within the observer. We obtain the fitted values from each model object using the `fitted` method. Each observation was repeated 4 times so there are 640 fitted values. The fixed effects coefficients estimated in the model are extracted from the model object with the `fixef` method.

```
> ( mfGab.fe2 <- fixef(mfGab.lmer2) )
(Intercept) StimStim2  StimStim3  StimStim4  StimStim5
    1.8210     0.1394     0.2422     0.2855     0.1710
  StimStim6 StimStim7  StimStim8  StimStim9 StimStim10
    0.0227    -0.2001    -0.5232    -0.8615    -1.2535
> mfGab.fe2[-1] <- mfGab.fe2[1] + mfGab.fe2[-1]
```

Since they are coded according to the treatment contrasts, we add the first coefficient to the subsequent coefficients to obtain the mean estimate at each spatial frequency.

The values of the the random coefficients, b in (2.9), evaluated at the parameter estimates are referred to either as the best linear unbiased predictors (BLUPS) or the conditional modes [11]. They are extracted from the model with the `ranef` method, which returns a list of data frames.

```
> str( mfGab.re2 <- ranef(mfGab.lmer2) )
List of 2
 $ Obs:Stim:'data.frame':   160 obs. of  1 variable:
  ..$ (Intercept): num [1:160] -0.02722 -0.00808
      0.02763 -..
 $ Obs      :'data.frame':   16 obs. of  1 variable:
  ..$ (Intercept): num [1:16] -0.0302 -0.0297 -0.0245
      -0.1..
 - attr(*, "class")= chr "ranef.mer"
```

The fixed effects coefficients represent the estimates for the population contrast sensitivity function whereas by combining the conditional modes with the fixed effects, we obtain predictions for the individuals in the sample.

```
> mfGab.pred2 <- mfGab.fe2 +
+    rep(mfGab.re2[[2]][, 1], each = length(mfGab.fe2)) +
+    ranef(mfGab.lmer2)[[1]][, 1]
```

The second term above corresponds to the observer specific effects and so its values are repeated for each spatial frequency, while the third term for the `Obs:Stim` effect is already of the correct length and in the correct order.

We can now compare predictions at different levels in a `lattice` plot with the mean contrast sensitivity values, calculated from the `mfGab` data frame (Fig. 2.4). We leave it as an exercise to make the same graph and comparisons for the simpler model, `mfGab.lmer`.

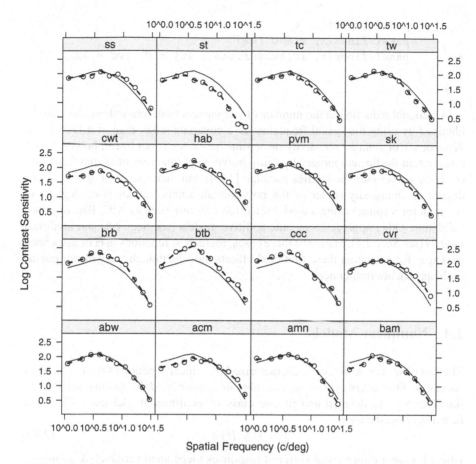

Fig. 2.4 Average contrast sensitivity from the ModelFest data set fitted with a mixed-effects model. The *solid lines* indicate the fixed effect prediction and the *dashed lines* have been adjusted for each observer by the estimates of the conditional modes

```
> Thr.mean <- t(with(mfGab, tapply(LContSens, list(Obs, Stim),
+   mean)))
> Thr.df <- stack(as.data.frame(Thr.mean))
> names(Thr.df) <- c("Mean", "Obs")
> Thr.df$SpatFreq <- SpatFreq
> Thr.df$pred2 <- mfGab.pred2
> print(
+ xyplot(Mean ~ SpatFreq | Obs, Thr.df, subscripts = TRUE,
+   id = Thr.df$pred2, scales = list(x = list(log = TRUE)),
+   xlab = "Spatial Frequency (c/deg)",
+   ylab = "Log Contrast Sensitivity",
+   panel = function(x, y, subscripts, id, ...){
```

```
+          panel.xyplot(x, y)
+          panel.lines(x, mfGab.fe2)
+          panel.lines(x, id[subscripts], lty = 2, lwd = 2)
+     } )
+ )
```

Additional reduction in the number of parameters to describe these data can be obtained by using the spatial frequency as a covariate rather than as a factor. The dependence of contrast sensitivity on spatial frequency is not linear, however. One could retain the linear approach by using polynomial functions of spatial frequency or spline curves (see the **splines** package [146]), but then one must decide on the degree of complexity (order of the polynomial, number of knots or degrees of freedom for a spline) using a model selection criterion such as AIC, BIC or cross-validation. Other approaches include additive models (see Sect. 2.5) and nonlinear models (see Sect. 2.4). In each of these cases, there exist functions in R or associated packages for extending these to mixed-effects models, though we will not pursue such approaches further here.

2.4   Nonlinear Models

The response cannot always be characterized as a linear function of the explanatory variables. One alternative is to consider *nonlinear models*. In this section we illustrate how to develop and fit one class of nonlinear model using R. These functions have the form

$$Y = f(\boldsymbol{X}; \beta) + \varepsilon \tag{2.10}$$

where Y is an n-dimensional vector of responses (dependent variable), $\boldsymbol{X}$ an n-by-p matrix of explanatory variables used in predicting the response, β is a p-dimensional vector of parameters, $f()$, a function and typically $\varepsilon \sim N(0, \sigma^2)$. When $f(\boldsymbol{X}; \beta) = \boldsymbol{X}\beta$, the nonlinear model simply reduces to the linear model. The R function nls ("*n*onlinear *l*east-*s*quares") determines estimates of parameters $\hat{\beta}$ that minimize the sum of the squared errors, $Y - f(\boldsymbol{X}; \hat{\beta})$, using by default a Gauss–Newton algorithm in an iterative search process. When $\varepsilon \sim N(0, \sigma^2)$, then the least-squares solution is also the maximum likelihood solution (Sect. B.4.2). R includes several functions that permit solving this type of problem, for example, the functions optim and nlm. Each of these requires the user to write a new function to calculate the sum of squared error for each new model considered. The function nls which we describe here is a convenient alternative that uses formula objects to specify the model to be fit.

2.4.1 Chromatic Sensitivity Across the Life Span

Knoblauch et al. [99] measured thresholds as a function of age for detecting equiluminant chromatic differences along three axes in the CIE xy chromaticity diagram. The ages in their sample ranged from 3 months to 86 years old, covering periods of the life span in which both developmental and aging factors would be expected to be operating. Pre-verbal participants were evaluated with a preferential-looking technique [170]. The data set Chromatic in the **MPDiR** package is obtained by

```
> data(Chromatic)
```

and contains four components: Log2Age, the base 2 logarithm of the age of each observer, Age, the age in years, Thresh, the threshold modulation for detection of a chromatic stimulus and Axis, a 3-level factor indicating along which of the three axes in the color space the observation was measured. The axes names—Protan, Deutan and Tritan—correspond to the color confusion axes for each of the three types of congenital dichromatic observer and, in theory, permit testing the sensitivity of each of the three classes of cone photoreceptor in the normal eye.

Figure 2.5 shows scatterplots of the thresholds as a function of age for each of the axes tested and was produced using the **ggplot2** package with the following code fragment.

```
> library(ggplot2)
> qplot(Age, Thresh, data = Chroscamatic,
+       facets = .~ Axis, col = "black",
+       geom = c("point"),
+       xlab = "Age (years)", ylab = "Chromatic
                Threshold") +
+       scale_x_continuous(trans = "log2",
+            limits = 2^c(-3.5, 7.5),
+            breaks = 2^seq(-2, 6, 2),
+            labels =  2^seq(-2, 6, 2) ) +
+       scale_y_log10(limits = 10^c(-3.5, -1.3),
+          breaks = c(0.001, 0.002, 0.005, 0.01, 0.02,
                0.05),
+     labels = c(0.001, 0.002, 0.005, 0.01, 0.02, 0.05))+
+       geom_smooth(colour = "black")
```

Both the ordinates and abscissas are logarithmically spaced as in the original publication (log base 10 on the ordinate and base 2 on the abscissa). A smooth loess curve has been added to each plot in black. This initial view of the data suggests that, in these coordinates, there are two trends in the data, an initial roughly linear decrease in threshold with age (the developmental trend), followed by a loss in sensitivity later in life (the aging trend). The graphs also suggest that the same functional form might adequately describe these trends along each axis, though the thresholds are higher along the tritan axis.

The authors fit these data with the following function,

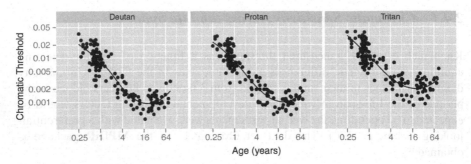

Fig. 2.5 Thresholds for detection of chromatic modulation as a function of age along three axes in color space. The *fitted curve* is obtained by local regression (loess)

$$T = aA^{-\alpha} + bA^{\alpha}, \tag{2.11}$$

where T is the threshold estimate, A the age, and a, b and α are parameters to be estimated. On log-log coordinates, the magnitude of the slopes of the two segments of the curve will approach α.

The nls function in the **stats** package [146] in R facilitates estimation of the free parameters in (2.11). The parameters are chosen to minimize the sum of the squared differences between data and predictions, a least squares criterion.

To use the function nls we need to specify the model as a formula object, just as we did for lm. The model (2.11) can be fit to the data by the following code:

```
> Chrom.nls <- nls(Thresh ~
+        a[Axis] * Age^-alph + b[Axis] * Age^alph,
+        data = Chromatic,
+        start = list(a = c(8, 8, 11) * 10^-3,
+                b = c(3, 3, 8) * 10^-5, alph = 1)
+                )
```

Three arguments are required. First, the model is specified as a formula. Unlike in the lm formulae, the asterisk, "*," here corresponds to multiplication and the circumflex, "^" to exponentiation. Because the terms a and b are subscripted by the factor Axis, different values of each will be estimated for each level of Axis. In this case, if an intercept term is desired, it must be included explicitly. In general, the interpretation of a formula object depends on the function that calls it, as we see here and have already seen with lm and xyplot. The "data" argument, as always, provides the frame for interpreting the terms in the formula object. The argument "start" is a list indicating initial values for the parameters. Because of the subscripting of the linear coefficients in the model, there are seven parameters in all for which to provide initial values and to estimate (three values of a, one for each level of Axis, three for b and one for alph).

The function nls returns an object of class "nls" containing information about the non-linear, least-squares fit. Over a dozen methods are defined for probing

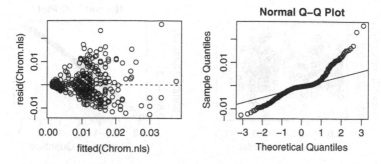

Fig. 2.6 Two diagnostic plots of the residuals obtained from fitting model (2.11) to the Chromatic data set with errors on a linear scale

objects of this class. The summary method prints out a table of the results in a format similar to that from a linear model.

```
> summary(Chrom.nls)
Formula: Thresh ~ a[Axis] * Age^-alph + b[Axis] *
          Age^alph

Parameters:
     Estimate Std. Error t value Pr(>|t|)
a1   8.36e-03   3.91e-04   21.40  <2e-16 ***
a2   8.88e-03   3.96e-04   22.43  <2e-16 ***
a3   1.21e-02   4.39e-04   27.65  <2e-16 ***
b1   3.42e-05   3.86e-05    0.89   0.376
b2   3.39e-05   3.86e-05    0.88   0.380
b3   8.61e-05   3.90e-05    2.21   0.028 *
alph 8.15e-01   4.14e-02   19.72  <2e-16 ***
---
Signif. codes:  0 `***' 0.001 `**' 0.01 `*' 0.05 `.'
                0.1 ` ' 1

Residual standard error: 0.00486 on 504 degrees of
    freedom

Number of iterations to convergence: 4
Achieved convergence tolerance: 4.44e-07
```

The diagnostic residuals vs. fitted values and Q–Q plots are displayed in Fig. 2.6 and were obtained with the following code fragment:

```
> par(mfrow = c(1, 2))
> plot(fitted(Chrom.nls), resid(Chrom.nls))
> abline(0, 0, lty = 2)
```

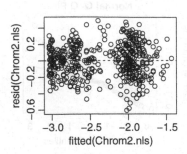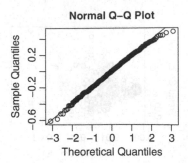

Fig. 2.7 Diagnostic plots of fit to `Chromatic` data set with the errors evaluated on a logarithmic scale

```
> qqnorm(resid(Chrom.nls))
> qqline(resid(Chrom.nls))
```

These plots reveal problems in the assumptions of the model. The left residuals plot displays heterogeneity of the residuals as a function of the fitted values. The Q–Q plot is sigmoidal, not linear. This pattern is consistent with a distribution that has too many extreme values relative to the Gaussian. The pattern of residuals leads us to minimize the logarithm of the errors rather than the errors. This change is easily implemented by taking logarithms of both sides of the formula object.

```
> Chrom2.nls <- update(Chrom.nls, log10(.) ~ log10(.))
```

The diagnostic plots for the new fit in Fig. 2.7 now appear reasonable.

The model includes the assumption that the same exponent, α can be used on each axis. We can consider a more complex model where we assign a different exponent for each axis. To do so, we add the subscript `Axis` to the term `alph`, and a model in which the exponent varies with chromatic axis is obtained. The single exponent model is nested in the multi-exponent model, and so we can readily test whether the added complexity of the multi-exponent model is needed to fit the data. The `anova` method takes a series of "nls" objects as arguments that correspond to nested models and performs a likelihood ratio test between each pair of models in the series. We first fit the more complex model. Notice that we had to update the start parameter list (the third argument) as well, since the new model has two more parameters.

```
> Chrom3.nls <- update(Chrom2.nls, . ~
+   log10(a[Axis] * Age^-alph[Axis] +
+   b[Axis] * Age^alph[Axis]),
+   start = list(a = c(8, 8, 11) * 10^-3,
+   b = c(3, 3, 8) * 10^-5, alph = c(1, 1, 1)))
> anova(Chrom2.nls, Chrom3.nls)
Analysis of Variance Table
```

```
Model 1: log10(Thresh) ~ log10(a[Axis] * Age^-alph +
   b[Axis] * Age^alph)
Model 2: log10(Thresh) ~ log10(a[Axis] *
  Age^-alph[Axis] +
    b[Axis] * Age^alph[Axis])
  Res.Df Res.Sum Sq  Df Sum Sq F value Pr(>F)
1    504      19.10
2    502      18.97   2  0.13    1.78   0.17
```

The results indicate that the simpler model with one exponent describes the data adequately. This conclusion is in agreement with that of the original article based on evaluation of the confidence intervals of the parameters.

The **MASS** package provides a confint method to compute confidence intervals for "nls" objects [178].

```
> library(MASS)
> confint(Chrom2.nls)
             2.5%    97.5%
a1     6.78e-03 7.94e-03
a2     7.26e-03 8.50e-03
a3     9.84e-03 1.16e-02
b1     2.29e-05 3.63e-05
b2     2.31e-05 3.68e-05
b3     5.66e-05 8.71e-05
alph   8.51e-01 9.40e-01
```

The reasonableness of the linear approximation involved can be evaluated by plotting the output of the profile method (left as an exercise). If the linear approximation is found to be inappropriate, bootstrap confidence intervals [55] can be obtained instead using the package boot [25].

The estimated parameters and fits to the data of the model Chrom2.nls shown in Fig. 2.8 match those reported by the authors.

2.5 Additive Models

Nonlinear models are particularly useful when the equation is motivated by an underlying theory and the estimated parameters can be given a meaningful interpretation, either with respect to the data or within the framework of the model. In the absence of theory, finding the best model to describe the data depends in part on the ability of the modeler to conjure up a "good" equation, i.e., one that fits the data well with a minimum number of parameters.

Additive models (AM) provide a flexible, nonparametric approach to describing data that vary nonlinearly as a function of a covariate in a manner that remains close to the framework of linear models [77, 198]. The additive model is of the form

$$Y = f_1(X_1) + f_2(X_2) + \cdots + \varepsilon, \tag{2.12}$$

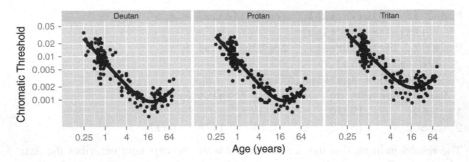

Fig. 2.8 Chromatic modulation thresholds as a function of age along three color axes. The *black curves* were obtained with the fitted values of model `Chrom2.nls` in which the exponent is the same in all color directions

where Y is the response, the f_i are smooth functions of the covariates, X_i and $\varepsilon \sim N(0, \sigma^2)$. The smooth functions are typically represented in terms of linear combinations of smooth basis functions, for example, splines, and, thus, can be added as columns to the model matrix. With a greater number of basis terms, more complex variations of the dependent variable can be described up to the limit of having the predicted curve pass through every point. Such a curve that describes the trend as well as the noise in the data will, in general, be poor at describing new samples from the same population. To control for this, a penalty for undersmoothing is introduced during the optimization. The penalty is usually implemented as a proportion, λ of the integrated square of the second derivative of f (related to its curvature). Larger contributions of the term result in less smooth models. The choice of degree of penalization (or smoothness) is controlled by minimizing a criterion related to prediction error (i.e., fitting some of the data and calculating the error on the remaining portion) called generalized cross validation (GCV).

2.5.1   Chromatic Sensitivity, Again

Additive models can be fit in R using the function `gam` in the **mgcv** package, one of the recommended packages that comes with R by default [198]. It works similarly to the other modeling functions that we have seen, requiring that the model be specified by a formula object. The novelty for `gam` is that smooth terms of a covariate are specified by including them as arguments to a function `s`. For example, the following code loads the package and fits an AM to the `Chromatic` data set and then prints out a summary of the fit.

```
> library(mgcv)
> Chrom.gam <- gam(log10(Thresh) ~ s(Age, by = Axis)
      + Axis,
```

```
+    data = Chromatic)
> summary(Chrom.gam)
Family: gaussian
Link function: identity
Formula:
log10(Thresh) ~ s(Age, by = Axis) + Axis

Parametric coefficients:
            Estimate Std. Error t value Pr(>|t|)
(Intercept)  -2.3830     0.0158 -150.50   <2e-16 ***
AxisProtan    0.0264     0.0224    1.18     0.24
AxisTritan    0.2040     0.0226    9.04   <2e-16 ***
---
Signif. codes:  0 `***' 0.001 `**' 0.01 `*' 0.05 `.'
   0.1 ` ' 1

Approximate significance of smooth terms:
                   edf Ref.df    F p-value
s(Age):AxisDeutan 7.75  8.59 94.5  <2e-16 ***
s(Age):AxisProtan 8.21  8.83 94.4  <2e-16 ***
s(Age):AxisTritan 7.80  8.63 63.9  <2e-16 ***
---
Signif. codes:  0 `***' 0.001 `**' 0.01 `*' 0.05 `.'
   0.1 ` ' 1

R-sq.(adj) =  0.817   Deviance explained = 82.6%
GCV score = 0.045504  Scale est. = 0.043121  n = 511
```

A smooth function of the covariate Age has been specified as an explanatory variable. The argument by = Axis indicates that a separate smooth function should be fit to the data for each level of the factor. Here, we include an additive contribution of the factor Axis for identifiability, since the smooth terms will be centered around 0.

Examining the parametric coefficients, the summary method indeed indicates a difference in the vertical height along the "Tritan" but not the "Protan" axis. The new aspect of the display is the inclusion of the results for the smooth terms. The cross-validation selection of the degree of smoothing produces a non-integer estimate of the degrees of freedom, the estimated (or effective) degrees of freedom (edf) for each smooth term.

We test the above model against the simpler model that the variation in chromatic sensitivity across the life span follows the same course along all three chromatic axes by fitting a single smooth for all three axes. The update method facilitates fitting the new model.

```
> Chrom2.gam <- update(Chrom.gam, . ~ s(Age) + Axis)
> anova(Chrom2.gam, Chrom.gam, test = "F")
```

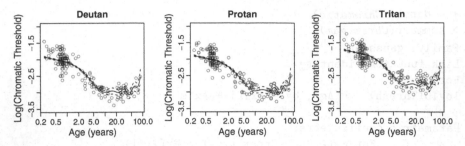

Fig. 2.9 Fits of the smooth function to the Chromatic data set for model Chrom2.gam, obtained using gam

```
Analysis of Deviance Table

Model 1: log10(Thresh) ~ s(Age) + Axis
Model 2: log10(Thresh) ~ s(Age, by = Axis) + Axis
  Resid. Df Resid. Dev  Df Deviance    F Pr(>F)
1       499      21.6
2       484      20.9 15.1   0.714  1.1   0.36
```

An approximate *p*-value from a likelihood test comparing the two models (see ?anova.gam and [198]) is obtained using the anova method and provides supporting evidence for a single curve describing the evolution of sensitivity with age. Unlike in the case of the nonlinear models, this inference does not depend on having chosen the correct equation to describe the data.

A plot of the simpler model with the data is shown in Fig. 2.9 for each of the three axes, using functions from the base **graphics** package. The format of the graphical output is different from that of "lm" objects. The plot method for "gam" objects generates a graph of the estimated smooth curve with standard errors of the estimated smooth indicated by dashed lines.[3] The graphs are output with a for loop. We use the estimated coefficients for the factor Axis to adjust the heights of the single smooth to the data along each axis. The points were plotted using transparency (the fourth parameter of the rgb function) to allow the curve to be visualized better behind the crowd of points. The fitted smooth function differs from the parametric curve used with nls at the youngest ages, providing evidence against (2.11) in this age range, though the oscillatory structure of the curve at the older ages suggests that additional smoothing might be appropriate. Such smoothing could conceivably lessen the differences between the two fits.

```
> opar <- par(mfrow = c(1, 3))
> cc <- coef(Chrom2.gam)[1:3] + c(0, coef(Chrom2.gam)[1],
```

---

[3]Use the gam.check function on the model object to visualize diagnostic plots that, in this case, raise no flags on the adequacy of the fit. We leave this verification as an exercise for the reader.

```
+          coef(Chrom2.gam)[1])
> names(cc) <- c("Deutan", "Protan", "Tritan")
> for (Ax in names(cc)) {
+   plot(Chrom2.gam, rug = FALSE,
+       shift = cc[Ax],
+       ylim = c(-3.5, -1) - cc[Ax],
+       main = Ax, log = "x",
+       xlab = "Age (years)",
+       ylab = "Log(Chromatic Threshold)")
+   points(log10(Thresh) ~ Age, Chromatic,
+       subset = Axis == Ax,
+       col = rgb(0, 0, 0, 0.5))
+ }
> par(opar)
```

2.6   Generalized Linear Models

GLM [127] extend linear models in two exceedingly useful ways. First, the
dependent variable is not necessarily assumed to be distributed as a Gaussian
random variable. The choice of possible distributional families now includes the
Gaussian but also the Bernoulli, binomial, Poisson and many others that are
members of what is called the *exponential family* (Sect. B.3.5). We can use the
GLM to model data whose dependent variable has any of these distributions. The
Bernoulli distribution (See Sect. B.2.2) is a particularly important case for us since
we will often use it to model binary responses ("Yes-No," etc.) in psychophysical
experiments. Warning: The term "exponential family" is potentially confusing.
The members of the exponential are themselves distributional families. Moreover,
the exponential distributional family (Sect. B.3.4) is just another member of the
exponential family; it has no special status.

Second, the response is modeled as a nonlinear transformation of a weighted
linear combination of the explanatory variables. The nonlinear transformation is
referred to as the *link function*. This sort of model is remarkably common in the
psychophysical literature.

Familiar linear models (such as ANOVA) can be used to develop analogous GLM
models (e.g. "GLM ANOVA") that use a very similar terminology in describing
them while accommodating nonlinear transformations of the response variable.

The linear predictor takes the form

$$\eta = \boldsymbol{X}\beta \tag{2.13}$$

where as with the ordinary linear model the design matrix $\boldsymbol{X}$ encapsulates all of
the information about the explanatory variables and β is a vector of unknown

parameters to be estimated. The relation of the linear predictor to the expected value of the response (or the mean function) is written as

$$g(\mathsf{E}[Y]) = \boldsymbol{X}\beta, \tag{2.14}$$

where g is the link function. The link function is chosen to be invertible.

Thus, in the Bernoulli case, we would have

$$g(\mathsf{E}[P(Y = 1)]) = \log\left(\frac{p}{1-p}\right) = \boldsymbol{X}\beta. \tag{2.15}$$

Note that the link function here is the logit transformation but as we will see, other functions can also be used. Inverting g, yields

$$\mathsf{E}[P(Y = 1)] = g^{-1}(\boldsymbol{X}\beta), \tag{2.16}$$

where we can think of the function $g^{-1}() = \psi()$ as a *psychometric function* that transforms an intensity variable into a probability of choosing a particular response in a binary task.

To summarize, a GLM is characterized by three components:

1. The responses Y_i are independently and identically distributed and sampled from the same member of the exponential family
2. The model contains a set of explanatory variables represented in a matrix $\boldsymbol{X}$ and coefficients in a vector β that together form a linear predictor
3. The linear predictor is related to the expected value of the μ by an invertible link function, g

2.6.1 Chromatic Sensitivity, Yet Again

The response variable for the previously analyzed Chromatic data set can also be modeled using GLM as shown next. Recall that fitting raw thresholds resulted in a distribution of residuals that violated the Gaussian assumption (Fig. 2.6). In Sect 2.4.1, we employed a log transform to stabilize the variance. The log transform requires the data values to be positive and, in fact, the diagnostic plots for the model fit to the transformed data in Fig. 2.7 are consistent with a log normal distribution of the raw threshold values.

It is plausible to consider modeling the raw data with other distributions defined only on positive values and that show a similar dependence of variance on the mean. For example, an interesting candidate within the family of distributions amenable to the GLM is the Gamma distribution. For a Gamma distributed variable, the standard deviation increases linearly with the mean, a pattern that could produce the fan-shaped pattern of residuals in Fig. 2.6. An advantage of the GLM approach is that we can analyze the data on their original scale.

In order to apply the GLM, we must be able to specify the model in the form of a linear predictor. We note that the exponent, α, estimated for (2.11) is close to unity. If we set it to this value, the model reduces to a 2-term polynomial that can be represented as a linear predictor.

```
> Chrom.glm <- glm(Thresh ~ Axis:(I(Age^-1) + Age),
+    family = Gamma(link = "identity"), data = Chromatic)
```

Recall that exponents in the formula object language are used to generate interaction terms in linear models. To protect against this interpretation and force the exponent to be interpreted as an arithmetic operation, it must be enclosed by the function I. We model the thresholds as an interaction with the sum of the age covariates so that there will be coefficients for the developmental and aging sections of the data and the possibility of different coefficients for each axis. The Gamma family is specified with an identity link so that the untransformed thresholds are analyzed.

As before, we can compare this model to one in which the coefficients are forced to be the same along all three axes though the heights of the curves can depend on the axis.

```
> Chrom2.glm <- update(Chrom.glm, . ~ Axis + (I(Age^-1)
+   + Age))
> anova(Chrom2.glm, Chrom.glm, test = "F")
Analysis of Deviance Table

Model 1: Thresh ~ Axis + I(Age^-1) + Age
Model 2: Thresh ~ Axis:(I(Age^-1) + Age)
  Resid. Df Resid. Dev Df Deviance     F Pr(>F)
1      506        101
2      504        100  2    0.506  1.23   0.29
```

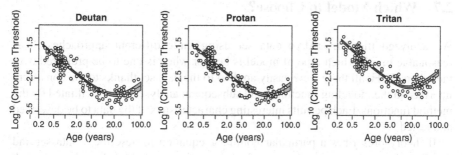

Fig. 2.10 GLM fits to the Chromatic data set for model Chrom2.glm

```
> opar <- par(mfrow = c(1, 3))
> xx <- 10^seq(-0.6, 2, len = 200)
> for (Ax in levels(Chromatic$Axis)){
+    plot(log10(Thresh) ~ Age, Chromatic,
+        subset = Axis == Ax,
+        xlab = "Age (years)",
+        ylab = expression(paste(plain(Log)[10], " (Chromatic
+            Threshold)")),
+        main = Ax, log = "x",
+        ylim = c(-3.5, -1))
+    Chrom.pred <- predict(Chrom.glm,
+        newdata = data.frame(Age = xx, Axis = Ax),
+        se.fit = TRUE)
+    polygon(c(xx, rev(xx)), with(Chrom.pred, log10(c(fit + 1.96
+        * se.fit, rev(fit - 1.96 * se.fit)))), col = "grey")
+    lines(xx, log10(Chrom.pred$fit))
+    points(log10(Thresh) ~ Age, Chromatic, subset = Axis == Ax,
+        pch = 21, bg = "white")
+ }
> par(opar)
```

Finally, we plot the predicted curves from the model Chrom2.glm with the data in Fig. 2.10. Even though the thresholds were fit in raw units, we scale the axes logarithmically to compare the data more easily with the previous plots and because it spreads out the data improving the visual display. By specifying the argument se.fit = TRUE in the predict method, standard error estimates are returned as well as the predicted values. We use these to plot a 95% confidence interval on the fit as a grey envelope around the predicted curve. The diagnostic plots should be examined systematically during the fitting process. In this case, they look reasonable. We leave verification of this as an exercise for the reader.

2.7 Which Model to Choose?

We analyzed the Chromatic data set using three different approaches. It is reasonable to ask which type of model is best or when is one to be preferred over the others. In R, all three were easily applied to the data set thanks to the consistent interface to the modeling functions, and subsequent analyses were facilitated by the method functions. Each permits interesting characteristics of the data to be described and tested.

If theory prescribes a particular nonlinear equation to describe a data set and the Gaussian assumption on the residual errors is reasonable, then the approach using nonlinear least squares with the nls function is recommended. If no obvious functional form is available for the data or the exact functional form is unimportant, then the use of gam is worth considering. In the event of doubt about the best choice of functional form to describe a data set, the possibility of drawing conclusions about similarities across data sets nonparametrically, as demonstrated above, certainly provides a powerful option. If the model can be expressed as a linear predictor, then lm or glm should certainly be considered. The choice of link function for the glm will allow certain nonlinear transformations of the dependent variable to be handled. In addition, the choice of exponential family member affords some flexibility in accounting for the distribution and variance of the dependent variable.

An advantage of glm is that the predicted responses of the mean of the response variable are on the untransformed scale, whereas if we were to fit a linear model to the transformed responses (e.g., to the logit transformation in (2.15)), then we would be estimating the mean logit value rather than the mean probability. We demonstrate this difference with a simple example. We generate five random numbers from a binomial distribution using rbinom. We assume that each is the number of successes observed out of 30 trials and we would like to estimate the probability of a success, assuming that it is constant across samples. For the example, we set the underlying value of p to 0.8. We use glm with a binomial family and the default logit link to estimate a model with only an intercept term and compare the estimated coefficient on the response scale with the mean proportion of successes of the sample and the mean of the logits.

```
> NumSuccess <- rbinom(5, 30, 0.8)
> pSuccess <- NumSuccess/30
> resp.mat <- cbind(NumSuccess, 30 - NumSuccess)
> pSuccess.glm <- glm(resp.mat ~ 1, binomial)
> # glm estimate
> plogis(coef( pSuccess.glm ))
(Intercept)
      0.787
> # mean proportion Success
> mean(pSuccess)
[1] 0.787
> # mean of logits on response scale
> plogis(mean(qlogis( pSuccess )))
[1] 0.801
```

The model returns the estimated coefficient on the linear predictor scale so we transform it to the response scale with the plogis function.

2.8 Exercises

2.1. Create two factors A and B, A having 3 levels as in Sect. 2.2 and B with only 2. Use expand.grid to make a data frame with two columns representing a factorial crossing of the levels of the two factors. Then, with model.matrix generate the design matrices for formula objects representing (a) a model containing only additive effects of the two factors and (b) also, including their interaction. Study the design matrices so that you understand them with respect to the levels of each factor and terms in the model. Repeat this exercise with the intercept removed from the formula object.

2.2. The design matrix indicates the contributions of the model terms to an estimated coefficient but it does not indicate the contrasts in a direct fashion. To visualize the contrasts, one may use the `ginv` function from the **MASS** package on the design matrix. The contrasts will be in the rows of the resulting matrix. Examine the contrasts for a model with (a) just factor A, (b) only additive effects of factors A and B, (c) additive effects of the two factors and their interaction. Using the `fractions` function from the same package will aid in the interpretation of the contrasts. Using the contrasts, interpret the relation of each estimated coefficient to the terms in the model.

2.3. Re-fit the model (2.8) for the `Motion` data set but rescale the values of `LnAge` to center them with respect to their mean, using the function `scale`. How does rescaling affect the estimated coefficients, their precision and their interpretation?

2.4. In the `Motion` data set, how would you test the significance of the 3-way interaction `LnAge:Mtype:Sex`?

2.5. Random effects can enter a model as covariates as well as factors. A random covariate x would be specified with respect to a grouping factor f as a term (x + 0 | f). Recall that the intercept term is implicit in a model formula so that it must be suppressed by adding a 0 or subtracting a 1 from the formula or else a random effect will be estimated for the intercept as well.

Refit the Gabor functions of the `ModelFest` data using a polynomial series of the covariate `log10(SpatFreq)` instead of the factor `Stim` as explanatory variables. Remember to isolate the terms with powers using the `I()` function. Evaluate how many terms of the polynomial are necessary to describe the average data and which terms need also be included as random effects to describe the individual contrast sensitivity curves.

2.6. Fit the `Chromatic` data set with `nls` using a different exponent on the developmental and aging segments (descending and ascending branches, respectively) and test whether the added parameters yield a significantly better fit.

2.7. In this chapter, the `Chromatic` data set was fit with nonlinear, generalized linear and additive models. Make a graph with the data for one chromatic axis and the predicted curves from each of the three models so as to compare visually the three fits.

2.8. The data set `CorticalCells` in the **MPDiR** package contains the mean responses in action potentials/second of six neurons recorded in visual cortical areas V1 and V2 of the macaque brain in response to increasing levels of stimulus contrast reported in a paper by Peirce [140]. Plot the contrast response data for each cell using either **lattice** or **ggplot2**. A typical model for the contrast response function of cortical cells is given by the Naka-Rushton equation,

$$R(c) = R_m \frac{c^n}{c^n + \sigma^n}, \tag{2.17}$$

a generalization of the Michaelis-Mention function for which $n = 1$. What is the interpretation of the parameters in this model? Fit this model to each data set using nls. Then, re-fit the models using the reciprocal of the SEM column of the data frame with the weights argument. How does this change the fits? Compare the fit to (2.17) and the same model with the exponent, n, constrained to equal 1, using a likelihood ratio test.

2.9. Peirce [140] considers a more general model that allows for the fall-off from saturation shown by some cells in this data set. His model is

$$R(c) = R_m \frac{c^{n_1}}{c^{n_2} + \sigma^{n_2}}. \tag{2.18}$$

Fit this model to the data sets and then for each data set perform a likelihood ratio test between the fits of the models (2.17) and (2.18). Repeat this exercise using the weights. Replot (or update) the plots from the beginning of this exercise to include predicted curves from each model.

2.10. Watson [182] models the average contrast sensitivity data from the ModelFest data set (the first ten stimuli) with a difference of hyperbolic secants using four parameters

$$CS(f; f_0, f_1, a, p) = \mathrm{sech}\left((f/f_0)^p\right) - a\,\mathrm{sech}\left(f/f_1\right), \tag{2.19}$$

where f is spatial frequency and

$$\mathrm{sech}(x) = \frac{2}{e^x + e^{-x}}. \tag{2.20}$$

Calculate the average of the log contrast sensitivities for the 16 observers in this data set and fit this model to the data. Refit the model using the reciprocal of the inter-observer standard deviations as weights (see the "weights" argument of nls). Refit the data using an additive model with a smooth function of spatial frequency.

2.11. The GLM can also be used for analyzing count data that would traditionally be modeled with a log-linear model [14]. Typically, one specifies a Poisson family and a log link. This is possible because the multinomial, on which log-linear models are based, and the Poisson distributions have the same sufficient statistics.

The Grue data set in the **MPDiR** package is taken from a study by Lindsey and Brown [113] in which they categorized the frequency of languages with respect to how they referred to the words "green" and "blue" (separate words, a combined word (so called grue languages) or the word " dark" for the two) as a function of environmental ultra-violet light exposure, UV_B. The data is in the format of a contingency table, wide format. Modify the data set so that it is data frame with three columns indicating the counts, the UV-B exposure and the type of language, the long format (see the `stack` function). Then, fit the data using `glm` and test the hypothesis that UV-B exposure is independent of language type.

Chapter 3
Signal Detection Theory

3.1 Introduction

Signal detection theory (SDT), developed in the 1950s, is a framework of statistical methods used to model how observers classify sensory events [72, 169]. In this chapter we describe commonly used signal detection models and methods for fitting them. A recurring theme in this book is the use of the generalized linear model (GLM) to fit psychophysical data, and here we describe the use of the GLM to fit signal detection data.

3.2 Sensory and Decision Spaces

The signal detection experiment is a classification task. On each trial the observer is presented with a stimulus and must classify it. The list of possible stimuli is the *sensory* or *stimulus* space and the list of possible classification responses is the *decision* or *response* space. For example, the sensory space could designate a large number of objects that are either vertical or horizontal and the decision space could contain the responses "vertical" and "horizontal." For each stimulus in the sensory space there is a correct classification known to the experimenter. Each of the objects in the vertical–horizontal stimulus space is designated as either "vertical" or "horizontal." In SDT, we seek to measure the observer's ability to classify correctly.

SDT can be applied to experiments where there are more than two possible stimuli and more than two possible responses [120, 189], but we will focus here on the important special case where there are only two of each.

One very important example of a signal detection task with two possible responses is the *Yes–No task*. On each trial, a stimulus is either presented or not presented at the whim of the experimenter. For example, one stimulus might be a very simple spatial pattern presented at low contrast against a homogeneous background and the second stimulus would be just the background alone. The observer's task is

K. Knoblauch and L.T. Maloney, *Modeling Psychophysical Data in R*, Use R! 32,
DOI 10.1007/978-1-4614-4475-6_3,
© Springer Science+Business Media New York 2012

to respond "Yes" (the stimulus was present) or "No" (it was not). The observer may know the probability that the experimenter will present the stimulus on each trial, the *prior probability* denoted π. The presence or absence of the stimulus on a given trial is independent of its presence or absence on other trials and independent of the observer's response on other trials.

There are four possible outcomes to a trial in a Yes–No task, depending on whether the signal is present or not and whether the observer responds "Yes" or "No." These are summarized together with the traditional terms used to describe them in the following table along with their abbreviations in parentheses: Intuitively,

	Signal	Not signal
Yes	Hit (H)	False alarm (FA)
No	Miss (M)	Correct rejection (CR)

H and CR are "desirable" outcomes and M and FA' are not. The full theory of signal detection allows us to assign rewards and punishments to each of the four outcomes [72, 120, 189]. In this chapter we will just consider models of observers who attempt to maximize the probability of a correct response on each trial, in effect assigning values of 1 to the correct classifications H and CR and a value of 0 to the incorrect classifications M and FA.

On any trial where the signal is presented, the probabilities of a hit (p_H) and of a miss (p_M) must sum to 1 (Why?). Similarly, the probabilities of a correct rejection (p_{CR}) and of a false alarm (p_{FA}) must also sum to 1. If we assume that the observer is not changing his behavior over the course of the experiment, these probabilities are constant and, to simulate an SDT experiment, we only need to know the two probabilities p_H and p_{FA}. It might seem more natural to specify the pair p_H and p_{CR}, the probabilities of the two correct responses, but the pair p_H and p_{FA}, the two probabilities associated with the response "Yes," is traditional. Given either pair, we can compute all four probabilities.

Suppose, for example, that the probability of a Hit p_H is 0.9 and that of a False Alarm p_{FA} is 0.1. We simulate a Yes–No experiment with 300 trials divided evenly between signal present and absent using the `rbinom` function as follows:

```
> N <- 150
> pH <- 0.9
> pFA <- 0.1
> H <- rbinom(1, N, pH)
> M <- N - H
> FA <- rbinom(1, N, pFA)
> CR <- N - FA
> (SDT_Resp.tab <-  as.table(matrix(c(H, M, FA, CR),
    nc = 2,
+     dimnames = list(Resp = c("Yes", "No"),
+     Stim = c("Signal", "Not Signal")))) )
```

```
        Stim
Resp  Signal  Not Signal
  Yes    136          15
  No      14         135
```

Note that we can reconstruct the number of signal trials as $n_S = 150 = 136 + 14$ and the number of non-signal trials as $n_N = 150 = 15 + 135$. The larger values in the H and CR cells compared to the M and FA cells, respectively, indicate that the observer is detecting the signal above chance. We can also compute the *prior probability* of a signal $\pi = n_S/(n_S + n_N)$. There is no requirement that $n_S = n_N$ but they often are chosen to be equal.

The SDT model assumes that, on each trial, the observer's information about the stimulus can be modeled as a random variable X, which is drawn from the *signal distribution* $f_S(x)$ on trials on which the "signal" is presented and, otherwise, it is drawn from the *noise distribution* $f_N(x)$. In the theory of SDT, there are no restrictions on how we choose the signal and noise distributions. In practice, though, they are typically restricted to particular distributional families, discrete (Sect. B.2.1) or continuous (Sect. B.2.4), most typically the Gaussian (Sect. B.2.5).

3.2.1 Optimal Detection

Now suppose that the observer is given the value of the random variable X, drawn from either the signal or noise distribution. How should he decide whether to respond "Yes" or "No?" The decision rule that maximizes the probability of the correct response is based on the *likelihood ratio*

$$\Lambda(X) = \frac{f_S(X)}{f_N(X)} \tag{3.1}$$

where X is the sensory information [72, 120, 189]. The likelihood ratio is just the ratio of the probability density of the signal and noise distributions at the point X. Intuitively, the greater this ratio is, the more likely it is that X came from the signal distribution. The decision rule we employ to classify X is to fix a *criterion* β and respond "Yes" when $\Lambda(X) \geq \beta$ and otherwise "No." Of course, we still need to select the criterion β to fully specify the rule. The value that maximizes the probability of a correct response is just $\beta = (1 - \pi)/\pi$.

We can express any probability p in odds form by the transformation $o(p) = p/(1 - p)$. This transformation is an increasing function that maps the interval $(0, 1)$ to the interval $(0, \infty)$ and is invertible as $p = o/(1 + o)$. Consequently, the odds ratio is an alternative notation for probability often written as $a : b$, interpreted as

$p = a/(a+b)$. For example, $1:1$ odds correspond to $p = 1/2$, $1:2$ odds to $p = 1/3$, $2:1$ odds to $2/3$.[1] For any probability p we can define the odds $p/(1-p)$ and the criterion we have selected is the reciprocal of the *prior odds*.

If π is small, β will be large, an intuitively satisfying outcome: if the occurrence of the signal is improbable, we set a high criterion and the observer is less inclined to say "Yes." Correspondingly, if π is small, we set a lower value of criterion. If π is 1 (the signal is present on all trials) then $\beta = 0$ and, as $\Lambda(X) \geq 0$, the observer always responds "Yes." We will sometimes refer to β as the *likelihood criterion* to avoid confusion with a second criterion, the *sensory criterion* which we introduce in the next section.

3.3   Case 1: Equal-Variance Gaussian SDT

The most commonly used model of signal detection in psychophysics is the *equal-variance Gaussian (or normal) SDT*. In this model, the signal distribution is normally distributed with mean $d' > 0$ and variance $\sigma^2 = 1$

$$f_S(x) = \frac{1}{\sqrt{2\pi}} e^{-\frac{(x-d')^2}{2}} \tag{3.2}$$

while the noise distribution is a second normal with mean 0 and variance $\sigma^2 = 1$

$$f_N(x) = \frac{1}{\sqrt{2\pi}} e^{-\frac{x^2}{2}} \tag{3.3}$$

The signal and noise distributions for the equal-variance signal detection model are shown in Fig. 3.1a (the code to plot the figure follows the figure). The only difference between X on noise and signal trials is that the constant d' has been added to the random variable on signal trials. For this reason, the signal distribution is often referred to as the *signal+noise distribution* and d' is thought of as the additive change in sensory state (or response) due to the presence of the signal.

We can derive a second form of the optimal decision rule that will prove useful. The logarithm is an increasing function and we can apply it to both sides of the optimal rule with only the mild restriction that $\beta \geq 0$. The left-hand side becomes the *log likelihood ratio*

$$\lambda(X) = \log \Lambda(X) = \log \frac{f_S(X)}{f_N(X)} \tag{3.4}$$

---

[1]The log of the odds ratio is the default link function, the "logit," used with a binomial GLM.

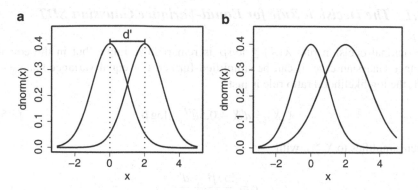

Fig. 3.1 (a) Equal-variance Gaussian distributions separated by a distance of d'. (b) Gaussian distributions of unequal variance with the same separation as in (a). The standard deviation of the signal distribution is 1.4 times greater than that of the noise distribution

```
> opar <- par(mfrow = c(1, 2), lwd = 3)
> x <- seq(-3, 5, len = 300)
> plot(x, dnorm(x), type = "l", ylim = c(0, 0.45),
+     cex.axis = 1.5, cex.lab = 1.3)
> lines(x, dnorm(x, mean = 2))
> arrows(0, dnorm(0) + 0.01, 2, dnorm(0) + 0.01, angle = 90,
+     code = 3, length = 0.05)
> text(1, dnorm(0) + 0.0375, "d'", cex = 1.8)
> segments(0, 0, 0, dnorm(0), lty = 3)
> segments(2, 0, 2, dnorm(0), lty = 3)
> mtext("a", side = 3, cex = 2, adj = 0, line = 0.5)
> plot(x, dnorm(x), type = "l", ylim = c(0, 0.45),
+     cex.axis = 1.5, cex.lab = 1.3)
> lines(x, dnorm(x, mean = 2, sd = 1.4) * 1.4)
> mtext("b", side = 3, cex = 2, adj = 0, line = 0.5)
> par(opar)
```

and the optimal decision rule is equivalent to the rule $\lambda(X) \geq \log\beta$. If we set β to be the reciprocal of the prior odds $(1 - \pi)/\pi$ then $\log\beta = \log((1 - \pi)/\pi)$. Since $\log(1/x) = -\log(x)$, the optimal rule in logarithmic form is just a comparison of the log likelihood ratio $\lambda(X)$ to the negative of the *log prior odds* $\log(\pi/(1 - \pi))$. In particular, when $\pi = 0.5$, $\log\beta = 0$.

If $d' = 0$ the signal and noise distributions coincide and $\lambda(X) = 0$. Since the signal is indistinguishable from the non-signal, the observer''s responses are independent of the presence or absence of the signal.

3.3.1 The Decision Rule for Equal-Variance Gaussian SDT

The optimal decision rule $\lambda(X) \geq \log \beta$ is remarkably simple, but in the equal-variance Gaussian case, it can be simplified further. For equal-variance Gaussian SDT, the log likelihood ratio rule is just

$$\lambda(X) = d'X - 0.5d'^2 \geq \log \beta \qquad (3.5)$$

which simplifies to $X \geq c$ with

$$c = \frac{\log \beta}{d'} + \frac{d'}{2} \qquad (3.6)$$

For equal-variance Gaussian SDT, then, the decision rule that optimizes probability correct is equivalent to a rule that compares the sensory information to a *sensory criterion*, c.

Warning: In elementary treatments of SDT based on the equal-variance Gaussian model, the optimal rule is often presented in this form, as a comparison of the sensory information X to a sensory criterion c. For other choices of signal and noise distributional families, the rule $\Lambda \geq \beta$ (or, equivalently, $\lambda \geq \log \beta$) is always the rule that maximizes the probability of a correct response, but it may not be possible to rewrite it as $X \geq c$ (see Sect. 3.4).

3.3.2 Simulating and Fitting Equal-Variance Gaussian SDT

The equal-variance Gaussian SDT observer is fixed by specifying the two parameters (d', c) or, equivalently, the two parameters (d', β). We also discovered above that we can specify any signal detection observer by specifying the probabilities p_H, p_{FA}. For the equal-variance Gaussian signal detection observer, we can translate from one parameterization to the other. This transformation is particularly easy in terms of the sensory criterion c. To begin with, the probability of a false alarm is just the probability that, when X is drawn from the noise distribution, $X \geq c$.

$$p_{FA} = 1 - \Phi(c) \qquad (3.7)$$

where $\Phi()$ is the cdf of the Gaussian with mean 0, variance 1 and $\Phi^{-1}()$ is its inverse. Similarly, a hit results when X, drawn form the signal distribution (with mean d'), is greater than the sensory criterion:

$$p_H = 1 - \Phi(c - d') \qquad (3.8)$$

The last two equations allow us to translate from (d', c) to (p_H, p_{FA}). We can go from (p_H, p_{FA}) to (d', c) by using the inverse of the cumulative distribution function (also referred to as the *quantile function*):

$$c = \Phi^{-1}(1 - p_{FA}) = -\Phi^{-1}(p_{FA}) \tag{3.9}$$

and

$$d' = c - \Phi^{-1}(1 - p_H) = \Phi^{-1}(p_H) - \Phi^{-1}(p_{FA}) \tag{3.10}$$

Consequently, we can readily simulate the performance of an equal-variance observer with specified parameters (d', c) in an SDT experiment with (n_S, n_N) signal and noise trials by translating (d', c) to (p_H, p_{FA}) and applying the simulation code above.

Suppose, for example, that we want to calculate the probabilities for the four outcomes for $d' = 1$, $c = 0.1$, and $n_S = n_N = 1,000$. Then, an experimental outcome is simulated by the following code fragment with the results indicated in the following table:

```
> dp <- 1 ; crit <- 0.1
> nS <- nN <- 1000
>  pFA  <- 1 - pnorm(crit)
>  pH <- 1 - pnorm(crit - dp)
>  FA <- rbinom(1,nN, pFA)
>  H <- rbinom(1, nS, pH)
>  CR <- nN - FA
>  M <- nS - H
```

	Signal	Not signal
Yes	0.829	0.46
No	0.171	0.54

To generate multiple simulations, it is more efficient to write a function that takes as input the values of d', c, and the numbers of signal and noise trials.

```
> simulate.SDT <- function(dp, crit, nS, nN = NULL) {
+     if(missing(nN)) nN <- nS
+     pFA <- 1 - pnorm(crit)
+     pH <- 1 - pnorm(crit - dp)
+     FA <- rbinom(1,nN, pFA)
+     H <- rbinom(1, nS, pH)
+     CR <- nN - FA
+     M <- nS - H
+     res <- matrix(c(H, M, FA, CR), nc = 2,
+         dimnames = list(c("Yes", "No"), c("S", "NS")))
+     as.table(res)
+ }
```

When the number of signal and noise trials is equal, only the number of signal trials need be specified. The function `missing` checks whether the formal argument nN has been specified and if not, sets the two equal. Our function returns an object of class "table." Multiple simulations are easily performed using the `lapply` function as shown below for three simulations.

```
> ( SDTsim.lst <- lapply(rep(1, 3), simulate.SDT,
+     crit = 0.5, nS = 1000) )
[[1]]
      S  NS
Yes 690 294
No  310 706

[[2]]
      S  NS
Yes 679 317
No  321 683

[[3]]
      S  NS
Yes 660 301
No  340 699
```

Although there is stochastic variation in the numbers of responses in each cell from one simulation to the next, the proportions of responses in each cell approximate the exact probabilities.

3.3.3   Maximum Likelihood Estimation of Equal-Variance Gaussian SDT Parameters

Given data such as those shown in the above tables, we would like to obtain maximum likelihood estimates (MLE) of (d', c). This turns out to be very simple for the equal-variance Gaussian observer. The values of FA are the results of a binomial random variable with probability p_{FA} and its maximum likelihood estimate is just $\hat{p}_{FA} = FA/n_N = FA/(FA + CR)$. The values of H are the results of a binomial random variable with probability p_H and its maximum likelihood estimate is just $\hat{p}_H = H/n_S = H/(H + M)$. The corresponding values from the table above are easily obtained with a second application of `lapply`

```
> ( PrSDTsim.lst <-  lapply(SDTsim.lst, "/", 1000) )
[[1]]
        S     NS
Yes 0.690 0.294
No  0.310 0.706
```

[[2]]

	S	NS
Yes	0.679	0.317
No	0.321	0.683

[[3]]

	S	NS
Yes	0.660	0.301
No	0.340	0.699

In the first example, the MLEs of p_{FA} are $\hat{p}_{FA} = 0.294$ and of p_H are $\hat{p}_H = 0.69$. But the equal-variance Gaussian SDT parameters (d', c) are an invertible transformation of (p_{FA}, p_H), and therefore we obtain the MLEs of d' and c by applying this transformation to the MLEs $\hat{p}_{FA}$ and $\hat{p}_H$:

```
> c(c = qnorm(1 - PrSDTsim.1st[[1]][1, 2]),
+     dp = qnorm(PrSDTsim.1st[[1]][1, 1]) -
+          qnorm( PrSDTsim.1st[[1]][1, 2]) )
     c      dp
0.542 1.038
```

See Sect. B.4.3 for justification.

This estimation method is commonly used without acknowledgment that the resulting estimates are maximum likelihood estimates. Of course, by applying a further invertible transformation, we can also compute the MLEs of the parameters (d', β), as well, if we prefer them to (d', c). The same justification applies.

3.3.4 Fisher's Tea Test

While the Yes–No experiment is often characterized as one of classifying trials as signal present and signal absent, the labels and nature of the judgment can be quite arbitrary. In the general case, the trials conform to two events that we may label A and B and the observer's task is to classify each trial. The response classifications then give rise to four possible outcomes, the response was A when the trial was of type A, the response was A when the trial was of type B, etc. In the absence of a clear logic for assigning terms such as Hit, False Alarm, etc., the choice can be made arbitrarily.

For example, the second chapter of Fisher's *The Design of Experiments* is entitled "The principles of experimentation illustrated by a psycho-physical experiment" [62]. In it, he describes an experiment in which eight cups of tea are prepared and to which milk is added before the tea in half and the reverse in the other half. The issue is whether a lady can, as she claims to be capable of, discriminate the order of mixture.[2] Fisher uses this example to make several points about the experimental method and the limits of interpretation of the results. The point that we emphasize here is that the structure of this experiment is equivalent to that of the Yes–No experiment described above.

Let the stimulus and response "tea followed by milk" be coded by B and "milk followed by tea" by A. We will arbitrarily label the cups in which tea was added first as the signal. Then, the probability of a Hit is $P(\text{Response} = B|\text{Stimulus} = B)$, a False Alarm, $P(\text{Response} = B|\text{Stimulus} = A)$, etc. and the analogy is complete. If sufficient numbers of trials are presented (eight is clearly not enough), one could from the outcome estimate the observer's sensitivity and bias for discriminating the order of mixture of milk and tea.

We will use this example to demonstrate how to explore the precision of the estimates from an SDT experiment. Let us suppose that there are $N = 1,000$ trials, half of which are of type A.

```
> N <- 1000
> Stim <- factor(rep(c("A", "B"), each = N/2))
```

In a real experiment, we would present them in randomized order (as discussed by Fisher), but since there are no dependencies between trials in our simulation, it is simpler to present first one event, then the other. We specify the sensitivity and criterion of the observer as a 2 element vector, where d' is the first element and c is the second. For example, the following codes for a biased observer with a difference in the distribution of underlying responses across A and B stimulus trials as

```
> dc <- c(1.3, 0.5)
```

The two equations relating d' and c to the normal quantiles for the Hit and False Alarm rates are both linear and we readily invert them to obtain the normal quantiles of the Hit and False Alarm rates for any pair of sensitivity and criterion values.[3]

$$d' = \Phi^{-1}(P_{\text{H}}) - \Phi^{-1}(P_{\text{FA}})$$
$$c = -\Phi^{-1}(P_{\text{FA}}) \tag{3.11}$$

```
> z2dc <- matrix(c(1, 0, -1, -1), 2, 2)
> dc2z <- solve(z2dc)
> p <- pnorm(dc2z %*% dc)
```

After transforming the quantiles to probabilities we can simulate the responses for the observer using the function rbinom to generate random binary events at the specified Hit and False Alarm rates. The simulated responses are stored in a factor Resp.

---

[2]The incident on which this anecdote is based is recounted at greater length in [155].

[3]Since the matrix of this system of equations is an involution (i.e., it is its own inverse), we do not need to use solve, but doing so so helps us keep track of which set of variables we are computing.

```
> Resp <- factor( ifelse(Stim == "B", rbinom(N, 1,p[1]),
+     rbinom(N, 1, p[2])), labels = LETTERS[1:2])
> Resp.tab <- table(Resp, Stim)[2:1, 2:1]
> z2dc %*% qnorm(Resp.tab[1, ]/(N/2))
        [,1]
[1,] 1.189
[2,] 0.423
```

The stimulus events, Stim, and the responses, Resp, are coded as factors. We tabulate the frequency of each type, extract the proportion of Hits and False Alarms, and transform them to normal quantiles with qnorm. The rows and columns were switched in the table Resp.tab to generate the table of responses in typical format. Then, the matrix from (3.11) maps the quantiles to d' and c.

Since the value of d' was not exactly the value that we specified, we might wonder how variable our estimation was and how it depends on the number of observations. We examine these questions by incorporating the above code into a function that we can use to explore systematically the parameters of this problem. We use some shortcuts in the function to make it run faster. In particular, we use the rbinom function to generate probabilities of Hits and False Alarms (PH and PF, respectively) for n experiments of N/2 trials, all at once.

```
> dpsim <- function(dc, N, n = 10000) {
+     z2dc <- matrix(c(1, -0.5, -1, -0.5), 2, 2)
+     dc2z <- solve(z2dc)
+     p <- pnorm(dc2z %*% dc)
+     PH <- rbinom(n, N/2, p[2])/N
+     PF <- rbinom(n, N/2, p[1])/N
+     apply(qnorm(cbind(PH, PF)), 1, diff)}
```

The function dpsim takes a two-element vector of d' and c as its first argument. N specifies the number of trials in an experiment, half being allocated to each type of stimulus event and n specifies the number of times the experiment will be repeated. The function returns the simulated values of d'. We leave as an exercise the modifications to simulate simultaneously the criterion value.

The following code fragment illustrates our simulation:

```
> dp.qu <- sapply(c(100, 1000, 10000), function(x) {
+     dpsim(c(0, 0), x, 10000)
+     })
> colnames(dp.qu) <- c(100, 1000, 10000)
> dhists <- data.frame(dp = as.vector(dp.qu),
+     Trials = rep(c(100, 1000, 10000), each = 10000))
```

Using sapply, we run the function for 10,000 experiments for each of three trial lengths, for an unbiased observer for whom the sensitivity is zero. The results are stored in a data frame to simplify plotting with the histogram function from the **lattice** package as shown in Fig. 3.2.

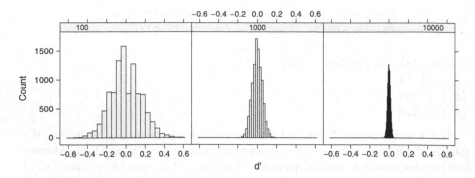

Fig. 3.2 Simulated values of d' for experiments of different numbers of trials. Each histogram is based on 10,000 simulated experiments with the number of trials per experiment indicated in the strips

Figure 3.2 illustrates how increasing the numbers of trials narrows the width of the distribution of simulated values. The 95% quantile values quantify this trend.

```
> apply(dp.qu, 2, quantile, c(0.025, 0.975))
           100    1000    10000
2.5%    -0.303 -0.0975 -0.0309
97.5%    0.311  0.0979  0.0303
```

A quick calculation on the quantiles suggests that the width decreases as the square root of the number of trials. General tools for the evaluation of power and sample size requirements for SDT analyses can be found in the **sensR** package [32].

```
> print(
+    histogram(~ dp | factor(Trials), data = dhists,
+          layout = c(3, 1), breaks = 25,
+       strip = strip.custom(style = 5),
+       xlab = expression("d'"), type = "count")
+ )
```

3.3.5   Fitting Equal-Variance SDT as a GLM

While normally overkill for this type of example, the SDT statistics can also be directly estimated using a GLM [48, 120]. The value of this exercise will become apparent when we proceed to analyze more complex psychophysical tasks. Consider, again, the formula for d'.

$$d' = \Phi^{-1}(p_H) - \Phi^{-1}(p_{FA}) = \Phi^{-1}(Pr(\text{‘Yes’}|P)) - \Phi^{-1}(Pr(\text{‘Yes’}|A)) \quad (3.12)$$

In the format of the equation on the right, the estimation of d' requires estimating the probabilities that an observer reports "Yes" conditional on the presence and absence of the signal and taking the difference of these values on a transformed scale. Viewed in this fashion, the expression for d' is equivalent to the contrast between the levels of a factor indicating the presence and absence of a signal as a linear predictor of the response for a GLM model. The model would be expressed as

$$g(\mathsf{E}[Pr(\text{Resp} = \text{"Yes"})]) = \beta_0 + \beta_1 X \tag{3.13}$$

where g is the link function and X indicates the presence/absence of the signal as a $0/1$ variable. Since the exponential family will be binomial, g will be the inverse of a sigmoidal function, typically mapping the interval $(0, 1)$ to $(-\infty, \infty)$. The default link function is the log of the odds ratio or logit function. To fit the equal-variance Gaussian SDT model, however, we need to specify that the link be an inverse Gaussian function, referred to as a "probit." When the signal is absent, $X = 0$ and β_0 provides an estimate of the normal quantile of the False Alarm rate. When the signal is present, $X = 1$ and the right-hand side provides an estimate of the normal quantile of the Hit rate. Then, β_1 must be the difference between the Hit and False Alarm rates on the probit scale, or d'. We express this relation for the variables in the tea problem as a model formula for a GLM as follows:[4]

```
> Resp ~ Stim
```

To summarize, the intercept in this model will be an estimate of the normal quantile of the False Alarm rate and, if the default "treatment" contrasts are used, the second coefficient in the model will correspond to the difference between the quantiles for the Hit rate and the intercept or d'.

```
> tea.glm <- glm(factor(Resp) ~ factor(Stim),
+    family = binomial("probit"))
> coef(summary(tea.glm))
              Estimate Std. Error z value Pr(>|z|)
(Intercept)    -0.423     0.0579   -7.31 2.65e-13
factor(Stim)B   1.189     0.0852   13.96 2.77e-44
```

Compare the coefficients with those obtained by manual calculation in Sect. 3.3.4, recalling the relation between the criterion and the False Alarm rate, (3.9). Here, we have modeled the binary responses while in Sect. 3.3.4, the estimates were based on average Hit and False Alarm rates. Both approaches lead to the same estimates when both can be applied to the same data set.

Obtaining SDT estimates via glm provides us with a model object, which can be explored with all of the methods and extractors available for analyzing this type of

---

[4]Note that this assumes that level A, the first level of the factor, corresponds to the signal absent case. See ?relevel for a simple way to reset the first level of a factor variable.

model. We can also extend this analysis easily to more sophisticated data sets by manipulating the formula object, but more on that later (see Sect. 3.3.6).

Finally, the estimates can be obtained by maximizing an expression for the likelihood directly using one of the optimization functions from R, such as optim. We demonstrate this here since understanding how to set this up is useful when additional constraints must be placed on the linear predictor or the predictor is nonlinear. In the code fragment below, we use the model.matrix function to create the model matrix, X, for the explanatory variable, Stim, a two-level factor. We then create a function, ll, to compute minus the log likelihood of the binomial model. The first argument is a vector whose values we wish to estimate and the second is the model matrix. The probabilities are related to a linear function of the explanatory variables by a Gaussian cdf. Finally the response variable, Resp, determines whether the probability p or 1 - p contributes to the total likelihood for each trial. The maximum likelihood estimates are obtained by an iterative procedure using the optim function. As its first two arguments, it takes initial estimates of the parameters and the function to minimize. A third argument giving the gradient of this function to be minimized is recommended generally but has not been used here. Additional arguments to the function to be minimized are supplied as named arguments.

```
> X <- model.matrix(~ Stim)
> ll <- function(b, X) {
+    p <- pnorm(X %*% b)
+    -sum(ifelse(Resp == "B", log(p), log(1 - p)))
+    }
> tea.opt <- optim(c(0, 1), ll, X = X)
> tea.opt$par

[1] -0.423  1.189
```

A five-element list is returned by optim with the estimates in the component par. The final value at optimization is given in the component value and a code indicating whether or not the convergence criterion was met is given in the component convergence.

3.3.6 Fitting with Multiple Conditions

One of the advantages of fitting SDT data with glm is that more complex experimental designs can be incorporated easily into the modeling using R's formula language. Consider the following data set, obtained from [120, p. 27]

```
> Recognition
  Stim Number Resp      Cond
1 Old      69  Yes    Normal
2 New      31  Yes    Normal
```

```
3  Old   31  No   Normal
4  New   69  No   Normal
5  Old   89  Yes  Hypnotized
6  New   59  Yes  Hypnotized
7  Old   11  No   Hypnotized
8  New   41  No   Hypnotized
```

The data are described as being from a representative observer performing a face recognition task. Prior to testing, the individual was presented with a series of images of faces to learn. In the experimental phase, the observer performed 100 trials with one face presented on every trial, in each of two conditions. In both conditions, half of the faces were new and the task was to identify the previously seen faces. The variable Stim indicates whether the face was "Old" or "New" and the variable Resp indicates how the observer responded to a question such as "Have you seen this face before?." The column Number indicates the number of responses within each response classification. The column Cond indicates, however, that the observer performed the task in two different states, "Normal" or "Hypnotized." This is like a Yes–No task in which a Hit corresponds to the response "Yes" when the face is "Old" and a False Alarm to the same response when the face is "New." The measure d' indicates the sensitivity of the observer to the category difference. Because there are two conditions, we can ask whether the sensitivity changes with the state.

One strategy is to run separate tests on each of the conditions, but then we would have no basis on which to compare the estimated values. Instead, we can obtain both values using glm with the formula specifying different nested models. The models can then be compared using a likelihood test.

The simplest model assumes that there is no effect of condition on any of the parameters. We fit this by leaving out the variable Cond.

```
> Recog.glm1 <- glm(Resp ~ Stim, binomial(probit),
+    Recognition, weights = Number)
> coef(summary(Recog.glm1))
            Estimate Std. Error z value Pr(>|z|)
(Intercept)   -0.126     0.0889   -1.41 1.57e-01
StimOld        0.932     0.1337    6.97 3.18e-12
```

Here, the formula only specifies a relation between the factors Resp and Stim. We could expand the frequencies of the responses and stimuli into long vectors of 1's and 0's for individual trials, but it is simpler to perform the specification by using the argument weights = Number. A single value of d' and bias is estimated for both conditions.

A second level of complexity is introduced by permitting the intercept term to vary across conditions, corresponding to a change in observer bias. We obtain this model by adding the factor Cond. We use the update method to fit the new model. It takes as first argument the model object from the first fit and as second the modified formula object. The "." on each side indicates to include the terms from the previous model.

Fig. 3.3 The estimated hit
and false alarm rates plotted
on the scale of the linear
predictor for model
`Recog.glm2`. The barred
segments indicate key
differences that correspond to
the fitted coefficients of the
model

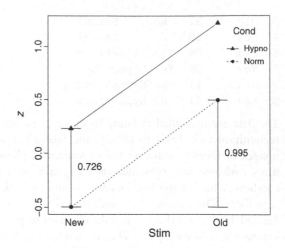

```
> Recog.glm2 <- update(Recog.glm1, . ~ . + Cond)
> coef(summary(Recog.glm2))
                Estimate Std. Error z value Pr(>|z|)
(Intercept)      -0.497        0.116   -4.30 1.74e-05
StimOld           0.995        0.139    7.17 7.28e-13
CondHypnotized    0.726        0.138    5.26 1.44e-07
```

Given the default treatment contrasts, the Intercept term corresponds to the bias
for the first level of `Cond`, "Normal" and the `StimOld`, the difference between the
two levels, which will be d'. The meaning of these coefficients is illustrated in the
interaction plot of Fig. 3.3 which shows the values of the estimated Hit and False
Alarm rates on the linear predictor scale. The term `CondHypnotized` gives the
difference between the two conditions for the first level of `Stim`, "Old." Since there
is no interaction, the difference between the values of `Stim` for the second level of
`Cond` is unchanged, i.e., there is no change in d' between conditions in this model.

To fit a model in which we allow for a difference in sensitivity between
conditions, we include the interaction term between `Stim` and `Cond`.

```
> Recog.glm3 <- update(Recog.glm2, . ~ . + Cond:Stim)
> coef(summary(Recog.glm3))
                Estimate Std. Error z value Pr(>|z|)
(Intercept)     -0.49585      0.131 -3.7824 1.55e-04
StimOld          0.99170      0.185  5.3491 8.84e-08
CondHypnotized   0.72340      0.182  3.9706 7.17e-05
StimOld:         0.00728      0.279  0.0261 9.79e-01
  CondHypnotized
```

The interaction term indicates the difference between the second levels of both the
`Stim` and `Code` terms that is not taken into account by the `CondHypnotized` term.
If the conditions produced a significant difference in sensitivity, this term would be

significantly different from 0 and the lines from an interaction plot, such as Fig. 3.3, would not be parallel. This hypothesis is not supported by the output of the summary method.

To formally compare the fitted models, we use the anova method to perform a likelihood ratio test. If we wanted to test that any parameters changed between conditions, we would compare the third model with the first one fit. Generally, however, when testing for sensitivity differences, we allow for differences in bias between conditions. In that case, it makes more sense to test the third model against the second. Because the models form a nested sequence, however, we can evaluate all three, with a single function call.

```
> anova(Recog.glm1, Recog.glm2, Recog.glm3,
    test = "Chisq")
Analysis of Deviance Table

Model 1: Resp ~ Stim
Model 2: Resp ~ Stim + Cond
Model 3: Resp ~ Stim + Cond + Stim:Cond
  Resid. Df Resid. Dev Df Deviance Pr(>Chi)
1         6        481
2         5        452  1     28.5 9.3e-08 ***
3         4        452  1      0.0    0.98
---
Signif. codes:  0 `***' 0.001 `**' 0.01 `*' 0.05 `.'
    0.1 ` ' 1
```

The results do not support a change in sensitivity between conditions, the same outcome supported, as well, by comparing AIC values.

```
>  AIC(Recog.glm1, Recog.glm2, Recog.glm3)
           df AIC
Recog.glm1  2 485
Recog.glm2  3 458
Recog.glm3  4 460
```

The approach is easily extended to encompass more experimental conditions by modifying the formula object appropriately. In the next chapter, for example, we will see how adding a covariate permits the modeling of psychometric functions.

3.3.7 Complete Separation

When using glm, you may see the following warning message displayed with the results for certain models.

```
Warning message:
```

```
glm.fit: fitted probabilities numerically 0 or 1
 occurred
```

This occurs when there are conditions in the data set for which the estimated probabilities are 0 or 1 and the situation is referred to as *complete separation*. For example, consider the following artificial data set:

```
> CmpltSep.df <- data.frame(Present = c(10000, 100),
+   Absent = c(0, 10000), Stim = factor(c(1, 0)))
```

Notice that for the first row there are 0 "Absent" responses, so that the observed proportion of responses "Present" is 1. If the estimated proportion in the model is 1 on the response scale then with the standard link functions, it will map into a value at infinity. Similarly, if the estimated proportion is 0, it will map into a value of minus infinity. When we model this data set with the default logit link, we obtain

```
> cmpsep.1 <- glm(cbind(Present, Absent) ~ Stim,
   binomial,
+   CmpltSep.df)
```

```
Warning message:
glm.fit: fitted probabilities numerically 0 or 1
occurred
```

That this situation is pathological is easily detected in the coefficient table from the summary method. In particular, note the standard error estimates.

```
> coef(summary(cmpsep.1))
            Estimate Std. Error   z value Pr(>|z|)
(Intercept)    -4.61        0.1 -4.58e+01        0
Stim1          36.51    671088.6  5.44e-05        1
```

In some situations, simply changing the link function is sufficient to eliminate this warning,

```
> cmpsep.2 <- glm(cbind(Present, Absent) ~ Stim,
+   binomial(probit), CmpltSep.df)
> coef(summary(cmpsep.2))
            Estimate Std. Error   z value Pr(>|z|)
(Intercept)    -2.33   3.73e-02 -62.49025    0.000
Stim1           9.85   4.83e+03   0.00204    0.998
```

though the standard error for the term Stim1 still appears unreasonable. This is, in fact, a demonstration of the Hauck–Donner phenomenon [78, 178] in which the standard errors for values far from the null hypothesis increase faster than the values themselves, thus rendering very large values insignificant.

Often, it is sufficient to gather more data to remedy this situation since it will likely be the case that the true proportions are not exactly 0 or 1 and with sufficient data, no zeros will occur. Consider how adding one "Absent" response to the above example modifies the fit.

Fig. 3.4 ROC contours for d' values of 0, 1, 2, and 3, indicated by each *solid contour*. The *point* indicates (d', c) from the example of Sect. 3.3.2 where the values of H, FA, and N were defined. The *dashed lines* indicate how d' changes when either the Hit (*vertical*) or False Alarm (*horizontal*) rate varies while the other stays fixed

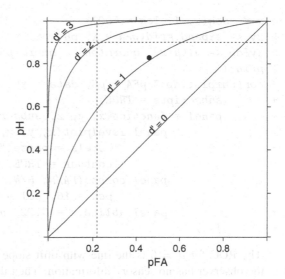

```
> cmpsep.3 <- glm(cbind(Present, Absent + 1) ~ Stim,
+    binomial(probit), CmpltSep.df)
> coef(summary(cmpsep.3))
             Estimate Std. Error  z value   Pr(>|z|)
(Intercept)     -2.33     0.0373    -62.5   0.00e+00
Stim1            6.05     0.2553     23.7   4.56e-124
```

With simple detection and discrimination tasks, the source of the complete separation is easily detected in the data set. With more sophisticated psychophysical tasks that involve more complex models, the underlying source of the problem may not be as evident. We will suggest some steps to take in later chapters.

3.3.8 The Receiver Operating Characteristic

As we noted above, the observer's performance in equal-variance Gaussian SDT is summarized by two parameters (d', c) and we freely translate back and forth between (d', c) and the equivalent parameterization (p_{FA}, p_H). Focusing on the first parameterization, we can interpret d' as the observer's fixed sensitivity in a particular task while c is a criterion that is under the observer's control and, for the ideal observer, depends on the prior odds of signal vs noise.

We can plot (p_{FA}, p_H) for a fixed value of d' and many values of c (or β) to get a receiver operating characteristic (ROC), as shown in Fig. 3.4 for $d` = 0, 1, 2, 3$ in the equal-variance normal case.

```
> x <- seq(0, 1, len = 500)
> xy <- expand.grid(pFA = x, pH = x)
> xy$dp <- with(xy, qnorm(pH) - qnorm(pFA))
> print(
+ contourplot(dp ~ pFA + pH, data = xy,
+         subscripts = TRUE,
+         panel = function(x, y, z, subscripts, ...){
+                 panel.levelplot(x, y, z, subscripts, at = 0:3,
+                         labels = paste("d' =", 0:3),
+                         contour = TRUE, region = FALSE)
+                 panel.points(FA/N, H/N,
+                         pch = 16, col = "black", cex = 1.25)
+                 panel.abline(v = 0.22, h = 0.9, lty = 2)
+         }) )
```

The ROC for $d' = 0$ is the line with unit slope where $p_{FA} = p_H$. When $d' = 0$, the observer has no sensory information. Then the observer is simply guessing on each trial and his rate of guessing "Yes" on signal trials must equal his rate of guessing "Yes" on noise trials, i.e., he has no basis for discriminating whether a trial is a signal trial or noise trial. As d' increases the ROC curves become more "bowed" toward the upper left corner of the plot. The observer is able to achieve lower FA rates for the same Hit rates (horizontal dashed line) or higher Hit rates for the same FA rates (vertical dashed line) and both represent better performance. We plot the maximum likelihood estimates $(\hat{p_{FA}}, \hat{p_H})$ for example from Sect. 3.3.2 on the ROC plot as a black point, and we see that the point is close to the ROC curve for $d' = 1$. In fact, it is precisely on the ROC curve (not shown) for $d' = 1.051$, the maximum likelihood estimate of d' for example. Proving that there is one and only one such curve is not difficult and is left as an exercise (Exercise 3.2). We use the contourplot function from the **lattice** package to draw contours of constant d' in Fig. 3.4. The ROC plot is a useful characterization of the levels of performance that the observer can achieve when d' is fixed and c or β is variable. We can also plot curves of fixed β for many different values of d' with the ROC curves previously plotted. These isocriterion curves are less frequently encountered but they provide additional insight into the structure of SDT. In particular, the tangent slope to the ROC curves at the intersections with any one isocriterion curve is all the same and, in fact, equal to β [72, pp. 36–38]. For example, they are all equal to 1 along the isocriterion curve that is a line running from top left to bottom right.

We can also plot a variant of the ROC curve where we replace p_{FA} by $\Phi^{-1}(p_{FA})$ and p_H by $\Phi^{-1}(p_H)$. The resulting zROC curve for the equal-variance Gaussian observer is a straight line of slope 1 offset from the diagonal $d' = 0$ by an extent that increases with d'. The criterion c or β determines the location along the line.

3.4 Case 2: Unequal-Variance Gaussian SDT

A second common model of signal detection is *unequal-variance Gaussian (or normal) SDT*. In this model, the signal distribution is normal again with mean $d' > 0$ but with variance σ^2,

$$f_S(x) = \frac{1}{\sqrt{2\pi}\sigma} e^{-\frac{(x-d')^2}{2\sigma^2}} \tag{3.14}$$

while the noise distribution is as in the equal-variance case, normal with mean 0 and variance 1,

$$f_S(x) = \frac{1}{\sqrt{2\pi}} e^{-\frac{x^2}{2}} \tag{3.15}$$

The signal and noise distributions for the unequal-variance signal detection model are shown in Fig. 3.1b with $\sigma^2 = 1.4$ so that the signal distribution is slightly wider (by a factor of $\sigma = 1.2$) than that of the noise. Here the difference between signal and noise distributions is not simply addition of a constant and the term the *signal+noise distribution* would be inappropriate.

3.4.1 The Decision Rule for Unequal-Variance Gaussian SDT

In this case, the log likelihood ratio is a quadratic function

$$\lambda(X) = -\frac{1}{2\sigma^2} \left[(1 - \sigma^2)X^2 - 2d'X + d'^2 + 2\sigma^2 \log\sigma \right] > \log\beta. \tag{3.16}$$

The equation in X is quadratic and is not equivalent to a simple inequality $X > c$. The following plot shows us why. We plot $\lambda(X)$ vs X in Fig. 3.5 when $d' = 1$ and $\sigma = 2$.

```
> dp <- 1 ; sig <- 2; logbeta <- 2
> X <- seq(-5, 5, len = 200)
> lamX <- function(X, dp, sig)
+    -((1 - sig^2) * X^2 - 2 * dp * X +
+          dp^2 + 2 * sig^2 * log(sig))/(2 * sig^2)
> plot(X, lamX(X, dp, sig), type = "n",
     ylab = "-log Likelihood")
> rts <- as.real( polyroot(
+    -c( dp * dp + 2 * sig^2 * log(sig) + logbeta * 2 * sig^2,
+    -2 * dp, 1 - sig^2) / (2 * sig^2) ))
> rect(-6, 2, rts[2], 11, col = "lightgrey", border = NA)
> rect(rts[1], 2, 6, 11, col = "lightgrey", border = NA)
> lines(X, lamX(X, dp, sig))
> abline(2, 0, lty = 2)
```

Fig. 3.5 A negative log
likelihood function for the
unequal-variance Gaussian
model. The *dashed line*
indicates a criterion of
$\log \beta = 2$ and the *grey regions*
indicate the two domains for
which this criterion is
satisfied

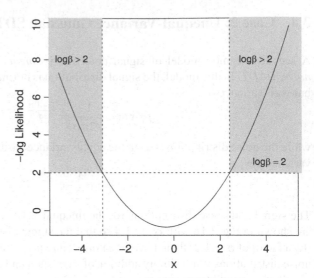

```
> abline(v =c (rts[2], rts[1]), lty = 2)
> text(3.5, 8, expression(paste(plain(log), beta > 2)))
> text(-4.5, 8, expression(paste(plain(log), beta > 2)))
> text(4, 2.3, expression(paste(plain(log), beta == 2)))
```

For $\log \beta = 2$, there are two intervals where $\lambda(X) > \log \beta$ and the ideal observer
should respond "Yes," and one where $\lambda(X) < \log \beta$ and the observer should respond
"No." We can therefore represent the decision rule by a pair of inequalities $c_1 < X <$
c_2 where the corresponding response is "Yes" if $\sigma^2 < 1$ and "No" if $\sigma^2 \geq 1$. If σ
increases with d', for example, then the resulting decision rule $c_1 < X < c_2$ is to
respond "No" to moderate values of X and "Yes" to larger *and* smaller values.

3.4.2   Simulating Unequal-Variance Gaussian SDT

The unequal-variance normal SDT observer is specified by three parameters
(d', σ, β) and we saw above that the optimal rule for responding to a trial where
the strength of the sensory variable X was

$$c_1 < X < c_2 \iff \text{Yes} \tag{3.17}$$

if $\sigma < 1$ and otherwise

$$c_1 < X < c_2 \iff \text{No} \tag{3.18}$$

if $\sigma > 1$. If $\sigma = 1$, of course, we are back to the equal-variance case. We can
readily simulate performance in the unequal-variance case by simply computing
the parameters (p_{FA}, p_H) given these rules. When $\sigma > 1$, for example, and the signal

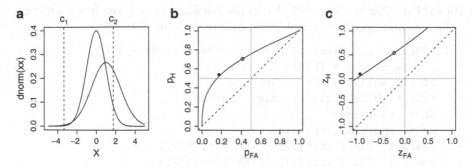

Fig. 3.6 (**a**) Signal absent and present distributions for the situation in which the $\sigma_p > \sigma_a$. The *dashed vertical lines* indicate the criterion points for example in the text. (**b**) Computed ROC curve for the distributions to the left with $d' = 1$. The *dashed line* corresponds to $d' = 0$. The *points* on this and the *graph* to the right are described in the text. (**c**) The zROC curve for the unequal-variance case is a line with slope equal to $1/\sigma$

distribution has greater variance than the noise distribution, then we need to solve for c_1 and c_2 and integrate the noise distribution from c_1 to c_2 to get p_{FA} and the signal distribution from c_1 to c_2 to get p_H. We illustrate these computations in drawing the ROC curve and zROC curve for unequal-variance Gaussian SDT when $d' = 1$, $\sigma = 1.5$, and $\log\beta = 1$. For convenience, we write a function for estimating the criterion points given these three values.

```
> UVGSDTcrit <- function(dp, sig, logbeta){
+     TwoSigSq <- 2 * sig^2
+     minLam <- optimize(function(X, dp, sig)
+        -((1 - sig^2) * X^2 - 2 * dp * X +
+           dp^2 + TwoSigSq * log(sig))/TwoSigSq, c(-10, 10),
+           dp = dp, sig = sig)$objective
+     if (logbeta < minLam) warning("complex roots")
+     cf <- -c(dp^2 + TwoSigSq * log(sig) + logbeta *
+        TwoSigSq,
+        -2 * dp, 1 - sig^2)/TwoSigSq
+     proot <- polyroot(cf)
+     sort(Re(proot))
+ }
```

In Fig. 3.6a, we draw the noise and signal distributions with the positions of the two criteria indicated by dashed vertical lines. Note that the ROC curve in Fig. 3.6b is no longer symmetric about the diagonal line that runs from the upper left to the lower right. Moreover, the zROC curve in Fig. 3.6c is a line but a line whose slope is no longer 1. In fact, it has slope $1/\sigma$ which in this case is $2/3$.

We can also simulate responses from an unequal-variance Gaussian observer. The following function generates Hit and False Alarm rates for the given parameter values and numbers of trials. We have included an argument `aggregate` as a logical variable that determines whether the values returned will be aggregated into average

Hit and False Alarm rates (TRUE, the default condition) or as a data frame with Nn
+ Ns rows of, respectively, signal absent and present trials, (FALSE).

```
> rUVGSDT <- function(dp, sig, logbeta, Nn, Ns = NULL,
+         aggregate = TRUE){
+    if (is.null(Ns)) Ns <- Nn
+    crit <- UVGSDTcrit(dp, sig, logbeta)
+    Rn <- rnorm(Nn)
+    Rs <- rnorm(Ns, dp, sig)
+    RespN <- (Rn < crit[1]) | (Rn > crit[2])
+    RespS <- (Rs < crit[1]) | (Rs > crit[2])
+    if (aggregate)
+        c(pFA = sum(RespN)/Nn, pH = sum(RespS)/Ns) else
+        data.frame(Resp = c(RespN, RespS),
+            Stim = factor(rep(0:1, c(Nn, Ns))))
+ }
```

We have plotted one simulation of unequal-variance Gaussian SDT with $n_S = 1,000$,
$n_N = 1,000$, and $\log\beta = 0$ as a solid point on the ROC plot (and the zROC plot) and
see that it is not far from the ROC or zROC curve. However, we cannot estimate
the parameters of the ideal observer employed in the simulation from this one point.
The observer is specified by three parameters and we have only two free parameters
in the data $(\hat{p}_{FA}, \hat{p}_H)$. Equivalently, there are many lines with many slopes that pass
through the single point. If we have a second point with the same values of d' and σ
and a different value of $\log\beta$, however, then we do have enough information to fit
the ideal observer to data. We show a second value of $(\hat{p}_{FA}, \hat{p}_H)$ as a hollow point
on both the ROC and zROC. This point is the result of simulating the ideal observer
with $d' = 1$, $\sigma = 1.5$, and $\log\beta = -0.5$, a simulation of an observer with the same
$d' = 1$, $\sigma = 1.5$ but with $\log\beta = -0.5$ instead of $\log\beta = 0$. We could obtain this
second point by just varying the prior odds of signal vs noise. It is evident in the
zROC that the two points determine a line and that the two conditions shown contain
enough information to pin down the ideal observer and estimate the parameters
$d' = 1$, $\sigma = 1.5$ and $\log\beta_1 = 0$, $\log\beta_2 = -0.5$ where now we explicitly subscript
the values of $\log\beta$ in the two experimental conditions. We leave as an exercise
modifying the above code to treat the case $\sigma < 1$.

```
> dp <- 1; sig <- 1.5; logbeta <- 1
> xx <- seq(-5, 5, len = 200)
> opar <- par(mfrow = c(1, 3), pty = "s" )
> plot(xx, dnorm(xx), type = "l", xlab = "X")
> lines(xx, dnorm(xx, mean = dp, sd = sig))
> c12 <-UVGSDTcrit(dp, sig, logbeta)
> abline(v = c12, lty = 2)
> mtext(c(expression(c[1]), expression(c[2])), 3, at = c12,
+           line = 0.3)
> mtext("a", side = 3, line = 1, adj = 0)
```

```
> minLam <- optimize(lamX, c(-10, 10),
+               dp = dp, sig = sig)$objective + 1e-6
> p <- t(sapply(seq(minLam, 10, len = 100), function(LBeta){
+         c12 <- UVGSDTcrit(dp, sig, LBeta)
+             1 - c(diff(pnorm(c12)), diff(pnorm(c12,dp,sig))) }
+ ) )
> SimPt1 <- rUVGSDT(dp, sig, 0, 1000)
> SimPt2 <- rUVGSDT(dp, sig, -0.5, 1000)
> plot(p, type = "l", xlim = c(0, 1), ylim = c(0, 1),
+           xlab = expression(p[FA]), ylab = expression(p[H]))
> points(c(SimPt1[1], SimPt2[1]), c(SimPt1[2], SimPt2[2]),
+           pch = c(16, 1), col = "black")
> abline(0, 1, lty = 2)
> abline(v = 0.5, h = 0.5, col = "grey")
> mtext("b", side = 3, line = 1, adj = 0)
> plot(qnorm(p), type = "l", xlim = c(-1, 1), ylim = c(-1, 1),
+           xlab = expression(z[FA]), ylab = expression(z[H]))
> points(qnorm(c(SimPt1[1], SimPt2[1])),
+           qnorm(c(SimPt1[2], SimPt2[2])),
+           pch = c(16, 1), col = "black")
> abline(0, 1, lty = 2)
> abline(v = 0, h = 0, col = "grey")
> mtext("c", side = 3, line = 1, adj = 0)
> par(opar)
```

One important point to note is that we need only induce the "not-so-ideal" observer to use two different values of $\log \beta$ in the two conditions. There is evidence that actual human observers tend not to select correct values of $\log \beta$ in signal detection tasks when $\beta \neq 1$, a phenomenon known as "conservatism" [72, pp. 91–94]. However, two points determine a line in the zROC even if they are not the points corresponding to the correct values of $\log \beta$ selected by the experimenter. Hence it is possible to fit data to the unequal-variance SDT ideal observer if the experimenter can plausibly induce the observer to alter $\log \beta$ without any alteration in d' and σ. Next we consider how to fit unequal-variance data by direct optimization.

3.4.3 Fitting Unequal-Variance Gaussian SDT by Direct Optimization

Because of the difference in variances between the signal absent and present conditions, the unequal-variance Gaussian SDT model cannot be fit using glm. It is, in effect, an example of a generalized nonlinear model. In this section, we will demonstrate how it can be fit by direct maximization of the likelihood, using some of the ideas and tools that we developed above. As noted above, a minimum condition is that data have been obtained for at least two criterion levels. Our first step will be to generate some simulated data. We will assume that $d' = 1$, $\sigma_S = 1.5$ and that $N = 5,000$ trials are run for each of the signal absent and present conditions. As minimum conditions, we will choose two levels of $\log \beta$ but will write the code so that an arbitrary number can be fit.

```
> dp <- 1; sig <- 1.5; N <- 5000
> lb <- c(1.5, 1)
```

We can use the rUVGSDT function to simulate 10,000 experimental trials (5,000 for each of signal absent and present conditions) for each criterion level. Setting aggregate = TRUE, the function returns a data frame with columns Resp and Stim as used in previous examples. We add a column, Cond, indicating the different criteria employed.

```
> UVGsim.df <- do.call(rbind, lapply(lb, function(LB)
+       rUVGSDT(dp, sig, LB, N, agg = FALSE)))
> UVGsim.df$Cond <- factor(rep(paste("C",
          seq(length(lb))),
+    sep = ""), each = 2 * N))
```

Note that we can abbreviate argument names if there is no possibility of confounding with abbreviated names of other formal arguments.

Next, we write a function, lUVG, to calculate the log likelihood for the unequal-variance Gaussian model. It takes two arguments, p and d. The first argument is a vector of length 2 + the number of criterion levels in the data set. The first two elements are the current estimates of d' and σ, respectively. The rest of the elements are the estimates of $\log \beta$ for each criterion level. The argument d is a data frame like the one that we just created. It returns the negative of the log likelihood for the data set given the parameters in p.

```
> lUVG <- function(p, d){ #p <-c(dp, sig, logbeta_1,_2,
    ...,_n)
+   sig <- p[2]
+   if (length(p[-(1:2)]) != length(levels(d[[3]])))
+       stop("Number of initial estimates of
+       log beta not equal to number of levels of
+       conditions in the data!")
+   Pr <- sapply(p[-(1:2)], function(lb, pr) {
+       crit <- UVGSDTcrit(pr[1], sig, lb)
+       1 - c(diff(pnorm(crit)),
+              diff(pnorm(crit, pr[1], pr[2])))
+       }, pr = c(p[1], sig))
+   ll <- sapply(seq_along(levels(d[[3]])), function(Cd)
    {
+           with(subset(d, Cond == levels(d$Cond)[Cd]),
+               ifelse(Stim == "1",
+                   Resp * log(Pr[2, Cd]) +
```

```
+                   (1 - Resp) * log(1 - Pr[2, Cd]),
+                   Resp * log(Pr[1, Cd]) +
+                   (1 - Resp) * log(1 - Pr[1, Cd])
+            ))
+        })
+    -sum(unlist(ll))
+ }
```

Once we have a function to calculate the log likelihood, we can use it with an optimization function to obtain the maximum likelihood estimates. We use optim with method = "L-BFGS-B." This method uses gradients and boxed-constraints. We specify only the lower limits so that $d' > 0$, $\sigma > 1$ and the values of $\log \beta$ are above the minimum of the parabola that defines the log likelihood ratio for this problem as in Fig. 3.5. These are precomputed using optimize for one-dimensional optimizations and assigned to minLam. We specify hession = TRUE so that we can compute standard errors of the estimated values. Finally, we assemble the estimates, standard errors, and z-values in a matrix with named rows and columns for pretty output.

```
> minLam <- optimize(function(X, dp, sig)
+        -((1 - sig^2) * X^2 - 2 * dp * X +
+          dp^2 + 2 * sig^2 * log(sig))/(2 * sig^2),
+          c(-20, 20), dp = dp, sig = sig)$objective
> UVG.opt <- optim(c(1.1, 1.6, 1.4, 1.1), lUVG,
    d = UVGsim.df,
+        method = "L-BFGS-B", hessian = TRUE,
+        lower = c(0, 1, rep(minLam, length(lb))))
> est <- UVG.opt$par
> names(est) <- c("dprime", "sigma",
+        paste("log beta", seq(length(lb)), sep = ""))
> est.se <- sqrt(diag(solve(UVG.opt$hessian)))
> cbind(Estimate = est, SE = est.se, z = est/est.se)
            Estimate    SE    z
dprime         1.04 0.193 5.39
sigma          1.44 0.309 4.65
log beta1      1.56 0.160 9.77
log beta2      1.08 0.241 4.49
```

An initial estimate for d' can be obtained by calculating Hit and False Alarm rates individually for each condition and averaging estimates. An initial estimate of σ can be obtained from the reciprocal of the line fit to the values in zROC space. Obtaining initial estimates of the $\log \beta$ values will be more challenging. We leave this as an exercise as well as modifying our code to handle the case $\sigma < 1$. The final values and standard errors indicate that the underlying values are reasonably estimated though the procedure should be repeated from multiple starting values to guard against the possibility that the algorithm located a local minimum.

3.5 Case 3: Exponential SDT

In principle, signal detection models can be constructed for any distribution that appropriately describes the distribution of the stimulus. For example, a less common model of signal detection is based on the exponential distribution. We think of the sensory information as a temporal delay where the signal distribution leads to shorter delays than the noise. For clarity, we denote the sensory information by T instead of X and replace x by t. Then the signal distribution for *exponential SDT* is an exponential distribution with rate parameter τ

$$f_S(t) = \tau e^{-\tau t} \tag{3.19}$$

while the noise distribution is a second exponential with $\tau = 1$.

$$f_N(t) = e^{-t} \tag{3.20}$$

The signal and noise distributions for the exponential signal detection model are shown in Fig. 3.7 with $\tau = 2$. We use the dexp function to generate the exponential pdfs. This is parameterized so that the second argument `rate` corresponds to τ and the distribution has mean $1/\tau$. When the signal is present, the time delay T will on average be shorter than it would be on noise trials.

```
> xx = seq(0, 5, len = 100)
> plot(xx, dexp(xx), type = "l",
+    xlab = "T", lwd = 2, ylim = c(0, 2))
> lines(xx, dexp(xx, rate = 2), lty = 2, lwd = 2)
```

For the exponential SDT above, the log likelihood ratio rule is just

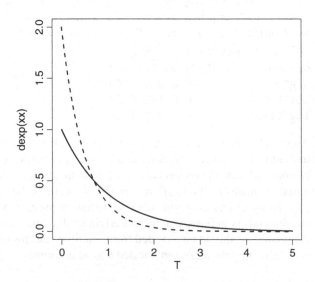

Fig. 3.7 Exponential distributions with $\tau = 1$ (*solid line*) and $\tau = 2$ (*dashed line*)

$$\lambda(T) = \log \tau + (1 - \tau)T > \log \beta \tag{3.21}$$

which simplifies to $T < c$ when $\tau > 1$ and $T > c$ when $\tau < 1$. When $\tau = 1$ the signal and noise distributions coincide. If we interpret the presence of the signal as a factor that speeds up the response, then $\tau > 1$ and the corresponding decision rule is of the form $T < c$. For equal-variance normal SDT and exponential then, the decision rule that optimizes probability correct is equivalent to a rule that compares the sensory information to a *sensory criterion c*. We leave setting up the estimation and fitting procedures as an exercise.

3.6   m-Alternative Forced-Choice and Other Paradigms

Suppose that in the tea example from Sect. 3.3.4, two cups of tea were presented on each trial in random order and the observer's task was to decide whether the cup with "tea followed by milk" was the first or the second cup. This corresponds to a two-alternative forced-choice paradigm. Calculating the value of d' when there are two alternatives in a trial can be simply performed using the same methods as for the Yes–No paradigm but on theoretical grounds the result must be divided by $\sqrt{2}$. The empirical justification for this correction, however, has been called into question under certain circumstances [202].

When there are more than two alternatives, there is no closed form solution, and the calculation involves solving for d' in the expression [72]

$$P_c = \int_{-\infty}^{\infty} \phi(x - d')\Phi(x)^{m-1}\,dx, \tag{3.22}$$

where P_c is the proportion of correct responses, ϕ and Φ are the Gaussian pdf and cdf, respectively, and m is an integer indicating the number of choices on each trial. This formulation is also valid for the case $m = 2$. Using P_c assumes that the observer is unbiased. While it is often assumed that m-alternative forced-choice (mAFC) tasks reduce bias in the observer's judgments, this assumption has come under fire, as well, recently [93, 202].

It is not difficult to calculate this integral in R using the function `integrate` and to solve for d' using the `uniroot` function, to estimate the root of $f(d') = 0$ with f defined as

$$f(d') = P_c - \int_{-\infty}^{\infty} \phi(x - d')\Phi(x)^{m-1}\,dx. \tag{3.23}$$

The function `dprime.mAFC`, available in the package **psyphy** [94], does this. It takes two arguments: `Pc` and `m`, the unbiased proportion of correct responses and the number of alternatives, respectively.

For example, suppose we take the variable `Stim` from the simulation above and relabel the values to indicate whether the event A was in the first or second interval.

```
> N <- 1000
> Stim <- factor(rep(Stim, 1000), labels = 1:2)
```

and we redefine the variable Resp with the same labels but this time specify an unbiased observer with a non–zero sensitivity.

```
> dc <- c(1.4, 0.7)
> p <- pnorm(dc2z %*% dc)
> Resp <- factor( ifelse(Stim == "1", rbinom(N, 1,p[1]),
+   rbinom(N, 1, p[2])))
> Resp.tab <- table(Resp, Stim)
```

We obtain the value of P_c from combining both Hits and Correct Rejections

```
> Pc <- sum(Resp.tab[2:3])/sum(Resp.tab)
> library(psyphy)
> dprime.mAFC(Pc, m = 2)
[1] 0.972
```

We compare this value with that obtained using the Yes–No formula with the correction

```
> (z2dc %*% qnorm(Resp.tab[c(2, 4)]/(2 * N)))[1]/sqrt(2)
[1] 0.972
```

noting that differences can arise in the two techniques when the observer is biased.

3.6.1 Letter Identification: 4AFC

As an example of mAFC with greater than two choices consider the data set ecc2 in the package **psyphy** [94]. These data are from a double-judgment task in which the observer's task was to choose one of four positions in which one of four letters was presented, Detection, and then to identify the letter, Identification [204]. The data set contains five columns: Contr, the luminance contrast of the letter, task, whether the task was detection or identification, Size, the height of the letter in minutes of visual angle and Correct and Incorrect indicating the number of correct and incorrect classifications, respectively.

To simplify, we only consider the subset of observations for the Identification of letters 12.4 min high.

```
> data(ecc2)
> ID12 <- subset(ecc2, task == "ID" & Size == 12.4)
> Pc <- with(ID12, Correct/(Correct + Incorrect))
```

dprime.mAFC does not take a vector as an input so we use sapply to run it for each value.

```
> sapply(Pc, dprime.mAFC, m = 4)
[1] 0.0206 0.3015 0.1542 0.6097 1.3325 2.1399
```

The function can be made to take a vector as an argument with the `Vectorize` function in the **base** package [146], which takes as arguments a function and its arguments to be vectorized and returns a vectorized version of the function. For example,

```
> dpvec.mAFC <- Vectorize(dprime.mAFC, "Pc")
> dpvec.mAFC(Pc, 4)
[1] 0.0206 0.3015 0.1542 0.6097 1.3325 2.1399
```

We note that d' increases with letter contrast.

An alternative for measuring d' in 2- and 3afc is provided by the **sensR** package [32]. The package defines family functions `twoAFC` and `threeAFC`, modifications of the binomial family for use with `glm` that return estimates of d' for formula objects in which unbiased probability correct is fit by a constant term.

3.6.2 d' for Other Paradigms

The **psyphy** package contains functions for calculating d' for three additional experimental paradigms.

`dprime.SD` In a Same–Different task, two stimuli are presented but the observer classifies trials with respect to whether the two stimuli are the same or different. In the tea example, Same trials would be *AA* (tea first on both trials) or *BB* (milk first on both trials) while Different trials would be *AB* (tea first, followed by milk first) or *BA* (milk first followed by tea first).

`dprime.ABX` In an ABX task, stimuli *A* and *B* are presented followed by either one of *A* or *B* and the observer judges whether the third stimulus was the same as either the first or second

`dprime.oddity` On each trial of the oddity task, three stimuli, two of which are the same, are presented in random order, and the observer classifies each trial as to which of the three was different.

The **sensR** package also contains functions for calculating d' in various paradigms including `samediff` for d' in Same–Different tasks and another family function, `triangle`, which can be used for d' estimates in oddity paradigms.

3.7 Rating Scales

As described in Sect. 3.3.8, more information is available when multiple points on the ROC curve are obtained. One way to obtain several points is by running several experiments in which the observer is nudged to alter his criterion, for example, by changing the payoff for a particular response and/or the a priori probability of the occurrence of a particular signal. A more efficient method entails having the observer employ several criteria simultaneously within the same experimental session. This can be done using rating scale procedures [72, 120, 189].

The basic idea of rating scales is that the observer responds to each trial by using an n-point scale indicating confidence that a signal trial occurred. For example, suppose that the observer is instructed to use a 5-point scale, responding "1" on trials on which he is most certain that the signal is absent and "5" on trials when he is most certain that it is present. Intermediate values correspond to intermediate values of certainty.

From such ratings, we can construct a table giving the frequency that each rating class was employed for each of the stimulus conditions, e.g., "Present," "Absent." We then divide the table into a series of 2×2 tables by setting a boundary between each adjacent pair of ratings and summing separately over the frequencies to either side of the boundary for each stimulus condition. If we use a 5-point scale, there are four such tables. We can treat each of these tables as an individual Yes–No experiment corresponding to a different criterion level and calculate Hit and False Alarm rates. Plotting the (P_{FA}, P_H) pairs gives the estimated values along the ROC curve.

Consider an example simulating an observer's behavior in such an experiment. In the example, we assume that there are 1,000 trials and that the signal distribution is Gaussian with a mean of 1. We draw 1,000 numbers from $\mathcal{N}(0, 1)$ using `rnorm` and add a value of $d' = 1$ to half of them to generate the internal responses for signal and noise trials. We choose the internal criterion values along a 6-point scale and assign them to a vector named `IntCrit`. Note that the name `Inf` is defined in R so that we can handle very extreme values. We, then, use the `cut` function to create a factor that has the internal responses assigned to a factor level on the basis of the category boundaries in `IntCrit`.

```
> N <- 1000
> dp <- 1
> Noise <- rnorm(N)
> IntResp <- rep(c(dp, 0), each = N/2) + Noise
> Stim <- factor(rep(1:0, each = N/2))
> IntCrit <- c(-Inf, -2:2, Inf)
> Ratings <- cut(IntResp, IntCrit)
> levels(Ratings) <- 1:6
```

To help visualize the situation, we plot the Noise and Signal + Noise distributions in the decision space with the category boundaries indicated as dashed lines in Fig. 3.8. For each category boundary, the frequencies of response are summed across categories to one side to obtain Hit and False Alarm rates.

```
> X <- seq(-3, 4, len = 1000)
> plot(X, dnorm(X), type = "l")
```

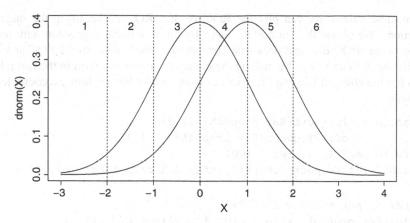

Fig. 3.8 Noise and Signal + Noise distributions with six category boundaries for a rating scale indicated by *dashed vertical lines*. Hit and false alarm rates are calculated by treating each category boundary as a criterion in turn and cumulative summing the numbers of responses to one side

```
> lines(X, dnorm(X, mean = 1))
> abline(v = -2:2, lty = 2)
> text(seq(-2.5, 2.5), 0.385, 1:6, cex = 1.3)
```

The rating frequency table is simply generated using `table`. We can then use the `apply` function to obtain cumulative sums over the frequencies to one side of each rating boundary using the `cumsum` function. We divide by the number of trials for each stimulus condition so that the values are rendered directly into Hit and False Alarm rates.

```
> ( Rat.tab <- table(Stim, Ratings) )
    Ratings
Stim   1    2    3    4    5    6
   0  10   56  163  185   76   10
   1   1   12   61  178  174   74
> ( Rat.proptab <- apply(Rat.tab[, 6:1], 1, cumsum)
    [6:1, ] /
+    (N/2) )
   Stim
          0     1
 1 1.000 1.000
 2 0.980 0.998
 3 0.868 0.974

 4 0.542 0.852
 5 0.172 0.496
 6 0.020 0.148
```

From these values, we can plot the ROC or the zROC curve, using the qnorm function. We obtain the fitted ROC curve by fitting a linear regression with unit slope to the zROC data and back-transforming the fitted values to the ROC space. We fix the fit to unit slope by making the covariate an offset term in the formula. This has the effect of keeping these values in the model but not fitting a coefficient to them.

```
> Rat.lm <- lm(qnorm(Rat.proptab[-1, 2]) ~
+               offset(qnorm(Rat.proptab[-1, 1]))))
> zfa <- seq(-4, 4, len = 100)
> RatROC.fit <- pnorm(cbind(1, zfa) %*% c(coef(Rat.lm), 1) )

> opar <- par(mfrow = c(1, 2))
> plot(Rat.proptab, xlim = c(0, 1), ylim = c(0, 1),
+       xlab = expression(P[FA]),
+       ylab = expression(P[H]))
> lines(pnorm(zfa), RatROC.fit)
> lines(pnorm(zfa), pnorm(zfa + coef(Rat.glm)[6]), lty = 2,
        lwd = 2)
> plot(qnorm(Rat.proptab), xlim = c(-3, 3), ylim = c(-3, 3),
+       xlab = expression(z[FA]),
+       ylab = expression(z[H]))
> abline(coef(Rat.lm), 1)
> abline(v = 0, h = 0, col = "grey")
> abline(coef(Rat.glm)[6], 1, lty = 2, lwd = 2)
> par(opar)
```

Naively, one might think that to test whether the zROC has a unit slope, i.e., the equal-variance assumption, that one would use lm with the covariate not as an offset, but this would be ignoring that there is error along both the abscissa and ordinate values. We will see how to approach this problem below, but first we need to look at the formal model underlying rating scale data and some functions available in R to fit it.

3.7.1   Ordinal Regression Models on Cumulative Probabilities

The rating scale responses are ordinal values; we can say that one value corresponds to a greater rating than another but the intervals between the values have no meaning. The method used to analyze the ratings above, based on cumulative frequencies of response, suggests the following ordinal regression model based on cumulative probabilities [53, 127, 178].

$$g(P(Y \le k \mid x)) = \theta_k - X\beta \tag{3.24}$$

where g is a link function, Y the rating whose probability being less than the kth category boundary is conditional on the explanatory variable x, $X\beta$ is a linear predictor, and θ is a category dependent intercept; it provides an estimate of the rating boundaries on the scale of the linear predictor. The minus sign attached to the linear predictor is a convention that results in probability increasing with increased rating and increase in the linear predictor.

We can fit a model like this using `glm` following a prescription described in McCullagh and Nelder [127, pp. 172–173] (also see [196]). First, the ratings are transformed into binary responses coding the cumulative frequency of response. For example, for each rating of an n-point scale, we create $n-1$ binary variables coding $P(Y \leq 1), P(Y \leq 2), \ldots, P(Y \leq n-1)$. Each binary response will be associated with a different level of an $n-1$-level factor in the explanatory variables, thereby coding a different intercept per rating boundary. The rest of the explanatory variables are packaged into the design matrix, X, as in other models with a linear predictor. In this way, the category boundaries are independent of the linear predictor. We could model the data that we simulated above, as follows:

```
> ordRat <- ordered(Ratings)
> cumRat <- as.vector(sapply(1:5, function(x)
    ordRat <= x))
> X <- matrix(
+ c(rep(c(1, 0, 0, 0, 0), each = N),
+ rep(c(0, 1, 0, 0, 0), each = N),
+ rep(c(0, 0, 1, 0, 0), each = N),
+ rep(c(0, 0, 0, 1, 0), each = N),
+ rep(c(0, 0, 0, 0, 1), each = N)), ncol = 5)
> X <- cbind(X, -rep(Stim == "1", 5))
> Rat.df <- data.frame(Resp = cumRat, X = X)
```

We first use the function `ordered` to coerce the factor `Ratings` to an object of class "ordered." Examination of its class reveals it to inherit from class "factor." Examination with `str`, however, shows that its levels include a coding of their order. We exploit the ordering to generate the vector of binary responses coding the membership in the cumulative proportions associated with each rating level, `cumRat`. Note that this vector is five times as long as `ordRat`. The first five columns of the matrix X code for the intercept per rating boundary and the last the binary factor indicating the presence/absence of the signal. Once the data frame containing the binary responses and the design matrix is created, we can use it with `glm` to fit the model. The "." on the right-hand side of the formula indicates the additive contribution of all columns not already included in the model, i.e., all but `Resp`, here.

```
> ( Rat.glm <- glm(Resp ~ . - 1, binomial(probit),
        Rat.df) )
Call:  glm(formula = Resp ~ . - 1, family =
        binomial(probit),
```

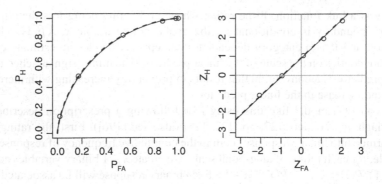

Fig. 3.9 The *solid lines* show the ROC and zROC curves estimated for the simulated rating scale experiment. The points are calculated for each rating scale boundary. The *dashed lines* were obtained using `glm` and described later in the text

```
    data = Rat.df)

Coefficients:
   X.1      X.2      X.3      X.4      X.5      X.6
 -2.040   -1.091   -0.110    0.942    1.992    0.929

Degrees of Freedom: 5000 Total (i.e. Null);   4994
   Residual
Null Deviance:        6930
Residual Deviance: 3400       AIC: 3410
```

The first five coefficients correspond to the estimated rating boundaries on the linear predictor scale. The last corresponds to the estimate of d'. The predicted ROC and zROC curves are shown in Fig. 3.9 as dashed lines.

The ordinal regression model of (3.24) is usually specified with the additional constraint that $\theta_1 \leq \theta_2 \leq \cdots \leq \theta_{n-1}$. Using `glm`, we cannot impose such a constraint although unless the data violate the model, the results will usually respect the requirement as above. To impose such a constraint or others, one must set up the likelihood to be maximized directly, as we have done elsewhere, for other problems. This approach is taken in several functions for fitting ordinal regression models available in R packages, for example, `polr` in package **MASS**, `lrm` in **rms** [76], `clm` in **ordinal** [31], and `vglm` in **VGAM** [201]. In addition, the likelihood is defined slightly differently, in terms of a multinomial model rather than a Bernoulli one, as here. Again, this difference will result in differences in the estimated values, but these are likely to be slight when the model is a valid description of the data. For example, compare the results from `polr` with those obtained using `glm` above.

```
> library(MASS)
> polr(ordRat ~ Stim, method = "probit")
Call:
polr(formula = ordRat ~ Stim, method = "probit")
```

```
Coefficients:
Stim1
0.933

Intercepts:
   1|2     2|3     3|4     4|5     5|6
-2.033  -1.093  -0.110   0.943   1.994

Residual Deviance: 2798.73
AIC: 2810.73
```

The `polr` function uses the formula interface common to many R functions. It can take an ordered factor as the response variable with the explanatory variables specified on the right-hand side. Its help page specifies that the intercept term must be left in the model. The link function is specified through the `method` argument.

The `clm` function in the **ordinal** package offers several interesting alternative models in which additional constraints can be placed on the criterion points to generate a set of models nested with respect to the default, unconstrained estimation of these points (i.e., other than that they form a monotonic, increasing sequence). The argument `threshold` can take the value "symmetric" to force the criteria to be placed symmetrically about the central one (for odd) or the central two (for even numbers of) points. If set to "equidistant," the criteria will be spaced equally along the linear predictor axis. Such constraints permit a model with potentially fewer parameters to be fit to the data.

```
> library(ordinal)
> Rat.clm0 <- clm(ordRat ~ Stim, link = "probit")
> Rat.clm1 <- update(Rat.clm0, threshold = "symmetric")
> Rat.clm2 <- update(Rat.clm1, threshold =
    "equidistant")
```

The numbers of fitted parameters for three models are:

```
> sapply(list(Rat.clm0, Rat.clm1, Rat.clm2), "[[",
    "edf")
[1] 6 4 3
```

A nested set of likelihood ratio tests indicates that the simplest model, equal spacing of the criteria, describes the data as well as the two more complex models in this case.

```
> anova(Rat.clm2, Rat.clm1, Rat.clm0)
Likelihood ratio tests of cumulative link models:

              formula:      link:   threshold:
Rat.clm2 ordRat ~ Stim probit equidistant
Rat.clm1 ordRat ~ Stim probit symmetric
```

```
Rat.clm0 ordRat ~ Stim probit flexible
```

```
           no.par  AIC logLik LR.stat df Pr(>Chisq)
Rat.clm2      3 2806  -1400
Rat.clm1      4 2808  -1400    0.01  1       0.92
Rat.clm0      6 2811  -1399    1.58  2       0.45
```

When the link function is specified as "probit," the clm function by default and the polr function fit an equal-variance Gaussian model to the rating scores. The slope of the zROC curve is constrained to unity. The clm function, however, offers the possibility of fitting the unequal-variance Gaussian model by exploiting its second formal argument, scale. This argument takes a one-sided formula which specifies the explanatory variables, factors or covariates, whose values will be scale dependent. The model fit is

$$g(P(Y \leq k \mid x) = \frac{\theta_k - \boldsymbol{X}\beta}{\exp(s_x)}, \tag{3.25}$$

where $\exp(s_x)$ codes the dependence of variance or scale on the explanatory variables. In the case of a Yes–No task for which the stimulus is coded by a two-level factor, the estimated value corresponds to the ratio of standard deviations for the Noise and Signal + Noise distributions. The exponential is used to constrain the estimated value to be positive.

For our simulated data set, we would fit the unequal-variance model by updating one of the model objects fit above with the equal-variance model. For example,

```
> Rat.clm.UV <- update(Rat.clm0, scale = ~ Stim)
> summary(Rat.clm.UV)
formula: ordRat ~ Stim
scale:    ~Stim
```

```
 link    threshold nobs logLik   AIC      niter max.grad
 probit flexible   1000 -1399.09 2812.19 7(0)   2.69e-11
 cond.H
 1.9e+01
```

```
Coefficients:
      Estimate Std. Error z value Pr(>|z|)
Stim1   0.9539     0.0771    12.4   <2e-16 ***
---
Signif. codes:  0 `***' 0.001 `**' 0.01 `*' 0.05 `.'
  0.1 ` ' 1
```

```
log-scale coefficients:
      Estimate Std. Error z value Pr(>|z|)
Stim1   0.0417     0.0568    0.73     0.46
```

```
Threshold coefficients:
     Estimate Std. Error z value
1|2  -2.0513     0.1234  -16.62
2|3  -1.1083     0.0681  -16.26
3|4  -0.1166     0.0534   -2.18
4|5   0.9577     0.0615   15.57
5|6   2.0447     0.1047   19.54
```

which shows the estimated scale coefficient. It is stored in the zeta component of the object. The ratio of standard deviations fit to the data is obtained by

```
> exp(Rat.clm.UV$zeta)
Stim1
 1.04
```

which is not far from unity, the value used to simulate the data. We suggest that the reader simulates some data where the ratio is not unity to see how it affects this parameter. A more formal test of the equal-variance assumption is obtained by a likelihood ratio test of nested models

```
> anova(Rat.clm.UV, Rat.clm0)
Likelihood ratio tests of cumulative link models:

             formula:      scale: link:  threshold:
Rat.clm0  ordRat ~ Stim ~1       probit flexible
Rat.clm.UV ordRat ~ Stim ~Stim   probit flexible

            no.par  AIC logLik LR.stat df Pr(>Chisq)
Rat.clm0         6 2811  -1399
Rat.clm.UV       7 2812  -1399    0.54  1       0.46
```

which provides no support for rejecting the equal-variance assumption.

3.7.2   Sensitivity to Family Resemblance

Maloney and Dal Martello [122] presented observers with pairs of images of children's faces in which the pairs were either siblings or not. The observers rated each pair as to their confidence that the pair was a sibling on a scale from 0 (lowest certainty that the pairs are related) to 10 (highest certainty that the pairs are related). The data set Faces in our package **MPDiR** contains the ratings of each of 32 observers on 30 pairs of faces. The ratings are in the column named SimRating. The column sibs contains a two-level factor indicating whether or not the pairs were siblings with levels "0" and "1" for not siblings and siblings, respectively. The data set also contains information about the age difference in months of the children,

agediff, a two-level factor indicating whether the children were of the same or
different gender, gendiff, a factor to identify each observer, Obs, and another to
identify each image pair, Image.

As in their analysis, we combine the ratings across observers to obtain cumulative
proportions higher than each rating level as a function of whether or not the pairs
were siblings.

```
> data(Faces)
> rat.tab <- with(Faces, table(sibs, SimRating))
> cum.tab <- apply(rat.tab, 1, cumsum)
> ( cum.prop <- 1 - cum.tab/rowSums(rat.tab) )
    sibs
            0      1
 0   0.7521 0.954
 1   0.6375 0.923
 2   0.5208 0.869
 3   0.4125 0.794
 4   0.3208 0.723
 5   0.2250 0.660
 6   0.1687 0.573
 7   0.1000 0.463
 8   0.0563 0.298
 9   0.0292 0.156
10   0.0000 0.000
```

The result in cum.prop is a two-column matrix with the proportions for each of the
two states of the image pairs. The two columns can be treated as False Alarm and
Hit rates and are plotted against each other in Fig. 3.10(a), (b) to obtain ROC and
zROC curves, respectively. The plot in (a) is in terms of the cumulative probabilities
and that in (b), transformed by qnorm. To avoid problems with infinite values, the
last pair of coordinates was excluded from the right plot.

A common method for estimating d' from the ROC curve is to average the d'
value for each pair of points plotted in Fig. 3.10 which gives a value of 1.07. This
is the estimate under the equal-variance Gaussian model, and the prediction with
unit slope and intercept d' is shown in (b) as a solid line. This line, transformed by
pnorm, gives the solid curve in Fig. 3.10(a).

The estimates of the criterion boundaries for the ratings are obtained by
transforming the False Alarm rates to the linear predictor scale.

```
> -qnorm(cum.prop[-11, 1])
       0       1       2       3       4       5       6
 -0.6811 -0.3518 -0.0522  0.2211  0.4654  0.7554 0.9591
       7       8       9
  1.2816  1.5871  1.8932
```

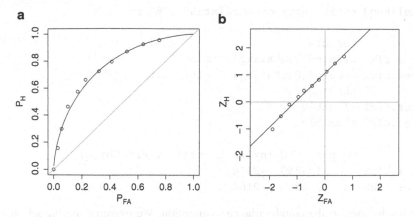

Fig. 3.10 (**a**) The hit and false alarm rates plotted against each other as an ROC curve for the Faces data set. The curve is obtained by back-transforming the line fit to the points in zROC space. (**b**) The hit and false alarm rates plotted as a zROC with the best-fitting straight line

```
> par(mfrow = c(1, 2))
> plot(cum.prop, xlim = c(0, 1), ylim = c(0, 1),
+    xlab = expression(P[FA]), ylab = expression(P[H]))
> abline(0, 1, col = "grey")
> dp <- mean(apply(qnorm(cum.prop[-11, ]), 1, diff))
> xx <- seq(-3, 3, len = 100)
> lines(pnorm(xx, lower.tail = FALSE),
+    pnorm(xx, mean = dp, lower.tail = FALSE))
> mtext("a", line = 0.5, cex = 2, adj = 0)
> plot(qnorm(cum.prop[-11, ]), xlim = c(-2.5, 2.5),
+    ylim = c(-2.5, 2.5), xlab = expression(z[FA]),
+    ylab = expression(z[H]))
> abline(v = 0, h = 0, col = "grey")
> lines(xx, xx + dp)
> mtext("b", line = 0.5, cex = 2, adj = 0)
```

Instead of the manual calculation performed above, we can use one of the modeling functions described in the previous section to analyze the data. For example, using clm we fit equal -and unequal-variance Gaussian models. Note that the ratings must be coerced to an ordered factor.

```
> Faces.clm1 <- clm(ordered(SimRating) ~ sibs,
+    data = Faces, link = "probit")
> Faces.clm2 <- update(Faces.clm1, scale = ~ sibs)
```

We test the equal-variance assumption using the anova method.

```
> anova(Faces.clm1, Faces.clm2)
```

```
Likelihood ratio tests of cumulative link models:

            formula:                      scale: link:
Faces.clm1 ordered(SimRating) ~ sibs ~1      probit
Faces.clm2 ordered(SimRating) ~ sibs ~sibs  probit
            threshold:
Faces.clm1 flexible
Faces.clm2 flexible


            no.par  AIC logLik LR.stat df Pr(>Chisq)
Faces.clm1     11 4350  -2164
Faces.clm2     12 4352  -2164       0  1       0.99
```

The results support the equal-variance assumption. We examine the model object for this case with the summary method.

```
> summary(Faces.clm1)
```

```
formula: ordered(SimRating) ~ sibs
data:    Faces

 link    threshold nobs logLik   AIC     niter max.grad
 probit flexible  960  -2163.83 4349.67 5(1)  2.28e-07
 cond.H
 1.4e+02


Coefficients:
       Estimate Std. Error z value Pr(>|z|)
sibs1   1.0737     0.0702    15.3   <2e-16 ***
---
Signif. codes:  0 `***' 0.001 `**' 0.01 `*' 0.05 `.'
  0.1 ` ' 1


Threshold coefficients:
     Estimate Std. Error z value
0|1   -0.6666     0.0579  -11.51
1|2   -0.3539     0.0546   -6.48
2|3   -0.0527     0.0534   -0.99
3|4    0.2312     0.0535    4.32
4|5    0.4678     0.0544    8.60
5|6    0.6994     0.0560   12.48
6|7    0.9173     0.0581   15.78
7|8    1.2094     0.0617   19.62
8|9    1.6092     0.0677   23.77
9|10   2.0499     0.0767   26.73
```

The location estimate corresponds to the estimation of d' in the model and agrees with the value calculated above by averaging values for the different criteria. The threshold coefficients are the criterion boundaries for the ratings. These also show broad agreement with the values obtained above.

One of the advantages of using a modeling function is that additional information is packed into the returned object that can be extracted with various method functions. For example, note the information returned by the summary method concerning the variability and significance of the estimated coefficients as well as information related to the overall fit of the model.

The other important advantage, however, is the possibility of fitting and testing more complex models to describe the rating judgments. For example, the Faces data frame contains other explanatory variables whose effects on the observer ratings can be evaluated. The factor gendiff indicates whether the genders of the individuals in the images were the same or different, agediff is a covariate indicating the age difference between the pair in months. To evaluate their influence we can test whether adding either of these terms leads to a significantly better description of the responses.

```
> addterm(Faces.clm1, scope = ~ sibs + gendiff +
    agediff,
+    test = "Chisq")
Single term additions

Model:
ordered(SimRating) ~ sibs
        Df   AIC   LRT Pr(Chi)
<none>       4350
gendiff  1 4352  0.04 0.84724
agediff  1 4337 14.63 0.00013 ***
---
Signif. codes:  0 `***' 0.001 `**' 0.01 `*' 0.05 `.'
   0.1 ` ' 1
```

The results do not indicate an influence of gender difference but they do suggest that the age difference may influence the ratings, so we fit a model with this term included.

```
> summary(update(Faces.clm1, location = . ~ . +
    agediff))
formula: ordered(SimRating) ~ sibs
data:    Faces

link    threshold nobs logLik    AIC      niter max.grad
probit flexible  960  -2163.83 4349.67 5(1)   2.28e-07
cond.H
1.1e+02
```

```
Coefficients:
      Estimate Std. Error z value Pr(>|z|)
sibs1   1.0737     0.0702    15.3   <2e-16 ***
---
Signif. codes:   0 `***' 0.001 `**' 0.01 `*' 0.05 `.'
  0.1 ` ' 1

Threshold coefficients:
      Estimate Std. Error z value
0|1    -0.6666     0.0579   -11.51
1|2    -0.3539     0.0546    -6.48
2|3    -0.0527     0.0534    -0.99
3|4     0.2312     0.0535     4.32
4|5     0.4678     0.0544     8.60
5|6     0.6994     0.0560    12.48
6|7     0.9173     0.0581    15.78
7|8     1.2094     0.0617    19.62
8|9     1.6092     0.0677    23.77
9|10    2.0499     0.0767    26.73
```

The results indicate that increasing age difference leads to a lower value of the linear predictor or a slightly lower internal response to the image pairs. The coefficient is significant but quite small, however, and its presence has little effect on the estimate of the influence of the factor sibs.

The objective of the above example was to demonstrate how more complex models describing the data can be explored. An important issue that we have overlooked is the random effects due to differences in sensitivity of observers and/or in difficulty of the images used in the study. We will revisit this data set to consider these effects in Chap. 9 when we examine mixed-effects psychophysical models.

Exercises

3.1. Derive a maximum likelihood fitting procedure for the parameters of the exponential model. Hint: just as in the Gaussian case, you begin with the maximum likelihood estimates of (p_{FA}, p_H) and then transform them to the desired parameterization. Plot ROC curves for three or four values of the parameter τ.

3.2. Prove that in the equal-variance Gaussian case, there is only one point on the ROC corresponding to any choice of (d', c). Hint: Consider YES responses.

3.3. In this exercise and the next we use data from a signal detection task reported by Devinck [51]. In Devinck's experiment, the signal was a large square patch that was presented at one of six levels of contrast and added onto a larger field in which the pixels were randomly modulated in luminance. The observer's task was to rate each trial on a scale from 1 to 5 as to how likely he thought that the signal was presented, where a rating of 1 indicates the least confidence and 5, the most. Each contrast level was presented 28 times, with the level chosen randomly from trial to trial, and interspersed with an equal number of catch-trials, in which no signal was presented. The data set for one observer is contained in the data frame ConfRates in our package **MPDiR** and consists of three components, the Contrast of the signal on each trial (a contrast of 0 indicates a catch-trial), the Rating given by the observer for that trial and a logical, and Stim indicating whether or not the signal was presented on a given trial. Fit these data with an equal-variance Gaussian model. Hint: See Sect. 3.7.

3.4. We continue to use the ConfRates data set from the previous exercise. In that exercise, you assumed that an equal-variance Gaussian model was appropriate. In this exercise, we ask you to test the equal-variance assumption for these data. Hint: See Sect. 3.7.

Chapter 4
The Psychometric Function: Introduction

4.1 Brief Historical Review

The psychometric function is a summary of the relation between performance in a classification task (such as the ability to detect or discriminate between stimuli) and *stimulus level* [59, 176]. Stimulus level is typically a measure of magnitude of a physical stimulus along a single physical dimension such as size, distance, light or sound intensity, concentration, or frequency. We will use the terms "stimulus level" and "stimulus intensity" interchangeably. We gave an example of a psychometric function in Chaps. 1 and 2, we discussed the close relationship between the psychometric function and the generalized linear model. While there is no need to use the GLM in fitting psychometric functions we will see that doing so makes it very convenient to apply advanced statistical methods to psychophysical data.

Psychophysicists use many experimental designs in evaluating human performance, many of which lead to data that can be summarized as a psychometric function [91]. For simplicity, we will focus on one of them, the venerable *method of constant stimuli*, one of the original methods proposed by Fechner [59]. In the method of constant stimuli, the experimenter first selects a physical stimulus dimension and then presents stimuli of different levels along that dimension in random order, typically repeating each level many times. At each presentation, the observer is limited to one of two responses, e.g., Present/Absent, Same/Different, etc. Again for simplicity, we will refer to such experiments as Yes–No experiments.

A Yes–No experiment is easily summarized. We need only specify the proportion of responses of one specific type (e.g., "Present," "Same," or "Correct") as a function of the stimulus level. We refer to this response as the "Yes" response and we will code the "Yes' response as 1, the alternative as 0. The psychometric function is just $P[\text{"Yes"} \mid I] = F(I)$ where I denotes stimulus intensity and F is the psychometric function. Stripped to its essentials, a method of constant stimulus experiment with N trials consists of a list of the stimulus levels $I = (I_1, I_2, \ldots, I_N)$ presented on each of N trials and the observer's response on each trial $r = (r_1, r_2, \ldots, r_N)$ where $r_i \in \{0, 1\}$.

K. Knoblauch and L.T. Maloney, *Modeling Psychophysical Data in R*, Use R! 32,
DOI 10.1007/978-1-4614-4475-6_4,
© Springer Science+Business Media New York 2012

Fig. 4.1 Simulated data (points) for a Yes–No experiment using the method of constant stimuli for $M = 7$ levels of intensity, with $n = 200$ presentations at each intensity level

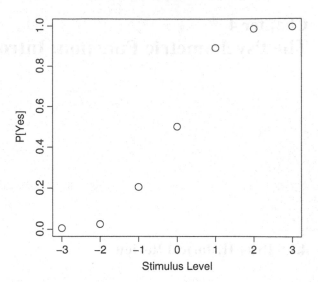

However, the experimenter will typically choose to work with a small number of distinct levels $\{I*_1 < I*_2 < \cdots < I*_M\}$ where level i occurs n_i times in the list of levels. The value M is often small (5–7) in typical experiments.

The observer's responses can then be summarized as the number of times $0 \le r_i \le n_i$ that he responded "Yes" when a stimulus of intensity $I*_i, i = 1, \ldots, M$ was presented. We have two ways of numbering the stimulus intensities $I = (I_1, I_2, \ldots, I_N)$ and $I* = \{I*_1 < I*_2 < \cdots < I*_m\}$. We will abandon the former in favor of the latter, allowing us to omit the $*$ in $I*_i$. Now I_i refers to a particular stimulus level that will be repeated n_i times over the course of the experiment. In collapsing the data in this way, we are excluding the possibility that the observer's performance changes during the experiment. Indeed, one justification for randomizing the order of presentation of the stimuli is to minimize the effect of any change in performance due to fatigue or practice. Accordingly, we collapse the data to the three vectors $I = (I_1, \ldots, I_M)$, $n = (n_1, \ldots, n_M)$, and $r = (r_1, \ldots, r_M)$. The total number of trials is, of course, the sum of the number of trials at each level: $N = n_1 + \cdots + n_M$.

In Fig. 4.1 we show a plot of the proportion of "Yes" responses vs intensity for a hypothetical experiment with $M = 7$ levels of intensity and $n_i = 200$ trials per level (a total of $N = 1,400$ trials). The key to this simulation is to specify the probability of a "Yes" response at each stimulus intensity level $p_i = F(I_i)$ and that we can do if we know the psychometric function $F()$. The n_i trials at level I_i are just the realization of a binomial variable with probability p_i (see Sect. B.3.2). With our notation, the observer in a method of constant stimuli experiment is reduced to M interleaved binomial variables, r_i, at previously chosen intensity levels I_i, arbitrarily set here to $-3, -2, \ldots, 3$.

The code used to generate the example is

```
> n <- 200
> M <- 7
> Intensity <- seq(-3, 3, len = M) # make a   vector
    of levels
> p <- pnorm(lntensity)
> nYes <- rbinom(M, n, p)
> pObs <- nYes/n
> plot(Intensity, pObs,
+    xlab = "Stimulus Level",
+    ylab = "P[Yes]")
```

To improve readability of the code, we replaced I by `Intensity`.[1]

To simulate these data, we assumed the psychometric function relating `level` to probability of a "Yes" response `p` is the cumulative distribution function of a Gaussian with $\mu = 0$ and $\sigma = 1$. We chose this psychometric function because it is often a good model of human data [57, 93]. The function `pnorm` accepts a vector of class "numeric" that corresponds to quantiles as its first argument and returns a vector of probabilities from a Gaussian cdf with values of `mean = 0` and `sd = 1`, as defaults.

The function `rbinom` takes as its first argument the number of random values to generate which here is $M = 7$. The second argument is a specification of the number of trials at each level $(n_1, \ldots, n_M)$ (here, `n = 200` for each level). The third argument is a vector of probabilities `p` with one for each level. The function `rbinom` returns a vector `r` of simulated counts of "Yes" responses at each level, the simulated data $r = (r_1, \ldots, r_M)$. We plot the proportion of "Yes" responses for each level.

The plot exhibits many of the characteristics of empirical Yes–No data. The probability of responding "Yes" increases from a value near 0 to almost 1 with increasing stimulus level. Of course the range of probabilities obtained depends on the range of stimulus levels chosen by the experimenter and also on the experimental design as we shall see below when we consider forced-choice designs.

The plot summarizes the observer's performance at the specified intensity levels but the experimenter would like to estimate how the probability of responding "Yes" varies continuously with stimulus level across all possible stimulus levels. He wishes to estimate the full psychometric function $p = \Phi(I)$.

Moreover, the experimenter typically wants a summary of the performance captured by a small number of readily interpretable parameters. The most common approach is to consider psychometric functions that are drawn from a *psychometric function family* that is typically the cumulative distribution function of a distributional family such as the Gaussian (Sect. B.2.5). Knowing the settings of the parameters completely specifies the psychometric function and also the model of the observer which it entails. If psychometric functions went to singles bars, they'd ask after each other's parameters rather than their zodiacal signs.

---

[1] Also, I is the name of a function in R, and it is best to avoid variables with the same names as functions.

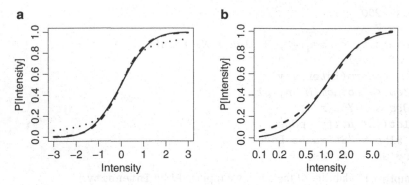

Fig. 4.2 (**a**) Psychometric functions based on the Gaussian (*solid*), logistic (*dashed*), and Cauchy (*dotted*) distributions. (**b**) Psychometric functions based on the log normal (*solid*) and Weibull (*dashed*) distributions. Note that the abscissa is logarithmic

The cumulative distribution functions (cdfs) of particular families of distributions (Gaussian, logistic, Weibull, etc.) have ranges running from 0 to 1 and are often used to specify psychometric function families (Appendix B). The p-functions associated with the built-in probability distribution families in R facilitate plotting and manipulating psychometric function families based on well-known cumulative distribution function families, notably the Gaussian and the Weibull. If, for example, we know or at least are willing to assume that the observer's psychometric function is drawn from the Gaussian family then we can write it as $p = \Phi(I; \mu, \sigma)$, where μ and σ are the location and scale parameters (Sect. B.2.6), respectively, of a Gaussian. We focus on estimating these parameters.

The earliest attempts at estimating models of the observer's psychometric function centered on least squares fits (possibly weighted) of polynomials to the response proportions correct transformed by a normal quantile function [73, 176]. As we saw in Sect. 2.7, this approach leads to estimates of the response variable and its variance on the quantile scale rather than in terms of correct proportion. In this chapter we focus on modern methods of fitting that maximize likelihood (Appendix B.4.2) and their implementation as GLMs [127]. Chapter 2 contains a brief introduction to GLMs.

We show examples from three different families of psychometric functions in Fig. 4.2a: the Gaussian, logistic, and Cauchy psychometric functions obtained, respectively, with the `pnorm`, `plogis`, and `pcauchy` functions with parameters approximately adjusted to match Φ (see Appendix B). The Gaussian psychometric function family is defined by

$$P(I) = \frac{1}{\sqrt{2\pi}\sigma} \int_{-\infty}^{I} e^{-\frac{(u-\mu)^2}{2\sigma^2}} \, du \qquad (4.1)$$

with free parameters μ and σ, the logistic psychometric function family is defined by

$$P(I) = \frac{\exp\left(-\frac{I-m}{s}\right)}{1 + \exp\left(-\frac{I-m}{s}\right)} \tag{4.2}$$

with free parameters m, s, and the Cauchy distribution is defined by

$$P(I) = \frac{1}{\pi} \arctan\left(\frac{I-m}{s}\right) + \frac{1}{2} \tag{4.3}$$

with free parameters m and s.

Just like the Gaussian, the other two distributions have two parameters, a *location* parameter and a *scale* parameter (see Sect. B.2.6). The three curves overlap but do not coincide. The Cauchy, in particular, does not asymptote at 0 and 1, respectively, on the left and right of the plot as rapidly as the other two. Of course, we can create a new family of psychometric functions based on any cumulative distribution function family. The fitting process, described in this chapter, intuitively corresponds to shifting and scaling these curves until they match as closely as possible the data points.

```
> opar <- par(mfrow = c(1, 2)) # set up a 1x2 array of plots
> # plot three location-scale psychometric function families
> Intensity <- seq(-3, 3, len = 100)
> plot(Intensity, pnorm(Intensity), type = "l", lwd = 2,
+            ylab = "P[Intensity]")
> lines(Intensity,plogis(Intensity, scale = sqrt(3)/pi),
     lty = 2, lwd = 3)
> lines(Intensity, pcauchy(Intensity, scale = 5/8), lty = 3,
     lwd = 3)
> mtext("a", adj = 0, cex = 2, line = 1)
> # plot two other psychometric function families
> #     (not location-scale families)
> Intensity2 <- seq(0.1, 10, len = 200)
> plot(Intensity2, plnorm(Intensity2), type = "l", lwd = 2,
     log = "x",
+          xlab = "Intensity", ylab = "P[Intensity]")
> lines(Intensity2, pweibull(Intensity2, shape = 1,
     scale = 1.5),
+ lty = 2, lwd = 3)
> mtext("b", adj = 0, cex = 2, line = 1)
> par(opar)
```

The three psychometric functions in Fig. 4.2a are nonzero for any choice of stimulus intensity: i.e., their use to model performance implies that there is some nonzero probability of getting a positive response at any stimulus level. When the physical stimulus intensity can only take on nonnegative values, (e.g., when light intensity, contrast, sound pressure or chemical concentration is the physical scale), it would make sense to consider functional forms that are 0 below some level of physical stimulus intensity.

The log normal and Weibull functions shown in Fig. 4.2b as solid and dashed curves, respectively, and plotted on a log abscissa are two examples of such functions. These are easily obtained with the `plnorm` and `pweibull` functions. The log normal psychometric function is defined as

$$P(I) = \frac{1}{\sqrt{2\pi}\sigma} \int_{-\infty}^{I} \frac{1}{u} e^{-\frac{(\log(u)-\mu)^2}{2\sigma^2}} \, du, \tag{4.4}$$

where μ and σ play the roles of location and scale, respectively, on a log axis. In the `plnorm` function, the arguments are called `meanlog` and `sdlog`, respectively. The Weibull psychometric function is defined as

$$P(I) = 1 - e^{\left(-\left(\frac{I}{\alpha}\right)^{\beta}\right)}, \tag{4.5}$$

where α and β play the roles of location and scale, respectively, but on a log axis. Note the corresponding arguments of `pweibull` are called `scale` and `shape`, respectively.

With appropriate choices of these parameters, the ordinate values are very similar over the range that most data are collected. When plotted on a log stimulus axis, these curves can shift without change of shape, just as the curves of Fig. 4.2a do on a linear abscissa. As noted above, on a log stimulus axis, the scale parameter acts as the location and the shape as the scale. A log normal curve can also be obtained by application of the `pnorm` function to the log of the stimulus dimension, as in the example of Chap. 1.

The Weibull function has seen extensive use in fitting psychometric function data in part because its cdf can be expressed as a closed form, unlike the Gaussian [144], and its parameters have a theoretical relation to a model of probability summation among independent, noisy mechanisms [21, 116]. The built-in probability functions of R allow the researcher to work with psychometric function families that have no analytic closed form.

Other choices are possible. Urban [176] suggested a generalized psychometric function based on the cumulative Gaussian, Φ, which we used in the example of Chap. 1 and plot as the solid curve in Fig. 4.2a using the `pnorm` function. Recently, researchers have considered using families of psychometric functions based on semi-parametric smooth curves [65]. We will illustrate these approaches in the next chapter.

In data from real experiments, the psychometric function may not asymptote at 0 for very low stimulus levels and may not asymptote at 1 for very high stimulus levels. Even at very low stimulus levels, the observer may continue to respond positively on a fixed proportion of trials. Similarly, occasional errors due to inattention, slips of the finger, or incorrectly entered data can result in less than perfect performance for stimuli that are clearly visible or discriminable.

To take into account these possibilities, a general form of the psychometric function, $P(x)$, can be written as

$$P(I) = \gamma + (1 - \gamma - \lambda)p(I; \alpha, \beta), \tag{4.6}$$

where $p()$ is a cumulative density function with location and scale parameters α and β on the scale I, γ is the lower asymptote, and λ is the distance from the upper asymptote.

Recent work has demonstrated the importance of including these additional (nuisance) parameters in the specification of the psychometric function to obtain the best fit to the data [188]. We will consider how to fit this more general psychometric function in the next chapter.

4.2 Fitting Psychometric Functions to Data

In this section we describe how to fit psychometric functions to data by maximum likelihood (see Sect. B.4.2). We assume that the experimenter has presented n_i trials at each of M intensities $I_i, i = 1, M$ and counts the number of "Yes" responses at each intensity $r_i, i = 1, m$. We will fit the data to a Gaussian psychometric function (4.1). The Gaussian family has two parameters, μ and σ. At any intensity I_i the probability of a "Yes" response depends on μ and σ and is $\Phi(I_i | \mu, \sigma)$ where $\Phi()$ denotes the cumulative distribution function of the Gaussian.

For each stimulus level, the count of "Yes" responses r_i is assumed (as above) to be a binomial random variable (Sect. B.3.2) with probability $\Phi(I_i | \mu, \sigma))$ of a "Yes" response and n_i trials. We can then compute that probability of the responses $(r_1, \ldots, r_M)$ at any one intensity I_i by the binomial formula

$$P[r_i | \mu, \sigma] = C(r_i, n_i)\Phi(I_i, \mu, \sigma)^{r_i}(1 - \Phi(I_i, \mu, \sigma))^{n_i - r_i}, \tag{4.7}$$

where $C(r_i, n_i)$ is the binomial coefficient for observing r_i successes out of n_i events. The overall probability of the data $r = [r_1, \ldots, r_M]$ at all levels is the product of the probabilities of each level

$$P[r | \mu, \sigma] = \prod_{i=1}^{M} P[r_i | \mu, \sigma] \tag{4.8}$$

which is also the likelihood of the data as a function of μ and σ. Estimation by the method of maximum likelihood (ML; Sect. B.4.2) is remarkably simple: we select $\hat{\mu}$ and $\hat{\sigma}$ to maximize this probability. The notation $\hat{\mu}$, etc. (read "mu hat") is traditional for ML estimators. Although the ML method is easy to define, it may appear somewhat arbitrary to the reader. Section B.4.3 describes some of the remarkable and desirable properties of ML estimates that motivate their use.

In particular, since the maximum of $f(x) > 0$ coincides with the maximum of $\log f(x)$, we get the same estimates $\hat{\mu}$ and $\hat{\sigma}$ if we instead maximize the logarithm of likelihood. Taking the logarithm of both sides of (4.8) gives the log likelihood, ℓ

$$\ell[\mu, \sigma | r] = \sum_{i=1}^{m} [r_i \log \Phi(I_i | \mu, \sigma) + (n_i - r_i) \log(1 - p_i(I_i | \mu, \sigma))] + C. \qquad (4.9)$$

The constant C contains the logarithm of all of the combinatoric terms $C(r_i, n_i)$ which do not depend on μ or σ. Since we are maximizing (4.9) and the maximum is not affected by C we can omit it from this point on.

The derivations in this section are readily extended to other choices of psychometric function (Weibull, logistic, etc.) including the four-parameter psychometric function family in (4.6) that allows for asymptotes other than 0 and 1. The only limit on the ML method is that our choice of likelihood function has a unique maximum.

4.2.1 Direct Maximization of Likelihood

In Chap. 1, we modeled the psychometric functions from the HSP data set using glm with the link set to "probit." This fits a Gaussian cdf to the data. With a bit more work, we could have performed the fit by maximizing the likelihood directly. Although, in general, it will be easier to fit data using glm as explained below, direct maximization offers more flexibility, in a number of situations, such as when the function defining the decision variable is nonlinear, when constraints need to be applied to the estimated coefficients, or when the link function is non-standard.

In order to perform a direct maximization for the HSP data set, we write a function, lnorm (shown below), that calculates the log likelihood according to (4.9). It takes two arguments. The first is a vector p with two entries that are possible estimates of the values of the parameters μ and σ. The second is a data frame d that includes three columns (and possibly others), named Q, nyes, and nno. Each is a vector of length M, corresponding to stimulus levels I, number of "Yes" responses, and number of "No" responses. Recall that HSP does not contain the latter two columns and that we had to calculate and add them to the data frame. We will do so again below. The data frame is just a convenient way to encapsulate the experimental data. Since different experiments will use different names for the three columns of the data frame, we cannot count on the terminology we are using so far in this chapter. With each data frame drawn from studies of actual human performance, we will likely encounter new names, illustrating MacLeod's shrewd observation that "...most scientists would rather share another person's toothbrush than adopt his terminology...." [119].

As in Chap. 1 we will transform the explanatory variable, Q, by its logarithm.

```
> lnorm <- function(p, d) {
+    mu <- p[1]
+    sigma <- p[2]
+    I <- log(d$Q);
+    pr <- pnorm(I, mean = mu, sd = sigma)   #
        Gaussian cdf
+    -sum( d$nyes * log(pr) +  d$nno * log( 1-pr ))
+    }
```

We use the pnorm function, again, to give us the normal cdf. The formulation can be simplified by using two additional arguments to pnorm: lower.tail and log.p. The former specifies on which side of the quantile the probability will be calculated and avoids having to write out terms of the form $1 - p$. The latter indicates that the logarithm of the probability should be returned directly by the function so that we can add the returned values directly.

```
> lnorm <- function(p, d) {
+    mu <- p[1]
+    sigma <- p[2]
+    I <- log(d$Q);
+    logpyes <- pnorm(I, mean = mu, sd = sigma,
+        lower.tail = TRUE, log.p = TRUE)
+    logpno  <- pnorm(I, mean = mu, sd = sigma,
+        lower.tail = FALSE, log.p = TRUE)
+    -sum( d$nyes * logpyes +  d$nno * logpno)
+    }
```

There is little reason to choose one version of lnorm over the other but the second illustrates some useful features of R. The version in our package **MPDiR** combines several of the steps above for compactness.

One last detail: the function that we will use to optimize, optim described below, will minimize not maximize the function it is given.[2] We added a minus sign so that the function lnorm returns the negative of log likelihood; minimizing this function will maximize likelihood.

The parameterization in terms of the mean and standard deviation of the Gaussian is different than that used in Chap. 1, but as we saw there, more directly related to the parameters of interest, the threshold and precision of the observations.

We calculate and add the response columns, nyes, and nno to the the data frame. We will estimate parameters for only one Observer/Run and leave fitting the other conditions as an exercise. The data are plotted in Fig. 4.3.

```
> HSP$nyes <- round(with(HSP, N * p/100))
> HSP$nno <- with(HSP, N - nyes)
> SHR2 <- subset(HSP, Obs == "SH" & Run == "R2")
```

---

[2]This is the default behavior of optim. It can be made to maximize by including the argument control = list(fnscale = -1).

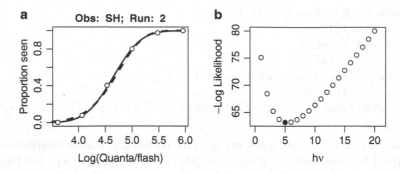

Fig. 4.3 (**a**) The probability of detecting a flash of light as a function of the log average number of quanta in the flash for the second run of observer SH. (**b**) The profile likelihood for the number of quanta/flash

Then, the optimization is performed with the following line of code.

```
> SHR2.norm <- optim(par = c(5, 0.35), lnorm, d = SHR2)
```

The function optim requires at least two arguments, par, a vector of initial estimates of the parameters to optimize, and a function, here lnorm, for which the parameters will be optimized. The argument par is a vector of parameter values that must correspond to the first formal argument of lnorm, which is p. Additional arguments to optim, such as d, are added as named parameters and are passed on to the function whose value is to be optimized. We can speed up the optimization process by providing not only the function lnorm but also a second function which returns the gradient of lnorm with respect to the parameters (the partial derivatives of lnorm with respect to each of the parameters). Providing such a gradient function will facilitate the fit and calculation of the standard errors of the parameters (see the example in [178]).

optim returns a list that includes the optimized parameters in a component par and the minimum of the function optimized (here the negative log likelihood) in a component value. Additional components include the number of function iterations, a code indicating whether the convergence criterion was reached and an optional message returned by the optimizer. If optim was called with the argument hessian = TRUE, an estimate of the Hessian matrix is also returned as a component of the list. The Hessian can be used to estimate confidence intervals for the parameters.

One disadvantage of using optim directly is that there are no built-in method functions to help us evaluate the results of the maximum likelihood fit.[3] Nevertheless, when the data are modeled in this fashion, it is not difficult to plot the data

---

[3]However, see the mle function in the recommended package **stats4** [146].

against the psychometric function defined by the parameter estimates as illustrated by the solid curve in Fig. 4.3. The code fragment generating this plot is

```
plot(I(p/100) ~ log10(Q), SHR2, xlab = "Log[Quant/flash",
    ylab = "Proportion seen")
> qq <- seq(3, 6, len = 200)
> pred.nrm <- pnorm(qq, mean = SHR2.norm$par[1],
+      sd = SHR2.norm$par[2])
> lines(qq, pred.nrm)
```

The reader can check that the values are very close to those obtained using `glm` in Chap. 1, taking into account its different parameterization.

Hecht et al. [79] used a psychometric function based on theory. They argued that variability in the detection of weak flashes was dictated principally by the quantum fluctuations of light. The variability of the emission from sources and the absorption by photopigments of integer numbers of photons was assumed to follow a Poisson density which, using their notation, is written as

$$P_n(a) = \frac{a^n e^{-a}}{n!}, \tag{4.10}$$

where P_n is the probability that a flash of light will provide at least the n quanta necessary for detection and a is the mean number of quanta per flash. If detection depends on the absorption of at least n quanta, the psychometric function should follow a curve of the probability of absorbing n photons or more, given by (4.10) summed from n to ∞ as a function of a. The exact shape of the psychometric function will be determined by n, the average number of photons required to generate a visual response.

Figure 4.4 shows a family of curves of the probability of absorbing at least n photons as a function of mean flash intensity, a, with each curve parameterized by n, the number of photons necessary to generate a visual sensation. The curve shape that best describes the frequency of seeing data gives the estimate, n, of the minimum number of quanta required for detection. Hecht et al. determined the best fit by sliding their data along the log abscissa, determining which curve was closest to the data, by eye, but it is straightforward to set up the problem in terms of maximizing the likelihood.[4]

To fit their model, we proceed as above and define a function, `lpois`, to calculate the negative log likelihood for a binomial probability that varies according to the cdf of 4.10, `ppois`. The function takes two arguments, a vector of parameters to adjust, p, and a data frame, d.

---

[4]They state, "No special statistical methods are necessary to determine which curve fits the data, since smaller and larger values of n are easily excluded by visual comparison." Such graphical short-cuts to fitting were common (and unavoidable) before availability of the enormous computational resources of the current era.

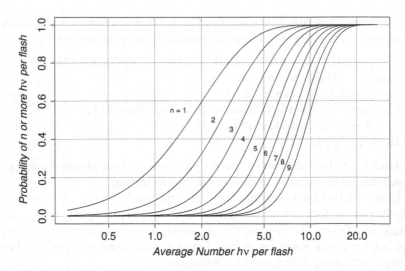

Fig. 4.4 The probability of a flash of light containing n or more quanta as a function of the average number of quanta in the flash. Each *curve* corresponds to a different average number of quanta per flash. The format of the graph was reproduced to closely follow that from [79] in which the notation $h\nu$, the product of Planck's constant and the frequency of light, giving the energy per photon, is used to denote a quantum of light

```
> lpois <- function(p, d)    { -sum(d$nyes * ppois(p[2],
    d$Q/p[1],
+     lower.tail = FALSE,  log.p = TRUE) +
+     d$nno * ppois(p[2], d$Q/p[1],
+        lower.tail = TRUE, log.p = TRUE))
+     }
```

The function ppois gives the sum. Its first argument, p[2], is the quantile, but here corresponds to the parameter of the curve, n, that we wish to estimate. The second argument is lambda which corresponds to the mean number of quanta in a flash, a. We scale this value by a second parameter, p[1], that implements the shifting of the curve along the logarithmically scaled abscissa. As detailed in [79], a number of factors intervene to attenuate the light intensity, measured at the corneal surface, before it reaches and is absorbed by the photopigments to trigger a visual event.

We again use the function optim to estimate the parameters for the psychometric function, giving starting values estimated from Fig. 4.4.

```
> SHR2.pois <- optim(c(18, 5), lpois, d = SHR2)
```

The predicted curve, obtained using ppois with the fitted parameters, is added to Fig. 4.3a as a dashed line and describes the data comparably to the Gaussian model.[5]

---

[5]The use of a Gaussian to describe these data is less heretical than one might think. For example, Crozier [42, 43] argued against a Poisson model and in favor of a Gaussian model to describe detection psychometric functions.

```
> pred.poi <- ppois(SHR2.pois$par[2],
+    exp(qq - log(SHR2.pois$par[1])), lower.tail = FALSE)
> lines(qq, pred.poi, lty = 2, lwd = 2)
```

The value of *n* is given by the second element of the par component.

```
> SHR2.pois$par
[1] 19.11  5.54
```

The Poisson distribution is discrete, however, and the density is only nonzero for integer values of the quantiles. ppois returns a constant value for quantiles in the half-open interval $[n, n + 1)$. Thus, we should truncate this value to 5. It is worth verifying that this is a minimum by profiling the likelihood for integer values around the minimum. The profile, shown in Fig. 4.3b, was obtained with the following code fragment:

```
> nn <- seq(1, 20)
> lpois1 <- function(q, p, d) lpois(c(q, p), d)
> SHR2.prof <- sapply(nn, function(iy) {
+    unlist(
+        optimize(lpois1, interval = c(1, 1000), p = iy,
+            d = SHR2) )
+ })
> plot(nn, SHR2.prof[2, ], xlim = c(0, 21),
+    xlab = expression(paste("h", nu)),
+    ylab = "-Log Likelihood")
> points(which.min(SHR2.prof[2, ]),
+    SHR2.prof[2, which.min(SHR2.prof[2, ])],
+    pch = 16, col = "black")
```

To calculate the profiled values, the problem reduces to estimating only one parameter for each value of *n*. The help page for optim suggests that it is better to use the function optimize in this case. We define lpois1 in terms of lpois but with the parameters separated. optimize requires an interval of values within which to search for the optimum that is provided as the second argument. We use the function sapply to loop through the values of *n* for each minimization. The values returned by the optimizer are in a list, however, so we unlist them, which allows sapply to return the profiled values and the minima of the objective function in a 2-row matrix, SHR2.prof. The minimum value is indicated by a black point at $n = 5$, slightly less than the value for 6.

Hecht, Shlaer, and Pirenne reported a value of 7 for this case. Comparison of the fits for values of *n* from 5 to 7 in Fig. 4.5 reveals the limits of the graphic method for determining the *best* fit to the data.

Fig. 4.5 Frequency of seeing data for observer SH for Run 2 from the HSP data set. The three *curves* indicate the maximum likelihood fits for the Poisson detection model for 5 (*solid*), 6 (*dashed*), and 7 (*dotted*) absorbed quanta per detection event

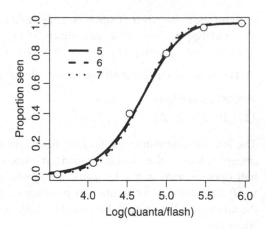

4.3 A Model Psychophysical Observer

In the next section, we describe fitting psychometric functions by glm. Here, we motivate its use by demonstrating that the relation between the variables of the linear predictor and the response variable in the binomial GLM is exactly that of the relation between the perceptual variables in the decision space and the observer's responses.

Psychophysical tasks typically concern a set of stimuli, $\{S_i\}$, that are presented in discrete trials to an observer. We use S here instead of I as previously, to emphasize that we could be dealing with representations beyond just intensity continua (see below). The observer's task is to classify each trial, yielding over the course of the experiment a discrete set of observable or behavioral responses, $\{R_j\}$. We will refer to these simply as the responses. We try to understand the observer's behavior by modeling the mechanisms that transform the stimuli to an internal representation and the processes of decision (or classification).

A general model of this process can be written

$$E[R = R_i] = g(f(S_i, \theta), \psi). \tag{4.11}$$

Here, S_i is some representation of the stimulus. It could be a detailed description of its distribution along physical continua (space, time, frequency, etc.), a feature or summary value of the stimulus (contrast, mean energy, position, etc.), or simply a nominal characterization of a particular stimulus from a specified set. The function f maps the stimulus variable to an internal representation as an *internal* response that we will refer to simply as the decision variable. When the decision variable takes a certain set (often a range) of values, the observer will respond R_i, otherwise an alternative choice from the set of behavioral responses. In general, the stimulus will include noise (some of which possibly added explicitly by the experimenter) so that the value of the decision variable for a given stimulus, S_i, is not deterministic. Thus, the expected value of the response will be a probability in the interval $(0, 1)$.

$$P(R = R_i) = g(f(S_i, \theta), \psi). \tag{4.12}$$

The function g maps the decision variable to this probability of emitting a particular response. It is, in effect, a psychometric function, as described in the preceding section. Both the functions f and g may require specification of additional parameters, θ (e.g., (α, β) of (4.6)) and ψ (e.g., (γ, λ) of (4.6)) for the model to fit the data adequately.

A further simplification can be introduced when the function f mapping the stimulus representation to the decision variable is taken to be linear. Then (4.12) becomes

$$P(R = R_i) = g(X_i \cdot \beta), \tag{4.13}$$

where for the moment we have ignored potential nuisance parameters, ψ. Now, all of the stimulus information is encoded in the values of the vector X and the coefficients β weight this information to produce the decision variable. All of the trials of an experiment can be taken into account by creating a matrix, $\mathbf{X}$, where each row indicates the stimulus information for that trial. Then, the model can be written as

$$\eta(P(R = R_i)) = \mathbf{X}\beta, \tag{4.14}$$

where $\eta = g^{-1}$. In this form, the model reveals itself to be a generalized linear model with link function η, the inverse of the psychometric function. We used this model in Chap. 3 to estimate the sensitivity of the observer, d'. In that case, the explanatory variable was a two-level factor indicating the presence or absence of the signal. In the case of fitting a psychometric function, the explanatory variables will often include covariates as well indicating the value of the stimulus along a physical continuum. This is how the intensity of the stimulus entered the model in the example in Chap. 1. Additional examples will be presented in succeeding sections to highlight this relation.

4.4 Fitting Psychometric Functions with `glm`

To summarize: in fitting psychometric functions we typically have in mind a psychometric function family such as the cdfs of Gaussian random variables. Each member of the family is indexed by the settings of a small number of parameters (for the Gaussian, the location parameter μ and the scale parameter σ). Fitting entails selecting the setting of the parameters to maximize the likelihood of the data observed. In the past few sections we did so by using optimization code to maximize likelihood directly.

In most experiments the experimenter is interested in measuring not just one psychometric function but several, in different experimental conditions, and he is testing whether performance differs with condition. If, for example, the experimenter systematically varies two experimental factors (e.g., the spatial frequency of a Gabor target and its spectral composition) in an ANOVA-like design, the resulting data can be organized by "row" and "column" with each row corresponding to an observation and the columns to the responses and the experimental factors and covariates for that observation (i.e., in a data frame). The optimization method is perfectly adequate to fitting psychometric functions in many conditions but then the experimenter must work out how to compare performance across conditions by nested-hypothesis tests (Sect. B.5), write the code to do so, and work out what it all means. He wishes to run down to the convenience store for a six-pack of beer but first he must build an automobile.

The glm function using a binomial family provides a powerful and simple method for fitting psychometric functions by maximum likelihood. The advantage of using glm is that we can then use it to analyze experimental designs with multiple conditions and multiple psychometric functions analogous to applications of the linear model (Chap. 2). For example, for the hypothetical situation proposed above with a response column named Resp and covariate columns for spatial frequency and spectral composition, named SF and SC, respectively, in a data frame named DataSet, we would write

```
glm(formula = Resp ~ SF + SC,
        family = binomial(link = probit),
        data = DataSet)
```

where we have included the formal argument names even though they would not be necessary here, because the arguments are in the same positions as in the function definition (see ?glm for the argument order and additional arguments that can be specified).

A formula object serves to define the relation between a response variable and the explanatory variables. For the binomial family, the response can be specified in three different ways: (1) as a vector of a two-level factor indicating the success/failure of individual trials, (2) as a two-column matrix indicating the numbers of successful and failed trials, (3) as a numeric vector indicating the proportion of successful trials. In the last case, the number of trials on which each proportion is based must also be indicated in the call, through the argument weights.

The explanatory variables typically include a covariate indicating the physical continua along which the stimuli vary. In its simplest form, the linear predictor is a linear function of this covariate. Additional factors, covariates, and their interactions can be included, however, to account for possible changes in the psychometric function with experimental condition. The formula language of R facilitates specifying and fitting models with and without specific terms. The significance of a term can be assessed using the anova method to compare a pair (or a sequence) of nested models, one of which does not contain the term of interest.

The functional form of the psychometric function (i.e., the psychometric function family) is chosen by specifying a link function as an argument of the binomial family function specified in the call to glm. The default link is the logit specifying a logistic psychometric function (4.2). A Gaussian cdf is obtained by specifying probit as the link. Two other links of general interest are the cauchit and cloglog, which allow specifying a Cauchy or Weibull (on a log intensity scale) cdf for the psychometric function, respectively. Users wishing to write their own link functions are advised to study the example on the help page for family and to study the function make.link in the **stats** package.

4.4.1 Alignment Judgments of Drifting Grating Pairs

The data set Vernier in the package **MPDiR** contains some results from an experiment measuring the sensitivity of an observer to differences in alignment between a pair of drifting, adjacent, horizontal gratings [167]. The capacity to detect the alignment of two features is referred to as Vernier acuity. The data set contains eight components

```
> data(Vernier)
> names(Vernier)
[1] "Phaseshift"  "WaveForm"   "TempFreq"   "Pc"
[5] "Direction"   "N"          "NumUpward"  "Num
                                           Downward"
```

Phaseshift is a numeric vector containing the vertical offsets between the two gratings in degrees of a grating cycle. For each condition, 11 phase shifts were tested between ±50°. The observer's task was to indicate which component of the pair appeared shifted upward. In this data set, the two gratings were of equal contrast. The component Pc indicates the proportion of trials in which a particular grating appeared shifted upward. From this proportion and the number of trials per condition, N, the numbers of upward and downward appearing trials were computed as components NumUpward and NumDownward. There were three experimental conditions, each specified by a two-level factor and indicated in components: WaveForm, whether the grating luminance profile was sinusoidal or square, TempFreq, the temporal frequency in cycles/sec (2 or 8 Hz) of the stimulus, and the Direction, whether the grating pair was drifting downward or upward.

All combinations of conditions were tested giving eight psychometric functions. As usual, it will be helpful to start by graphing the data to get a feel for what conditions might influence the results. We use the xyplot function from the **lattice** package to obtain an overview of the data. It would help to visualize the trends if we could include an initial fitted psychometric function for each condition. xyplot provides for connecting the points by line segments, plotting a regression line or loess curve through each panel using either the type argument or by providing panel functions in the panel argument. These depend on linear models, but here we have a generalized linear model.

We can write a custom panel function that will add such a curve. It takes the proportion correct and the number of trials as the second and third arguments from

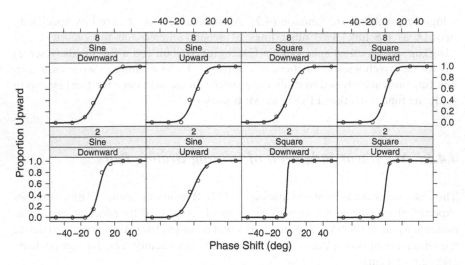

Fig. 4.6 Psychometric functions from the Vernier data set [167]. Each panel is fit with a Gaussian cdf using the custom panel function, panel.psyfun, described in the text

which it calculates the numbers of successful and unsuccessful trials. We allow, as well, for a user specification of the link function. glm is used to obtain a model for each panel from which predicted values are obtained and displayed with the panel.lines function. A more sophisticated version might permit the user to control a number of other features, such as the number of points used to draw the smooth curves and whether standard errors of the fit should be included.

```
> panel.psyfun <- function(x, y, n, lnk = "logit", ...) {
+    xy.glm <- glm(cbind(n * y, n * (1 - y)) ~ x,
+           binomial(lnk))
+    rr <- current.panel.limits()$xlim
+    xx <- seq(rr[1], rr[2], len = 100)
+    yy <- predict(xy.glm, data.frame(x = xx),
+           type = "response")
+    panel.lines(xx, yy,  ...)
+ }
```

The panel function once defined can be exploited in a user-specified function assigned to the panel argument of xyplot as shown in Fig. 4.6. The strips above each plot identify the conditions. We use the argument aspect = xy that tries to set the aspect ratio according to the 45° banking rule. This is claimed to facilitate comparison of slope differences between curves. At a glance, the curves are quite similar though the rising portion of the 2 Hz square-wave condition appears steeper than the others. There might be a slight difference in the steepness between upward and downward conditions for both waveforms of the 2 Hz condition, though this is more subtle and less evident by eye.

```
> print(
+           xyplot(Pc ~ Phaseshift | Direction * WaveForm *
+              TempFreq,
+           data = Vernier, layout = c(4, 2), aspect = "xy",
+           xlab = "Phase Shift (deg)", ylab = "Proportion Upward",
+           panel = function(x, y, ...){
+                    panel.xyplot(x, y, ...)
+                    panel.psyfun(x, y, 20, lnk = "probit",
+                       lwd = 2, ...)
+                    }
+           ))
```

We begin then by fitting a model including all possible effects. We use the default link function, `logit` and obtain fits for `probit` and `cauchit` links with the `update` method. Note that the link must be specified by updating the family argument of `glm` through which it is passed as an argument.

```
> v <- list()
> v[["logit"]] <- glm(cbind(NumUpward, NumDownward) ~
+    Direction * WaveForm * TempFreq * Phaseshift,
+    binomial, Vernier)
> v[["probit"]] <- update(v[[1]], family
  = binomial(probit))
> v[["cauchit"]] <- update(v[[1]], family
  = binomial(cauchit))
> v[["weibull"]] <- update(v[[1]], family
  = binomial(cloglog),
+    . ~ Direction * WaveForm * TempFreq *
+       log(Phaseshift + 50.1))
```

Sun et al. [167] fit their data with a Weibull cdf. This is achieved here by taking the log of the covariate `Phaseshift` and using the `cloglog` link (see p. 154).

As the values must be greater than 0, we added a constant to translate the values, as did the original authors.[6] We can compare the different forms of the psychometric function using the AIC criterion. We have stored each model object as a named component in a list, v, to easily extract the four AIC values for comparison, using the `sapply` function.

```
> sapply(v, AIC)
   logit   probit  cauchit  weibull
     113      110      158      116
```

The lower AIC for the probit link suggests that the Gaussian cdf provides a better description of the data, so we will continue the example with the probit link.

---

[6]H. Sun, personal communication.

4.4.2 Selecting Models and Analysis of Deviance

The models that we fit above to the `Vernier` data set include all of the possible
interactions of the terms in the formula object and generate 16 coefficients to
describe the variation of psychometric functions across the eight conditions. R contains a number of tools for testing the significance of these terms to find the most
parsimonious description of the data. The functions `dropterm` and `addterm` in the
MASS package test the significance of dropping and adding terms, given a model
object, while respecting the marginality conditions of the model.[7] We have started
with the most complex model, so that we will proceed with repeated applications of
`dropterm`. For models with a large number of factors, **MASS** contains a function
`stepAIC` which will hunt systematically through the hierarchy of models for the
one with the lowest AIC value. However, when the model is not too unwieldy, it can
be informative to perform the search interactively as we will here.

The first application of `dropterm` indicates that the highest order interaction is
not significant, so we update the model without that term.

```
> library(MASS)
> dropterm(v[["probit"]], test = "Chisq")
Single term deletions

Model:
cbind(NumUpward, NumDownward) ~ Direction * WaveForm *
    TempFreq * Phaseshift
                                           Df Deviance AIC    LRT
<none>                                         11.3 110
Direction:WaveForm:TempFreq:Phaseshift  1     11.6 109 0.356
                                           Pr(Chi)
<none>
Direction:WaveForm:TempFreq:Phaseshift     0.55
> v.prob2 <- update(v[["probit"]], . ~ . -
+     Direction:WaveForm:TempFreq:Phaseshift)
```

Examining the analysis of deviance table with `anova` suggests that the third-
order terms including the `Direction:WaveForm` interaction are not significant,
subsequently verified using `dropterm`. The `anova` method with a single model
object as argument performs a sequential decomposition, so the order of the terms
may matter. The `dropterm` function evaluates the effect of eliminating each non-

---

[7]Lower-order terms are marginal to higher-order terms. For example, main effects are marginal
to interactions and simpler interactions are marginal to more complex ones. It is advised to test
higher-order interactions without removing marginal effects that include the same terms as the
higher-order effects. Conversely, one should not test marginal effects in the presence of significant
non-marginal effects.

marginal term, in turn, and gives a result that does not depend upon the order of the
terms. We update the model without the non-significant, non-marginal terms.

```
> v.prob2.anova <- anova(v.prob2, test = "Chisq")
> data.frame(terms = rownames(v.prob2.anova),
+    P = v.prob2.anova[, 5] )
> dropterm(v.prob2, test = "Chisq")
```

```
                            terms          P
1                            NULL         NA
2                       Direction   9.11e-01
3                        WaveForm   7.37e-01
4                        TempFreq   8.23e-01
5                      Phaseshift   9.34e-280
6               Direction:WaveForm  4.89e-01
7               Direction:TempFreq  6.06e-01
8                WaveForm:TempFreq  6.88e-01
9             Direction:Phaseshift  8.38e-01
10             WaveForm:Phaseshift  4.98e-04
11             TempFreq:Phaseshift  4.55e-04
12     Direction:WaveForm:TempFreq  6.06e-01
13  Direction:WaveForm:Phaseshift  1.33e-01
14  Direction:TempFreq:Phaseshift  1.45e-03
15   WaveForm:TempFreq:Phaseshift  9.34e-06
```

```
> dropterm(v.prob2, test = "Chisq")
Single term deletions

Model:
cbind(NumUpward, NumDownward) ~ Direction + WaveForm +
    TempFreq + Phaseshift + Direction:WaveForm +
    Direction:TempFreq + WaveForm:TempFreq +
    Direction:Phaseshift + WaveForm:Phaseshift +
    TempFreq:Phaseshift + Direction:WaveForm:TempFreq +
    Direction:WaveForm:Phaseshift +
    Direction:TempFreq:Phaseshift + WaveForm:TempFreq:Phaseshift
                                 Df Deviance  AIC   LRT  Pr(Chi)
<none>                                 11.6  109
Direction:WaveForm:TempFreq       1    12.6  108  0.95  0.3309
Direction:WaveForm:Phaseshift     1    11.7  107  0.06  0.8138
Direction:TempFreq:Phaseshift     1    19.0  114  7.34  0.0067 **
WaveForm:TempFreq:Phaseshift      1    31.3  126 19.64 9.3e-06 ***
---
Signif. codes:  0
```

```
> v.prob3 <- update(v.prob2, . ~ . -
+   Direction:WaveForm:(TempFreq + Phaseshift))
```

Continued application of dropterm suggests dropping one second-order interaction that is not included in the retained third-order ones, but a further application results in no additional terms to eliminate.

```
> dropterm(v.prob3, test = "Chisq")

Single term deletions

Model:
cbind(NumUpward, NumDownward) ~ Direction + WaveForm +
   TempFreq + Phaseshift + Direction:WaveForm +
   Direction:TempFreq + WaveForm:TempFreq +
   Direction:Phaseshift + WaveForm:Phaseshift +
   TempFreq:Phaseshift + Direction:TempFreq:Phaseshift +
   WaveForm:TempFreq:Phaseshift
                                Df Deviance AIC   LRT Pr(Chi)
<none>                             12.7 106
Direction:WaveForm              1  13.1 104  0.44  0.5082
Direction:TempFreq:Phaseshift   1  20.9 112  8.19  0.0042 **
WaveForm:TempFreq:Phaseshift    1  34.1 125 21.34 3.8e-06 ***
---

> v.prob4 <- update(v.prob3, . ~ . - Direction:WaveForm)
```

The two significant third-order interaction terms correspond actually to second-order interactions of factors with the steepness of the psychometric function, indicated by the slope of the covariate term, Phaseshift. The significant interaction WaveForm:TempFreq:Phaseshift corresponds to the greater steepness that we remarked in Fig. 4.6 for the 2 Hz, square-wave condition. The Direction:TempFreq:Phaseshift term corresponds to the slight difference in the slopes that we noticed for different levels of Direction in the 2 Hz condition.

Ordinarily, we would leave the analysis here and proceed to plotting the model, but there is an additional simplification that is reasonable to consider in the case of these data. As noted, the interactions of the factors with the covariate term in the model influence differences in steepness of the psychometric function across conditions and concern the precision or sensitivity of the observer to differences along the physical dimension of the covariate. The factor terms, being marginal to these interactions, are more difficult to interpret. In this case, however, the marginal terms correspond to biases in the judgment, possibly induced by the experimental condition. It is possible, for example, that one of the experimental conditions induces the observer to detect a downward shift when in fact the phase shift is in the upward direction. This is, in fact, what occurs when the contrasts of the two gratings are unequal [167]. When the intercept term of the linear predictor is 0, the Gaussian psychometric function passes through the 50% point at 0 phase shift.

If a factor is significant, it suggests that the experimental condition introduces a bias away from 50% in the observer's judgments. This reasoning leads us to examine a model without main effects of the factors as a test of the observer's bias and to test it against the model that includes them.

```
> v.prob5 <- update(v.prob4, . ~
+    Phaseshift:((TempFreq + Direction + WaveForm) +
+     + TempFreq:(Direction + WaveForm)) - 1)
> anova(v.prob5, v.prob4, test = "Chisq")

Analysis of Deviance Table

Model 1: cbind(NumUpward, NumDownward) ~ Phaseshift:TempFreq +
   Phaseshift:Direction + Phaseshift:WaveForm +
   Phaseshift:TempFreq:Direction +
   Phaseshift:TempFreq:WaveForm - 1
Model 2: cbind(NumUpward, NumDownward) ~ Direction + WaveForm +
   TempFreq + Phaseshift + Direction:TempFreq +
   WaveForm:TempFreq + Direction:Phaseshift +
   WaveForm:Phaseshift + TempFreq:Phaseshift +
   Direction:TempFreq:Phaseshift + WaveForm:TempFreq:Phaseshift
  Resid. Df Resid. Dev Df Deviance P(>|Chi|)
1        58       14.8
2        52       13.2  6     1.67       0.95
```

Examination of the analysis of deviance table indicates that we can retain the simpler model without main effects in which the experimental conditions do not lead to biases in the offset detection task. The final model is described as

```
> Pc ~ (WaveForm + Direction):TempFreq:PhaseShift
```

4.5 Diagnostic Plots

Now, that we have chosen a model, we should examine diagnostic plots to verify that assumptions underlying the model are reasonably met by the data. Plotting an object of class "glm" produces the same diagnostic plots as plotting an object of class "lm." In fact, there is no plot.glm function and the plot.lm method serves for both. The function class applied to objects returned by glm returns a vector of two elements.

```
> class(v.prob5)
[1] "glm" "lm"
```

These objects inherit from class "lm" so that when a method for "glm" does not exist, the call is dispatched to the method for the next element, here "lm." If a method for "lm" does not exist, then the default method is called. If there is no default method, an error will occur.

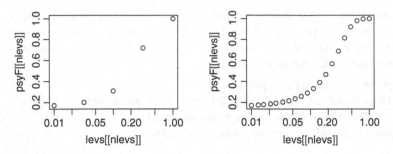

Fig. 4.7 Underlying probabilities of response for 5 and 20 stimulus levels

There is a problem, however, with interpreting the standard diagnostic plots from binomial models that is especially exacerbated in the case of binary data. We will illustrate this with a simulated example. Let's consider two experiments, one with 5 stimulus levels (a typical number used in a real experiment) and one with 20 levels (an atypically large number). The sampled points from the true underlying psychometric functions are illustrated in Fig. 4.7 with the code to generate the values.

```
> GrpResp <- Grp.glm <- vector("list", 2)
> names(GrpResp) <- names(Grp.glm) <- names(levs)
> for (nlevs in names(levs)) {

> opar <- par(mfrow = c(1, 2))
> levs <- psyF <- vector("list", 2)
> names(psyF) <- names(levs) <- c("5", "20")
> for (nlevs in names(levs)) {
+          levs[[nlevs]] <- round(10^seq(-2, 0,
              len = as.integer(nlevs)), 3)
+          psyF[[nlevs]] <- pnorm(levs[[nlevs]], mean = 0.2,
              sd = 0.2)
+          plot(levs[[nlevs]], psyF[[nlevs]], log = "x")
+ }
> par(opar)
```

Next, we simulate the binary responses of individual responses of the individual trials. We assume that each level has been repeated 100 times. Then, we create a data frame with the responses and the levels. Finally, we fit the model to each set of responses using `glm` and store the results in a component of the list `indiv.glm`.

```
> Ntrials <- 100
> indiv.lst <- indiv.glm <-  vector("list", 2)
> names(indiv.lst) <- names(indiv.glm) <- names(levs)
> for (nlevs in names(levs)){
```

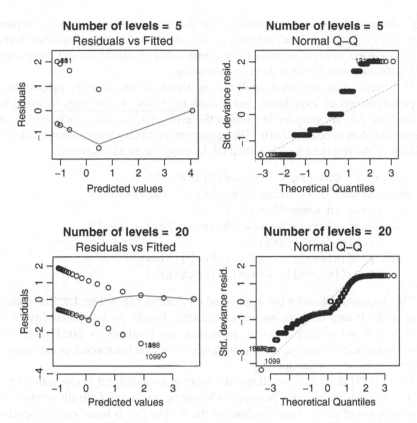

Fig. 4.8 Diagnostic plots for glm fitted to simulated binary responses with 5 (*top*) and 20 (*bottom*) stimulus levels. The *left-hand plots* indicate the deviance residuals plotted against the fitted values (on the linear predictor scale) and the *right-hand plots* indicate the ordered residuals plotted against theoretical Gaussian quantiles

```
+    Resp <- rbinom(Ntrials * length(psyF[[nlevs]]), 1,
+        psyF[[nlevs]])
+    indiv.lst[[nlevs]] <- data.frame(resp = Resp,
+        levs = levs[[nlevs]])
+    indiv.glm[[nlevs]] <- glm(resp ~ levs,
+        binomial(probit), indiv.lst[[nlevs]])
+ }
```

We next examine the first two diagnostic plots for both data sets, the residuals vs the fitted values and the QQ plots (Fig. 4.8).

The simulated data with 20 levels clearly indicate the typical patterns obtained for these two diagnostic plots, but the results are evident, as well, in the data with only 5 levels. First, consider the residual vs fitted value plots. There are only two responses $(0, 1)$ so that the residuals appear to fall along two contours. At the extremes, the residuals approach 0 as the expected proportions of response approach the only responses available to the observer. The QQ plots also deviate in a systematic fashion, seeming to be constructed of two sigmoidal components with a discontinuity between them in the middle.

Observers' data are often analyzed in terms of the average proportion of responses across an experiment and not on the basis of responses to individual trials. If the data are analyzed in terms of the average responses at each level, do the diagnostic plots appear similarly? We aggregate the binary responses into a format suitable as the response for a binomial GLM using the `table` function.

```
> GrpResp <- Grp.glm <- vector("list", 2)
> names(GrpResp) <- names(Grp.glm) <- names(levs)
> for (nlevs in names(levs)) {
+    GrpResp[[nlevs]] <- t(with(indiv.lst[[nlevs]],
+        table(resp, levs)))
+    Grp.glm[[nlevs]] <- glm(GrpResp[[nlevs]] ~
+        levs[[nlevs]], binomial(probit))}
```

The diagnostic plots for the aggregated data are shown in Fig. 4.9. The patterns seen in the binary data are no longer evident, though the interpretability of the plots is still not entirely lucid. In particular, the residuals vs fitted value plots show systematic features in the residuals that would be interpreted as evidence for heterogeneity under other models.

Wood [198] describes two diagnostic plots that are helpful in evaluating fits to binary data. These are each based on a bootstrap analysis of the distribution and independence of the deviance residuals of the fit. The first is based on a comparison of the empirical distribution of the residuals with a confidence envelope about them, generated by a bootstrap method. The second evaluates dependencies among the residuals by comparing the number of runs of positive and negative values in the sorted deviance residuals with the distribution of runs from bootstrapped samples. We have included in our package **MPDiR** a function, `binom.diagnostics`, for running these analyses along with a `plot` method for visualizing them. We run the diagnostics for the fits to the simulated binary data for 5 and 20 stimulus levels using `lapply` which returns a list with the output of the two runs as components. We used 10,000 simulations which takes 10–20 min on the platforms that we used.

```
> indiv.diags <- lapply(names(levs),
+    function(x) binom.diagnostics(indiv.glm[[x]],
+        nsim = 10000))

> names(indiv.diags) <- c("5", "20")
```

The results for each run are returned in a list of class "binom.diag" that contains five components, for example, for the five stimulus level data set.

```
> str(indiv.diags[[1]])
```

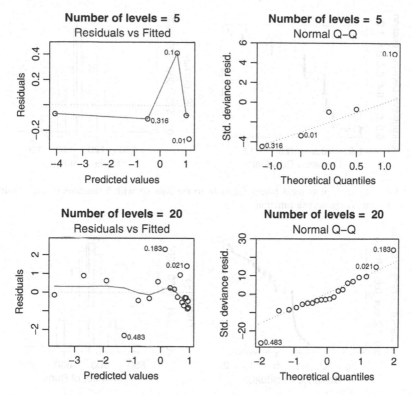

Fig. 4.9 Diagnostic plots for `glm` fitted to the simulated binary data after aggregation into average responses with 5 (*top*) and 20 (*bottom*) stimulus levels. The *left-hand plots* indicate the deviance residuals plotted against the fitted values (on the linear predictor scale) and the *right-hand plots* indicate the ordered residuals plotted against theoretical Gaussian quantiles

```
List of 5
 $ NumRuns  : int [1:10000] 134 119 139 132 123 136 133 154 159..
 $ resid    : num [1:10000, 1:500] -4.52 -4.44 -4.38 -4.33 -4.3..
 $ Obs.resid: Named num [1:500] -0.53002 1.9238 -0.77231 0.8698..
  ..- attr(*, "names")= chr [1:500] "1" "2" "3" "4" ...
 $ ObsRuns  : int 133
 $ p        : num 0.595
 - attr(*, "class")= chr [1:2] "binom.diag" "list"
```

NumRuns is a vector indicating the number of runs of positive and negative values in the ordered deviance residuals for each bootstrap fit. The second component, resid, is a matrix with each column the deviance residuals for one bootstrap fit. Obs.resid indicates the deviance residuals from the fit to the original data, and ObsRuns gives the number of runs for the original data. Finally, p is the proportion of times the number of runs from the bootstraps is less than or equal to the observed number.

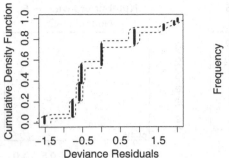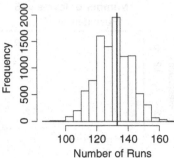

Fig. 4.10 Diagnostic plots for a binary GLM fit to the data set with 5 stimulus levels, obtained with the `binom.diagnostics` function

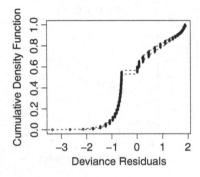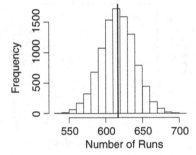

Fig. 4.11 Diagnostic plots for a binary GLM fit to the data set with 20 stimulus levels, obtained with the `binom.diagnostics` function

```
> plot(indiv.diags[[2]], cex = 0.5)
```

The `plot` method generates a pair of graphs, shown for the simulated data sets in Figs. 4.10 and 4.11. The graphs on the left display the distribution of residuals as points with the bootstrap estimate of the 95% confidence intervals as dashed lines. For better visualization, we reduced the point size in half using the `cex` argument. A suspect fit to the data would show points falling systematically outside the limits of the dashed lines. The right-hand graphs display histograms of `NumRuns`, the number of runs of the same sign residuals from the boostrap fits. The vertical line indicates the observed number of runs. A vertical line too far to the left would indicate that the fit to the obtained data contained too few runs, a sign of a systematic error in the fit. We extract the observed proportions for the fits to the two simulated data sets from the `p` component of the object.

```
> sapply(indiv.diags, "[[", "p")
    5    20
0.595 0.520
```

Versions of `binom.diagnostics` can also be found in the **MLDS** and **MLCM** packages that have been modified for the objects output by the modeling functions therein. We will exploit these in Chaps. 7 and 8 when we analyze the fitted data structures to binary judgments associated with difference scaling and conjoint measurement, respectively. We leave as an exercise for the reader the modification of `binom.diagnostics` to handle binomial fits, in general, and the evaluation of the model selected to fit the `Vernier` data.

Additional useful tools for analyzing binomial models can be found in the **binom-Tools** package [33].

4.6 Complete Separation

In Sect. 3.3.7, we discussed the warning messages emitted by the `glm` function that signal evidence for complete separation. That discussion was in the context a of Yes–No experiment for a single intensity value. It is not infrequent, however, to obtain such a warning message when fitting psychometric functions with `glm`, especially when simulating a large number of data sets, as in the preceding section.

An extreme example that would generate the warning is obtained by setting all responses to "No" to one side of a criterion point and all responses to "Yes" on the other, but less extreme conditions can also generate the warning as in the following simulated data set:

```
> x <- 10^seq(-2.5, -0.5, len = 6)
> pr <- pweibull(x, 3, 0.075)
> Trials <- 30
> set.seed(16121952)
> ny <- rbinom(length(pr), 30, pr)
> nn <- 30 - ny
> res <- glm(cbind(ny, nn) ~ log10(x), binomial(cloglog))
```

```
Warning message:
glm.fit: fitted probabilities numerically 0 or 1 occurred
```

We chose six points along a psychometric function following a Weibull cdf. The stimulus levels are separated by a constant ratio. Thirty trials were generated for each stimulus level. Since the example is based on the generation of random samples, the phenomenon that we wish to demonstrate might not occur on every run of the simulation. Therefore, we initialized the default random number generator to a state from which we have already found the phenomenon to occur using the `set.seed` function. Of course by doing this, the results and the outcome are no longer random. Plotting the data and the fitted function can help us understand why the warning is generated.

Figure 4.12 reveals that only one point falls on the rising section of the psychometric function, so there is very little information about its true shape in this range of stimulus values. Several strategies are possible to improve the situation. The best approach would be to repeat the experiment with more stimulus levels in the range over which the

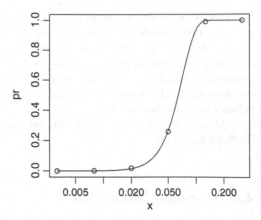

Fig. 4.12 Simulated data set illustrating conditions that lead to warning message of complete separation when the data are fit with `glm`

function is increasing and fewer where it is not. Adaptive psychophysical methods, discussed in the next chapter, will concentrate the sampling at stimulus levels in this region. In some situations, increasing the number of trials for each stimulus level will help as it will be less likely to obtain all "No" or "Yes" responses at stimulus values near the rapidly changing portion of the curve, thus, generating more points in this region. Another alternative is to fit the data using a penalized likelihood function, which introduces regularization in the fitting process. GAM and loess approaches to fitting psychometric functions use penalized likelihoods and are discussed at the end of the next chapter. Another approach is provided by the function `brglm` in the package of the same name [103]. This function uses a similar interface to `glm` and implements a method of obtaining a bias-reduced fit developed by Firth [61] that performs better than maximum likelihood methods in the presence of complete separation.

In later sections, we illustrate how to fit data from a variety of psychophysical experiments using `glm`. In more complex situations, the psychometric function is not visualized explicitly, yet similar constraints apply for avoiding a warning from `glm` of complete separation, i.e., increasing the number of trials and modifying the number and spacing of the stimuli sampled.

4.7   The Fitted Model

Finally, we can verify how well the predicted model describes the data graphically. To use the `predict` method to generate sufficient points to draw a smooth curve through the data, we must create a new data frame with the same variables as in the model. This is conveniently accomplished using the function `expand.grid`.

```
> nd <- expand.grid(Phaseshift = seq(-50, 50, len = 200),
+    WaveForm = c("Sine", "Square"),
+    TempFreq = factor(c(2, 8)),
+    Direction = c("Downward", "Upward"))
> v.pred <- predict(v.prob5, newdata = nd, type = "response")
```

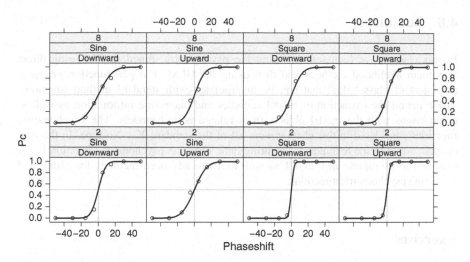

Fig. 4.13 Psychometric functions from the Vernier data set with the predicted curves from model v.prob5 described in the text. This model includes no intercept so that all of the functions attain their half maximal point at 0, as indicated by the intersection of the *grey vertical* and *horizontal lines*

```
> nd$pred <- v.pred
> nd$ID <- with(nd, WaveForm:TempFreq:Direction)
> Vernier$ID <- with(Vernier, WaveForm:TempFreq:Direction)
```

We add the predicted values as a column pred to the data frame. We will need to create a unique identifier for each condition in both the original data frame and the new one in order to relate the predicted values to the correct conditions in the data set. We do this by creating a new factor, ID, from the interaction of the three factor variables. The code displayed in Fig. 4.13 illustrates how to pass the predicted values to the panel function using the identifier and the subscripts argument. The reduced model with only six parameters fits the data remarkably well.

```
> print(
+ xyplot(Pc ~ Phaseshift | Direction + WaveForm + TempFreq,
+        subscripts = TRUE, ID = Vernier$ID, aspect = "xy",
+        data = Vernier, layout = c(4, 2),
+        panel = function(x, y, subscripts, ID, ...){
+                panel.xyplot(x, y, ...)
+                panel.abline(v = 0, h = 0.5, col = "grey", ...)
+                which <- unique(ID[subscripts])
+                llines(nd$Phaseshift[nd$ID == which],
+ +}

+))
```

4.8 Summary

In this chapter, we focussed on fitting the psychometric function, first using direct maximum likelihood methods and then using the GLM. The `glm` function produces an object of class "glm" that can be manipulated with standard method functions. These permit the extraction of model statistics and diagnostic information as well as comparisons with the model objects from related nested models. The psychometric functions considered in this chapter were all of the standard Yes–No type. In the next chapter, we examine techniques for fitting data when the psychometric function is not expected to asymptote at 0 and 1 as well as more advanced methods for fitting and analyzing psychometric functions.

Exercises

4.1. For fitting binomial models with `glm`, the response variable can be specified in several ways. In Sect. 4.4, we specified the response variable as a two-column matrix with columns indicating the integer counts of the two responses. In other places, we code the response as a logical or two-level factor for a Bernoulli variable indicating the outcome of individual trials. A third possibility is not only to use a vector of the proportion of successes as the response but also to specify the number of trials in the weights argument. Refit the psychometric functions for the HSP data set using this last method and compare the results from the `summary` and `anova` methods to those obtained in the text.

4.2. The two aggregated response formats, (p, N) or $(Np, N - Np)$, can each be transformed into a vector of Bernoulli responses, i.e., a vector of Np 1's and $N - Np$ 0's. Create a data set from the HSP data set in this format and refit the data with `glm`. Compare the results from the `summary` and `anova` methods and note which characteristics of the model have changed with the data format and which have remained invariant.

4.3. Modify the `panel.psyfun` function from Sect. 4.4.1 so that the number of points used in the smooth curves can be modified and so that the standard error of the fit can be added to the graphs as an envelope as in Fig. 1.3. You will need to pay attention to the order in which the components within each panel are drawn. What other options might be interesting to add to this panel function?

4.4. The **MASS** package contains a function `stepAIC` that performs an automatic evaluation of a range of nested models by default on the basis of the AIC model selection criterion. Read the help page for this function and study the examples. Then apply it to one of the models fit to the Vernier data set. Does the final model agree with that found in the text using likelihood ratio tests? How would you change the arguments so that the model evaluation is based on BIC? How does this change the final selected model?

4.5. Modify the `binom.diagnostics` function from the MPDiR package to handle binomial data that are specified in terms of average proportion of successes? With this modified function, evaluate the final model that we selected for the Vernier data set in Sect. 4.4.1.

4.6. The data set `GlassPatterns` in the **MPDiR** package contains observations from a single observer who was presented on each trial with a pattern of hundreds of dot pairs. The pairs could be oriented to fall along a set of virtual concentric circles. If all of the dot pairs were so aligned, a concentric pattern, called a Glass Pattern after its discover [69], would be readily visible. In this experiment, only a varying proportion of the dot pairs were concentrically organized and the rest were oriented randomly. The proportion of concentrically oriented dot pairs is indicated by the component `Coherence` in the data frame and varied between 0 (all dot pairs randomly oriented) and 0.6 (60% of the dot pairs concentrically organized). The observer's task on each trial was to decide if a concentric pattern was present or not (one-sixth of the trials contained images of zero coherence). The column `Pyes` gives the proportion of trials for each condition on which the observer responded that a pattern was present, and the column `N` indicates the total number of trials for a condition. Besides the covariate `Coherence`, there is also a factor `Direction` with levels `A0`, `A45`, ..., `A315` that indicates the color of the dots. The color is specified as an angle (or direction) in a plane of constant luminance in a color space. The angles are also specified as a numeric variable in the column `Angle`.

Plot the data as a function of coherence for each color direction.

4.7. Continuing with the `GlassPatterns` data set, fit three models to the data in which (1) the intercept and slope of the covariate `Coherence` in the linear predictor are independent of color direction, (2) only the slope of the covariate is independent of color direction, and (3) both the intercept and slope depend on color direction. Perform a likelihood ratio test to evaluate which model best describes the data.

4.8. Add the predicted psychometric functions for each of the three models fitted in Exercise 4.7 to the graphs created in Exercise 4.6. Can you evaluate what the simplest model that best fits the data is?

4.9. Calculate and plot the coherence thresholds as a function of color direction on the same graph, for criterion values of 0.5 and 0.75. Calculate the coherence values that would yield $d' = 1$ as a function of color direction and add these to the graph, too. Why is it so low (hint: consider the false alarm rates)?

Chapter 5
The Psychometric Function: Continuation

In the previous chapter we showed how to use direct optimization methods and
the generalized linear model (GLM) to fit psychometric functions to Yes–No data.
In an extended example using GLM, we illustrated the procedure for selecting
a model to fit multiple psychometric functions, across a series of experimental
conditions and how to evaluate the goodness of fit of the model. In this chapter, we
continue the exploration of fitting psychometric functions and demonstrate methods
required when the observer's task is to select one among many alternatives, a type
of experiment referred to as m-alternative forced choice (mAFC). In addition, we
illustrate methods for assigning standard errors and confidence limits to estimated
parameters. Finally, we end with a short discussion of non- (or semi-)parametric
methods for fitting psychometric functions.

5.1 Fitting Multiple-Alternative Forced Choice Data Sets

With many common experimental designs, the observer's response at chance will
not asymptote at 0. For example,

2-Alternative Temporal, Forced-Choice: The observer is presented with two
intervals delimited in time only one of which contains a signal. As in the Yes–No
task in the previous chapter, the stimulus level varies from trial to trial. However,
even if the observer is reduced to guessing at low stimulus levels, the expected
rate of correct response (denoted γ) is 0.5, not 0.

2-Alternative Spatial, Forced-Choice: The choice on a given trial is between two
positions on a display. The signal is at only one location and chance performance
is again $\gamma = 0.5$.

K. Knoblauch and L.T. Maloney, *Modeling Psychophysical Data in R*, Use R! 32,
DOI 10.1007/978-1-4614-4475-6__5,
© Springer Science+Business Media New York 2012

m-Alternative Identification: On each trial, the observer is required to identify which of a set of m items was presented on a trial and expected chance performance will be at $\gamma = 1/m$.

When the observer must choose between m stimuli presented in the same trial, as in the first two cases above, the paradigm is referred to as multiple-alternative forced choice or $mAFC$. The first two examples are examples of 2AFC. When stimuli are presented (or not presented) in successive intervals of time, 2AFC is also commonly referred to as "two-*interval* forced choice" (2IFC) and mAFC as mIFC.

In fitting such data, one would like the estimation process to take into account the discrepancy between the chance performance level and 0. We will consider two methods for achieving this end using GLM. In the first, we use the standard built-in link functions and establish the asymptote using an *offset term* in the model; in the second we use *customized link functions* that describe (4.6) with $\lambda = 0$ and lower asymptotic value given by γ.

5.1.1 Using an Offset to Set the Lower Asymptote

The `glm` function can include an offset term when modeling a data set. According to its help page, this corresponds to an "a priori known component to be included in the linear predictor during fitting." If we fit a model that is *linear* in the stimulus intensity and from which we *exclude* the intercept term, then the linear predictor will equal the offset when the stimulus is 0. The link function applied to the asymptotic chance probability gives the appropriate value to which to set the offset. For example, assume a model

$$\eta(Y) = b_0 + \beta_1 X, \tag{5.1}$$

where X is a covariate indicating the stimulus intensity, Y is the probability of a correct response, and b_0 is the offset. If we want the psychometric function to asymptote at a value of γ, then we choose $b_0 = \eta(\gamma)$. This is more clearly illustrated by an example.

The **psyphy** package contains a data set, ecc2, from an observer who performed a 4AFC experiment reported in [204]. One of four letters from the set {b, d, p, q} was presented to the observer at low contrasts at one of four positions at $2°$ eccentricity from central fixation. The observer performed a double judgment in which first the position at which the letter was presented was specified (Detection) and then the letter was named (Identification). The data frame contains five variables

```
> library(psyphy)
> names(ecc2)
[1] "Contr"      "task"      "Size"      "Correct"      "Incorrect"
```

indicating the luminance contrast of the stimulus, the task performed (DET or ID), the height of the letter in minutes of arc, and the numbers of correct and incorrect

Fig. 5.1 Psychometric data from the `ecc2` data set for the 4AFC detection task of 20.6 min high letters. The *fitted curves* were obtained by fitting models with an offset term and linear predictors as described in the text

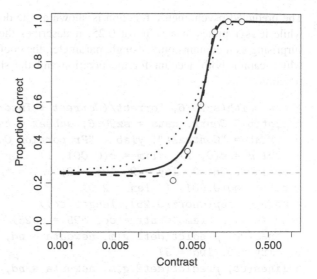

judgments. We will extract data from both tasks for the letter size 20.6 min but will initially fit psychometric functions only to the detection data. We begin by extracting the subset of the data.[1]

```
> sz20.6 <- subset(ecc2, Size == 20.6)
```

The probability of a correct detection as a function of the stimulus contrast is plotted in Fig. 5.1. Since on each trial the stimulus appeared in one of four locations, the expected chance performance is 0.25, indicated by the dashed horizontal line in grey. If we model these data with the `probit` link, then the offset should be set to

```
> qnorm(0.25)
[1] -0.674
```

The simplest model that we might try is to suppose that the linear predictor includes only the stimulus covariate, `Contr`.

```
> q25 <- rep(qnorm(0.25), nrow(sz20.6))
> det1.glm <- glm(cbind(Correct, Incorrect) ~ Contr - 1,
+   binomial(probit), sz20.6, subset = task == "DET",
+   offset = q25)
```

The offset is included as a separate argument, but we could have instead included it in the formula object as follows:

```
> ~ Contr + offset(q25) - 1
```

---

[1]It is generally not good practice to test for equality between floating point numbers but here the size value 20.6 is effectively a constant specifying a condition within the larger experiment.

The predicted psychometric function is shown as the dotted curve in Fig. 5.1 and while it asymptotes at a value of 0.25, it describes the data poorly. This is not surprising as it contains only a single parameter, the coefficient of the term `Contr`, which cannot both account for the position and the shape of the psychometric function.

```
> p <- with(sz20.6, Correct/(Correct + Incorrect))
> plot(p ~ Contr, data = sz20.6, subset = task == "DET",
+    xlab = "Contrast", ylab = "Proportion Correct",
+    ylim = c(0, 1), xlim = c(0.001, 1), log = "x", type = "n")
> abline(0.25, 0, lty = 2, lwd = 2, col = "grey")
> cc <- seq(0.001, 1, len = 200)
> nq25 <- rep(qnorm(0.25), length(cc))
> nd <- data.frame(Contr = cc, q25 = nq25)
> lines(cc, predict(det1.glm, newdata = nd, type = "response"),
+    lty = 3, lwd = 3)
> lines(cc, predict(det2.glm, newdata = nd, type = "response"),
+    lty = 2, lwd = 3)
> lines(cc, predict(det3.glm, newdata = nd, type = "response"),
+    lwd = 3)
> points(p ~ Contr, data = sz20.6, subset = task == "DET",
+    pch = 21, bg = "white", cex = 1.5)
```

We can do better by adding a second term to the linear predictor that depends on the square of the contrast. It has been argued that the necessity of a quadratic term in contrast detection reflects a nonlinear transducer function that relates stimulus intensity to d' [93].

```
> det2.glm <- update(det1.glm, . ~ . + I(Contr^2))
```

As seen previously, we must isolate the quadratic term with the function `I` so that the square is interpreted as a mathematical operation rather than as a formula operation. The fitted curve shown in dashes describes the data very well but has the inconvenience that it dips slightly below chance for a range of values outside that of the data set.

In the spirit of exploration, we consider dropping the term that is linear in contrast.

```
> det3.glm <- update(det2.glm, . ~ . - Contr)
```

The predicted fit is shown as the solid curve. This curve does not dip below the chance level, but it also describes the two lowest points less well. Usually, thresholds are estimated from the middle or upper levels of the psychometric function. In that case, this model seems to do as well as the previous one. It is possible to continue with higher degree terms, but this approach is evidently ad hoc. In addition, high-order polynomial models are apt to fluctuate widely outside the range of the data or even between data points [13, 185].

When the False Alarm rate, given by the expected level of chance performance, is taken into account as implemented here with an offset, then the linear predictor values are in units of d' [93]. For example, consider a Yes–No experiment analyzed with the equal-variance Gaussian model. This corresponds to using a probit link. Then, in the case of a single covariate, x, such as the stimulus level, we have

$$d' = z(P(H)) - z(P(FA)) = z(\Phi(\beta_0 + \beta_1 x)) - z(\Phi(\beta_0)) \qquad (5.2)$$

Recalling that $z = \Phi^{-1}$,

$$d' = \beta_1 x. \qquad (5.3)$$

From this, we can define a criterion-free measure of threshold, $1/\beta_1$, as the stimulus level that generates a performance of $d' = 1$ [93, 166]. If instead, a better fit is obtained with the square of the covariate, as in the example above, then the criterion-free measure would be $1/\sqrt{\beta_1}$. In the case of a model with both linear and quadratic terms, the solution corresponds to one of the roots of a quadratic equation.

5.1.2 Using Specialized Link Functions to Set the Lower Asymptote

The **psyphy** package [94] contains specialized link functions that can be used with glm to fit (4.6) directly to the data. Each of these links has a two part name separated by a '.' in which the prefix is mafc and the suffix gives the name of a built-in link for the binomial family. Each takes one argument, an integer, .m > 1, specifying the number of alternatives used to generate the experimental data (default, .m = 2). For example, a probit link for a 3afc experiment would be specified as

```
> mafc.probit(.m = 3)
```

Continuing with the data set for the letter size of 20.6 min, we will fit psychometric functions for both tasks and test whether the intercepts and slopes of the two linear predictors are equal, again using the Gaussian cdf to define the shape of the psychometric function. These two terms affect the location and the slope of the psychometric function. The location will determine the threshold and the slope the steepness with which the function rises from chance to perfect performance. In the original data, the contrasts were chosen to be roughly equi-spaced on a log axis. Unlike when using an offset, we will model the logarithm of the stimulus intensity.[2] Thus, we begin by adding a column to the data frame for the logarithm of the contrasts. Then, we fit three models of increasing complexity to the data and compare them using a likelihood ratio test.

---

[2]Using an offset with the logarithm of contrast would have forced the psychometric function to the offset detection level at log contrast = 0 or contrast = 1, not the desired behavior. Here, the log transform serves to facilitate distinguishing a shift along the log contrast axis from a change in slope.

```
> sz20.6$LContr <- log10(sz20.6$Contr)
> m1 <- glm(cbind(Correct, Incorrect) ~ LContr,
+    binomial(mafc.probit(.m = 4)), data = sz20.6)
> m2 <- update(m1, . ~ . + task)
> m3 <- update(m2, . ~ . + task:LContr)
> anova(m1, m2, m3,  test = "Chisq")
```

```
Analysis of Deviance Table
```

```
Model 1: cbind(Correct, Incorrect) ~ LContr
Model 2: cbind(Correct, Incorrect) ~ LContr + task
Model 3: cbind(Correct, Incorrect) ~ LContr + task + LContr:task
  Resid. Df Resid. Dev Df Deviance Pr(>Chi)
1       10      181.6
2        9        7.5  1      174   <2e-16 ***
3        8        6.5  1        1     0.32
---
Signif. codes:  0 `***' 0.001 `**' 0.01 `*' 0.05 `.' 0.1 ` ' 1
```

The specialized link function, `mafc.probit`, is used with the `binomial` family function just as the built-in links, except that it cannot be quoted. In the first model, a single psychometric function was fit to all of the data; the threshold and steepness of the psychometric functions for both tasks were constrained to be the same. The linear predictor in this model contains only an intercept and the covariate `LContr`. The model object was assigned to the name `m1`. In the second model, the factor `task` was added as a main effect. In this case, the thresholds may differ between tasks but the curves display the same steepness. Finally, in the third model, the interaction term `task:LContr` allows a difference in steepness between the curves to be estimated. Comparing the three models with the `anova` method indicates that we can reject that the thresholds are the same but not that the steepness varies between curves; thus, we retain model `m2`. This model also has the lowest AIC value of the three.

The data and selected model are shown in Fig. 5.2 with the tasks being indicated by different symbols. The expected level of chance performance is indicated as a dashed line. Note that the `predict` method uses the link function in the model object so that the predicted curves asymptote correctly. In this example, the factor `task` had only two levels, but the method generalizes without alteration to cases with more.

```
> p <- with(sz20.6, Correct/(Correct + Incorrect))
> pch <- 1:2
> plot(p~Contr, data = sz20.6, pch = pch[unclass(sz20.6$task)],
+    xlab = "Contrast", ylab = "Proportion Correct",
+    ylim = c(0, 1), xlim = c(0.02, 0.3), log = "x", type = "n")
> xx <- seq(-1.7, -0.5, len = 100)
> nd <- data.frame(LContr = rep(xx , 2),
+    task = rep(c("DET", "ID"), each = 100))
```

Fig. 5.2 Detection (*circles*) and identification (*squares*) data for 20.6 min letters from the ecc2 data set. The *curves* indicate the model fit with different thresholds but the same steepness parameter for both tasks

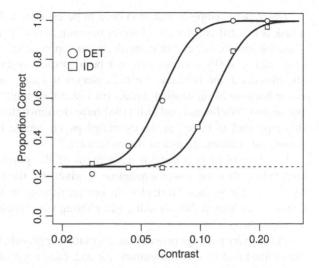

```
> pred <- predict(m2, newdata = nd, type = "response")
> abline(0.25, 0, lty = 2)
> lines(10^xx, pred[1:100], lwd = 2)
> lines(10^xx, pred[-(1:100)], lwd = 2, pch =20 +
    pch[unclass(sz20.6$task)],
+   bg = "white")
> points(p ~ Contr, data = sz20.6, pch = 20 +
    pch[unclass(sz20.6$task)],
+   bg = "white")
> legend(0.02, 0.9, legend = c("DET", "ID"), pch = c(21, 22),
+   pt.cex = 1.8, bty = "n")
```

5.1.3   Estimating Upper and Lower Asymptotes

In the above example, the lower asymptote of the psychometric functions was fixed at the expected level of chance performance. Situations arise, however, in which it is desirable to estimate this value. For example, in Yes–No detection tasks, the proportion of correctly classified trials may not fall to zero at low signal levels, due to bias or internal noise. The observer displays a nonzero "False Alarm" rate in the terminology of signal detection theory (Chap. 3). Even with an mAFC task, however, the limited number of trials in an experiment can result in data that would be fit better if the lower asymptote were estimated rather than fixed at the expected value.

The upper asymptote may also need to be estimated. This situation arises when
a task is too difficult for the observer to reach perfect performance no matter how
great the stimulus level. For example, as one approaches the resolution limit, letters
presented at 100% contrast may not be identified on every trial. Even when the
stimulus condition is "easy" for the observer to judge, however, occasional errors
occur because of inattention, blinks (in vision), or finger slips in responding on a
button-box. Wichmann and Hill [188] have demonstrated that occasional errors of
this type lead to biases in the estimated psychometric function unless the upper
asymptote is estimated as a nuisance factor.

Equation (4.6) is a general description of the psychometric function parameterized so that the lower asymptote is given by the variable γ and the upper
by $1 - \lambda$. Since these variables do not participate in the linear predictor, they
cannot be estimated directly using glm although they could be included using direct
maximization.

The **psyphy** package provides an alternative approach. Specialized link functions
are defined that take two arguments, g and lam. Each of these link functions has
a two part name in which the suffix is 2asym and the prefix is given by a name
of a built-in link for the binomial family. These provide links for fixed values
of the arguments. The function psyfun.2asym estimates a best fit to the data by
alternating between calls to glm for current best estimates of arguments to the link
function and calls to optim to estimate the values of g and lam, until the change in
log likelihood for successive alternations falls below a tolerance.

For example, we will simulate data from a Weibull-shaped psychometric function. In our simulation there are 100 trials at each of eight intensity levels, cnt. We
use the rbinom function to generate random numbers of successes with probabilities
specified by the simulated psychometric function at the specified intensity levels.
The parameters and levels were chosen so that the simulated observer performs at
100% success over several levels. Then, we artificially simulate an upper asymptote
error (e.g., due to inattention) by subtracting 1 correct response at the upper
asymptote.[3]

```
> b <- 3        # steepness parameter
> g <- 0.02 # lower asymptote
> d <- 0        # 1 - upper asymptote
> a <- 0.027 # threshold
> num.tr <- 100
> cnt <- 10^seq(-2, -1, length = 8)
> truep <- g + (1 - g - d) * pweibull(cnt, b, a)
> ny <- rbinom(length(cnt), num.tr, truep)
> nn <- num.tr - ny
> ny.mod <- ny
> ny.mod[8] <- ny[8] - 1
> nn.mod <- num.tr - ny.mod
```

---

[3]Note that if ny[8] is 0, improbable but possible, the code fails. The simplest correction would be
to test for ny[8] == 0 and resample if it is but we omit it from the example for simplicity.

Fig. 5.3 Simulated frequency of seeing data following a Weibull cdf. The *solid curve* is fit to the *white points*. The *dashed* and *dotted curves* have been fit to the data with the *grey point* (almost invisible under the *rightmost white point*), representing one erroneous judgment, substituted for the *underlying white point*. In the case of the *dotted curve*, the fit permitted an estimate of the lower and upper asymptotes, whereas in the other two fits, these values were constrained to 0 and 1, respectively

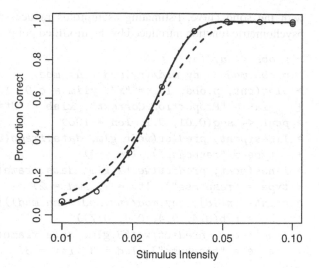

```
> weib.sim <- data.frame(ny = ny, nn = nn, cnt = cnt)
> weib.sim.mod <- data.frame(ny = ny.mod, nn = nn.mod,
+    cnt = cnt)
```

We fit models to both data sets and compare the fitted curves in Fig. 5.3.

```
> weib.glm <- glm(cbind(ny, nn) ~ log10(cnt),
+    binomial("cloglog"), data = weib.sim)
> weib2.glm <- update(weib.glm, data = weib.sim.mod)
```

The modified point is indicated in grey. The fit to the original data is indicated by the solid curve and the fit to the data with the modified point is indicated by the dashed curve. It is remarkable how sensitive the fit is to one erroneous judgment. The reduction of the proportion correct by 0.01 at this point reduces the steepness of the estimated psychometric function and biases the threshold estimate. The curve fit to the modified data shows systematic departures from the data at both high and low-intensity levels.

Using the link function `weib.2asym` as an argument in the function `psyfun.2asym` permits us to fit a curve with an estimate of the upper asymptote.

```
> weib3.glm <- psyfun.2asym(cbind(ny.mod, nn.mod) ~ log10(cnt),
+    link = weib.2asym, init.g = 0.05, init.lam = 0.025)
lambda =      0.0034      gamma =  0.0281
+/-SE(lambda) =    ( 0.00127 0.00907 )
+/-SE(gamma) =  ( 0.0138 0.0564 )
```

Initial estimates for the lower and upper asymptotes can be passed through the arguments `init.g` and `init.lam`, respectively. The final estimates are returned ± estimates of the standard errors based on the Hessian at the final iteration of `optim`. The fitted curve is shown as the dotted curve in Fig. 5.3. It is almost coincident with the solid curve. Estimating asymptotes reduces the bias of the estimated psychometric function introduced by the modified point.

```
> p.obs <- ny/(ny + nn)
> p.obs.mod <- ny.mod/(ny.mod + nn.mod)
> plot(cnt, p.obs, log = "x", ylim = c(0, 1),
+    ylab = "Proportion Correct", xlab = "Stimulus Intensity")
> pcnt <- seq(0.01, 0.1, len = 100)
> lines(pcnt, predict(weib.glm, data.frame(cnt = pcnt),
+    type = "response"), lwd = 2)
> lines(pcnt, predict(weib2.glm, data.frame(cnt = pcnt),
+ type = "response"), lty = 2, lwd = 2)
> points(cnt[8], (ny.mod/(ny.mod + nn.mod))[8], pch = 21,
+    bg = rgb(0.4, 0.4, 0.4, 0.7))
> lines(pcnt, predict(weib3.glm, data.frame(cnt = pcnt),
+    type = "response"), lwd = 3, lty = 3)
```

In the above example, the upper asymptote is treated as a nuisance parameter. If fitting several psychometric functions at once as with the Vernier and ecc2 data sets above, the question arises as to whether to obtain an estimate of λ for each curve. Doing so adds one parameter for each psychometric function. The current version of psyfun.2asym is not capable of fitting multiple values of λ when data from multiple conditions are analyzed. Yssaad-Fesselier and Knoblauch demonstrate fitting multiple λ's using functions for performing a generalized nonlinear model [204].

When the data for several psychometric functions would be expected to be subject to the same error at the asymptote, for example, when obtained from data in the same session or simply from the same observer, it may be convenient to estimate a common value of λ for all conditions. This can be accomplished with psyfun.2asym.

An alternative to fitting a single asymptote or multiple asymptotes is to treat the upper asymptote as a random parameter [204]. This is not an easy exercise, however. Unless one is interested specifically in estimating the variance at the asymptote, rather than just accounting for its fluctuations to stabilize threshold and steepness estimates, the gain in generality of this approach may not be worth the effort.

Kuss et al. use a binomial mixture model to model the upper asymptote behavior in which, in addition to the observer's binary decision on each trial, there is a binary probability of a lapse [105]. They use a Bayesian inference procedure to estimate the parameters with the functions provided in package **PsychoFun** available at http://www.kyb.tuebingen.mpg.de/~kuss, at the time of this writing.

Foster and Żychaluk propose a nonparametric method for fitting psychometric functions using local regression techniques and cross-validation to optimize the choice of bandwidth [65, 126, 208]. They present examples in which the flexibility of the local regression curve automatically takes into account uncertainties in both asymptotes without having to introduce additional parameters. We will examine this in Sect. 5.6.

5.2   Staircase Methods

The data discussed so far in this chapter were collected using the method of constant stimuli. A series of fixed stimulus levels are chosen ahead of time and each presented randomly a fixed number of times over the course of the experiment. This gives a balanced and complete data set but does not necessarily represent the most efficient method for estimating thresholds, the primary use of psychometric functions. Very often, adaptive procedures that attempt to home-in on the threshold value over the course of an experiment are used. A staircase procedure is, perhaps, the simplest of such procedures, wherein the stimulus intensity is increased or decreased according to the response of the observer. Over the course of a sufficient number of trials, the intensity values will fluctuate up and down around the targeted threshold. Various rules for how many successes or failures will be required for changing the stimulus level permit the experimenter to adjust the criterion value estimated. More sophisticated procedures have been developed, such as the QUEST [184] algorithm, wherein all preceding trials are used as a priori information on the underlying psychometric function to obtain an a posteriori estimate from which the optimal intensity level of the next stimulus is chosen.

The data set `StairCase` in the **MPDiR** package is from a 2AFC detection experiment that employed a 3-up, 1-down staircase. In this procedure, three successive correct responses are required to decrement the contrast of the stimulus but one incorrect response is sufficient to increment it. To limit the influence of the dependence of the stimulus levels on the observer's decisions, two randomly interleaved staircases were run, though the data set has not retained the random shifting between the two. In addition, the stimulus for each staircase started at a high value and was reduced by a factor of 2 until the first error and thereafter varied by factors of 1.26. Each of the staircases was 48 trials long. Experimenters who use staircase methods will exclude the initial trials of a staircase up to the first reversal from analysis. There is no reason to do so and we will include them in fitting the data. The two staircases are plotted as a function of trial in Fig. 5.4, rendering the source of the procedure's name self-evident.

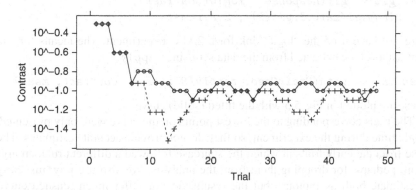

Fig. 5.4 Data from the `StairCase` data set indicating the stimulus contrast across trials for two randomly interleaved staircases, in a 2AFC detection experiment

Fig. 5.5 Proportion correct estimated from the staircase trials and the predicted psychometric function fit to the data

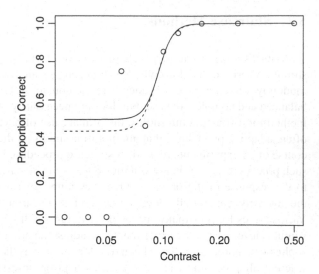

The data set contains four variables

```
> names(StairCase)
[1] "Trial"     "Contrast"  "Response"  "StairCase"
```

indicating the trial number of the staircase, the contrast level, the response (1—correct, 0—incorrect), and a factor indicating to which of the two staircases the trial belonged. Thresholds are commonly obtained by averaging the stimulus level at the turnaround points or all of the stimulus levels after a fixed number of turnarounds. Over the course of the experiment, however, sufficient repetitions occur at most intensities permitting the fitting of a psychometric function to the data.

To fit a psychometric function to the data set, we can exploit the same functions that we used above. The only difference is that the response is binary and not a proportion. However, the glm function works equally well for this limiting case of a binomial variable.

```
> sc.glm <- glm(Response ~ log10(Contrast),
+    binomial(mafc.logit(2)), StairCase)
```

Here, we have used the "logit" link for a 2AFC experiment. The means for each contrast level are extracted from the data set using tapply.

```
> sc.mns <- with(StairCase, tapply(Response, Contrast, mean))
```

These are plotted in Fig. 5.5 with the fitted (solid) curve.

The trials corresponding to the lowest points on the curve were only presented a single time during the experiment, so they do not provide accurate estimates. They arise from the early trials in which the staircase followed a different rule, an argument, perhaps, for dropping them from the analysis. We also use psyfun.2asym to estimate both asymptotes, but the results do not differ much (dashed curve).

This function will only accommodate the response in the formula formatted as a two-column matrix of successes and failures, which explains the modification of the formula argument. The data from the two staircases can be analyzed as independent runs and the homogeneity of results tested by comparing with the fit above. We leave this as a an exercise for the reader (Exercise 5.3).

```
> plot(as.numeric(names(sc.mns)), sc.mns, log = "x",
+           xlab = "Contrast", ylab = "Proportion Correct")
> cnt <- seq(0.03, 0.5, len = 200)
> lines(cnt, predict(sc.glm, newdata = data.frame
    (Contrast = cnt),
+           type = "response"))
> sc2.glm <- psyfun.2asym(cbind(Response, 1 - Response) ~
+           log10(Contrast), data = StairCase, link = logit.2asym,
+           init.g = 0.5)
lambda = 7.46e-12 gamma = 0.439
+/-SE(lambda) = ( NA NA )
+/-SE(gamma) = ( 0.349 0.534 )
> lines(cnt, predict(sc2.glm, newdata = data.frame
    (Contrast = cnt),
+           type = "response"), lty = 2)
```

5.3 Extracting Fitted Parameters

The previous sections concentrated on fitting a psychometric function and testing the significance of variables that explained its shift and shape by model comparisons. Once the best model is chosen, the experimenter is often interested in determining either a stimulus value that produces a criterion value of performance, referred to as a *threshold*, or a range of stimulus values that span a fixed criterion of performance, termed a *discrimination limen*. A graphical illustration of the determination of the threshold for two levels of performance is shown in Fig. 5.6 by the dotted lines with the arrows pointing to the threshold stimulus value for each level. The line segment below the abscissa axis labels is the distance between the two thresholds and corresponds to a difference limen.

In R the p-functions (such as pnorm) define probability cdfs that are convenient models of the psychometric function. For each p-function there is a q- or *quantile function* that is just the inverse of the corresponding cdf. The quantile functions can be used to map probabilities back to values of the stimulus level and they permit us to calculate thresholds from the linear predictor model coefficients. We will use them to estimate the stimulus level needed to achieve any specified level of probability in the psychometric task.

Two steps are necessary. The first is to translate the model coefficients into the parameters of the probability function. When the probability distribution is specified

Fig. 5.6 Psychometric
function with thresholds
indicated by the *arrows* at the
ends of the *dotted lines* for
performance criteria of
$p = 0.5$ and $p = 0.75$. The
distance on the stimulus axis
between these two points,
indicated by the segment
below the abscissa axis
labels, is a difference limen

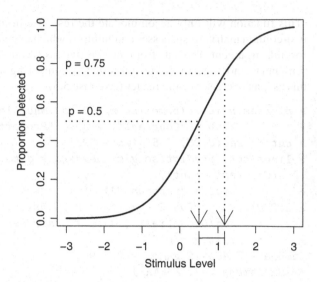

in terms of a location and a scale parameter on a linear stimulus continuum, such
as μ and σ of the normal distribution, then fitted coefficients—corresponding to the
slope and intercept, (β_0, β_1)—are related to the cdf parameters by

$$\text{location} = -\frac{\beta_0}{\beta_1} \tag{5.4}$$

$$\text{scale} = \frac{1}{\beta_1}. \tag{5.5}$$

This parameterization applies to the probit, logit, and cauchit links.

In the case of functions that display a location and scale parameter on a log axis,
the situation is bit more complex. We consider the Weibull psychometric function
for which the explanatory covariate in the linear predictor is the logarithm of the
stimulus level. The relation of the Weibull parameters (α, β) to the complementary
log–log link is given by

$$\log(-\log(1-p)) = \beta \log(x) - \beta \log(\alpha). \tag{5.6}$$

The right-hand side is linear in $\log(x)$ so we can equate the terms to the fitted
coefficients from a Weibull fit by

$$\alpha = e^{-\frac{\beta_0}{\beta_1}} \tag{5.7}$$

$$\beta = \beta_1. \tag{5.8}$$

The second step is to take into account any differences between the lower
asymptote of the fitted function and 0 or the upper and 1. From (4.6), we can solve
for a probability of correct response, p, as

$$P = \frac{p - \gamma}{1 - \gamma - \lambda}, \tag{5.9}$$

where P is the corrected value of probability to use for a performance level at p and γ and λ are the parameters defining the lower and upper asymptotes, respectively.

Examples of threshold and difference limen calculations for a normal distribution were presented in Chap. 1, so we will present an example here for the Weibull function fit obtained by using the "cloglog" link in Sect. 5.1.3, which includes a few subtle traps of which to beware. For example, (5.6) uses the natural (Napierian) logarithm while the stimulus levels in the example were transformed by common (decadic) logarithms. The two are related by a factor of $\log(10)$ which is introduced in the calculation below.

For the Weibull function, the parameter α is often used directly as the estimate of threshold. When asymptotes are 0 and 1, then this value corresponds to a probability of detection of $1 - e^{-1} \approx 0.632$. We estimate the threshold for this level of performance for the model weib3.glm, in which the asymptotes were also estimated.

```
> ln10 <- log(10)
> palpha <- 1 - exp(-1)
> gam <- weib3.glm$gam
> lam <- weib3.glm$lambda
> cc <- coef(weib3.glm)
> P <- (palpha - gam)/(1 - gam - lam)
> thresh_alpha <- qweibull( P, shape = cc[2]/ln10,
+     scale = exp(-ln10 * cc[1]/cc[2]) )
> weibfit.params <- c(thresh_alpha, cc[2]/ln10)
> names(weibfit.params) <- c("alpha", "beta")
> weibfit.params
 alpha    beta
0.0263  3.3278
```

Note that the argument shape corresponds to the steepness (scale) parameter and that the argument scale determines the location of the psychometric function on a logarithmically scaled abscissa. Compare these estimated values with those used to simulate the data. To obtain a difference limen, one would specify two probability levels and difference the resulting quantiles (see Sect. 1.3.2).

5.4 Methods for Standard Errors and Confidence Intervals

Standard errors for the linear predictor coefficients are given by the summary method and are simply returned as the second column of a matrix from the summary object by passing it to the coef method. For example,

```
> coef(summary(weib.glm))
```

```
              Estimate Std. Error z value Pr(>|z|)
(Intercept)      11.11        0.820     13.6 7.61e-42
log10(cnt)        7.02        0.513     13.7 1.12e-42
```

These are calculated from the diagonal elements of the variance–covariance matrix, as can easily be verified.

```
> sqrt(diag(vcov(weib.glm)))
```

```
(Intercept)  log10(cnt)
      0.820       0.513
```

The variance–covariance matrix is also useful for obtaining standard errors of linear combinations of the coefficients that may be of interest. For example, suppose we wanted to know the standard error of the linear predictor for a contrast of 0.026, which is near the value of $\hat{\alpha}$ for the fit to these data. The standard error calculated with vcov is

```
> sqrt(c(1, log10(0.026)) %*% vcov(weib.glm) %*%
+     c(1, log10(0.026)))
          [,1]
[1,]  0.0788
```

It is much easier (and more in the spirit of R) to obtain this value using the predict method with the argument se.fit = TRUE.

```
> predict(weib.glm, newdata = list(cnt = 0.026),
+     se.fit = TRUE)$se.fit
[1]  0.0788
```

The predict method is versatile, however. By additionally specifying method = "response", standard errors for the psychometric function are returned. These are useful in illustrating the precision of the fitted psychometric function graphically. We illustrate this in Fig. 5.7 using the polygon function to create a grey envelope that spans ±2 SE's around the curve.

```
> cnt <- 10^seq(-2, -1, length = 8)
> plot(cnt, p.obs, log = "x", type = "n", ylim = c(0, 1),
+          ylab = "Proportion Correct", xlab = "Stimulus
                   Intensity")
> pcnt <- seq(0.01, 0.1, len = 100)
> weib.pred <- predict(weib.glm, newdata = list(cnt = pcnt),
+          type = "response", se.fit = TRUE)
> polygon(c(pcnt, rev(pcnt)),
+          with(weib.pred, c(fit + 2 * se.fit, rev(fit - 2
                   * se.fit))),
+          col = "grey", border = "white" )
> lines(pcnt, weib.pred$fit, lwd = 2)
```

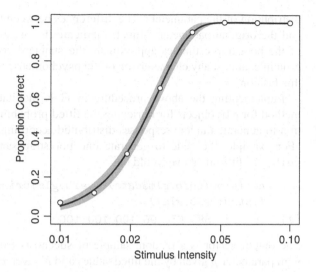

Fig. 5.7 Psychometric data from a simulated psychophysical observer based on a Weibull cdf. The best fit curve is indicated with ±2 standard errors of the fit as the *grey* envelope

```
> points(cnt, (ny/(ny + nn)), pch = 21,
+          bg = "white")
```

A `confint` method is also available, which uses profiling to calculate 95% confidence intervals by default, though other widths can be specified through the argument `level`.

```
> confint(weib.glm, level = 0.99)
              0.5 %  99.5 %
(Intercept)   9.11   13.35
log10(cnt)    5.77    8.42
```

5.5 Bootstrap Standard Error Estimates

Several investigators have proposed using bootstrap methods [55] to evaluate standard errors and confidence intervals related to psychometric function fits [64, 121, 175, 188]. These have the advantage of relying on fewer assumptions about the distribution of errors and permitting estimates of bias in the parameter estimates, as well.

For example, the procedure described by Maloney [121] is to generate a new sample of responses, r_1^*, given the fitted probabilities for each stimulus condition. This sample is called a bootstrap sample. The bootstrap sample is then fit with the same model used on the original data. This yields the vector of bootstrap estimated parameters from the first bootstrap sample, β_1^*. The above procedure is repeated a large number of times and the vector of estimates from the fits to each bootstrap sample is stored. The histograms of elements of β_i^* yield information about the distribution of the parameters. The differences between the means of the bootstrap and the original parameter estimates indicate the biases. The standard deviations of the bootstrap estimates approximate the standard errors of the parameters. In principle, almost any characteristic of the psychometric function can be assessed in this fashion.

Implementing the above procedure in R is facilitated by using the `fitted` method for `glm` objects for retrieving the fitted probabilities and the `rbinom` function to generate random responses distributed according to the fitted probabilities. For example, the code to generate one bootstrap sample of responses for the `weib.glm` fit from above would be

```
> rbinom(nrow(weib.glm$data), weib.glm$prior.weights,
+     fitted(weib.glm))
[1]    6   23   38   67   96  100  100  100
```

The output vector is a random sample of successes from a binomial distribution with parameter p given by the fitted values and N given by the number of trials. Try running it several times to see how the numbers of successes vary. We can now write a function incorporating this line of code that will compute a specified number of bootstrap estimates based on a model object.

```
> psyfun.boot <- function(obj, N = 100){
+     n <- obj$prior.weights
+     f <- fitted(obj)
+     resp.bt <- matrix(rbinom(N * length(n), n, f), ncol = N)
+     bt.res <- sapply(seq_len(N), function(x) {
+         r <- resp.bt[, x]
+         res.bt <- glm(cbind(r, n - r) ~ model.matrix(obj) - 1,
+             binomial(obj$family$link))
+         cc <- coef(res.bt)
+         names(cc) <- names(coef(obj))
+     cc
+     })
+     bt.res
+ }
```

The function returns a $p \times N$ matrix, each row of which contains the parameter estimates of the N bootstrap samples. While the above function is quite serviceable and illustrates the ease with which such an operation can be coded, it is more practical to take advantage of tools that are already available from within R. The package **boot** [25] contains functions for performing bootstrap analyses based on the book by Davison and Hinkley [46] and is a recommended package, i.e., usually installed with R by default. Besides functions for generating several types of bootstraps, it contains `print` and `plot` methods for objects of class "boot" and a function for calculating several types of bootstrap confidence intervals.

The main function for performing bootstraps in the package is named `boot`. It requires minimally a data set and a function indicating what statistic to calculate

from the data. For example, the function psyfun.stat takes a data set as an
argument and returns the coefficients from glm with a binomial family and the
"cloglog" link. It expects the first two columns of the data set to contain the number
of successes and failures and a covariate to be in the third column. In the fourth
column, it expects to find the number of successes from a bootstrapped sample.

```
> psyfun.stat <- function(d){
+     nn <- rowSums(d[, 1:2])
+     mm <- model.matrix(as.matrix(d[, 1:2]) ~ log10(d[, 3]))
+     t.glm <- glm(cbind(d[, 4], nn - d[, 4]) ~ mm - 1,
+       binomial("cloglog"))
+         as.vector(coef(t.glm))
+ }
```

The function psyfun.gen will generate the bootstrapped number of successes
added to column 4, above.

```
> psyfun.gen <- function(d, mle) {
+    nn <- with(d, ny + nn)
+    d$resp <- rbinom(nrow(d), nn, mle)
+    d
+ }
```

It takes a data set as an argument and a second argument, mle, that gives the fitted
probabilities from the model.

As an example, we evaluate the parameters estimates from the model in
weib.glm using the boot function and the helper functions defined above. We
begin by adding to the previously defined data set, weib.sim, an additional column
indicating the actual number of successes. The function psyfun.gen will replace
this with the bootstrapped values. We then run the bootstrap for 1,000 trials,
specified by the argument R.

```
> library(boot)
> weib.d <- cbind(weib.sim, resp = weib.sim$ny)
> weib.boot <- boot(weib.d, statistic = psyfun.stat,
+    R = 1000, sim = "parametric",
+    ran.gen = psyfun.gen,
+    mle = fitted(weib.glm))
```

Note that the arguments to psyfun.gen are entered as named parameters.[4] The
output is an object of class "boot," and the print method displays basic results

```
> weib.boot
```

PARAMETRIC BOOTSTRAP

---

[4]This requires about 15 s on the laptop computer used at the time of this writing.

```
Call:
boot(data = weib.d, statistic = psyfun.stat, R = 1000,
     sim = "parametric", ran.gen = psyfun.gen,
     mle = fitted(weib.glm))
Bootstrap Statistics :
     original   bias      std. error
t1*    11.11    0.173        0.854
t2*     7.02    0.105        0.532
```

The two rows correspond to the intercept and slope parameters, respectively, of the model. The first column indicates their original values from the object weib.glm. Not all estimators are unbiased and we can use the bootstrap to check whether a particular estimator is biased and estimate the magnitude of bias. The column labeled "bias" gives the difference between the mean of the bootstrap parameter estimates and the original values. Finally, the last column gives the SD of the bootstrap parameter estimates. Note that these are slightly less conservative than those given by the summary method for the glm object. The 1,000 bootstrap estimates of the two parameters are contained in the t component of the "boot" object.

We can examine the distribution of the bootstrap samples with the plot method which displays their histogram and a QQ plot, either against standard normal or χ^2 quantiles. In the latter case, the degrees of freedom must be supplied. The top two pairs of plots in Fig. 5.8 were generated by the commands

```
> plot(weib.boot, index = 1)
> plot(weib.boot, index = 2)
```

The index argument specifies the variable for which the plots are generated. A vertical dashed line on the histogram indicates the value of the original parameter. A transformation of the bootstrap values may be displayed via the t argument. For example, the distribution of the Weibull parameter α shown in the bottom pair of graphs in Fig. 5.8 was obtained with the following code:

```
> alpha <- with(weib.boot, exp(-log(10) * t[, 1]/t[, 2]))
> cc <- coef(weib.glm)
> plot(weib.boot, index = 1, t0 = exp(-log(10) * cc[1]/cc[2]),
+     t = alpha)
```

When transformed parameters are specified, the vertical line added to the histogram is specified through the argument t0.

The function boot.ci calculates several types of bootstrap confidence intervals. Continuing with our example,

```
> boot.ci(weib.boot, type = c("norm", "basic", "perc", "bca"),
+     L = influence(weib.glm)$hat, index = 1)
BOOTSTRAP CONFIDENCE INTERVAL CALCULATIONS
Based on 1000 bootstrap replicates
```

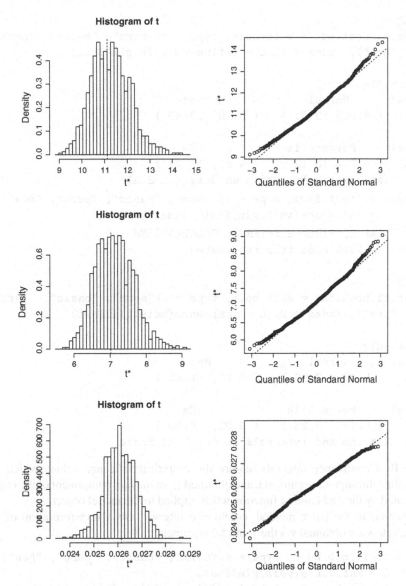

Fig. 5.8 Each row of plots gives the output of the plot method for an object of class "boot." The *graph* on the *left* shows a histogram of bootstrap estimates for a parameter. The *right graph* displays a QQ plot of the estimates with respect to a standard normal distribution. The *top pair* was generated from the estimates of the intercept term in the model weib.glm. The *middle pair* is based on the slope of the covariate term in the model. The *bottom pair* was generated for the parameter α of a Weibull cdf fit to the data, obtained from a combination of the intercept and slope terms

```
CALL :
boot.ci(boot.out = weib.boot, type = c("norm", "basic", "perc",
    "bca"), index = 1, L = influence(weib.glm)$hat)
```

```
Intervals :
Level       Normal                  Basic
95%    ( 9.26, 12.61 )      ( 9.10, 12.45 )
```

```
Level      Percentile                BCa
95%    ( 9.77, 13.12 )      ( 9.80, 13.22 )
Calculations and Intervals on Original Scale
> boot.ci(weib.boot, type = c("norm", "basic", "perc", "bca"),
+    L = influence(weib.glm)$hat, index = 2)
BOOTSTRAP CONFIDENCE INTERVAL CALCULATIONS
Based on 1000 bootstrap replicates
```

```
CALL :
boot.ci(boot.out = weib.boot, type = c("norm", "basic", "perc",
    "bca"), index = 2, L = influence(weib.glm)$hat)
```

```
Intervals :
Level       Normal                  Basic
95%    ( 5.87,  7.96 )      ( 5.75,  7.87 )
```

```
Level      Percentile                BCa
95%    ( 6.17,  8.28 )      ( 6.22,  8.32 )
Calculations and Intervals on Original Scale
```

The BCa confidence intervals require the empirical influence values which are specified through the argument L and obtained from the hat component of the object returned by the influence function when applied to the model object.

Similar to the plot method, confidence intervals for a transformation of the parameters are obtained via the t and t0 arguments.

```
> boot.ci(weib.boot, type = c("norm", "basic", "perc", "bca"),
+    L = influence(weib.glm)$hat,
+    t = alpha, t0 = exp(-log(10) * cc[1]/cc[2]))
```

```
BOOTSTRAP CONFIDENCE INTERVAL CALCULATIONS
Based on 1000 bootstrap replicates
```

```
CALL :
boot.ci(boot.out = weib.boot, type = c("norm", "basic", "perc",
    "bca"), t0 = exp(-log(10) * cc[1]/cc[2]), t = alpha,
    L = influence(weib.glm)$hat)
```

Fig. 5.9 The fitted psychometric functions from the 1,000 bootstrap estimates of the parameters in the model weib.glm are plotted using a *transparent black line color*. The *points* indicate the proportion of correct responses from the data set weib.sim

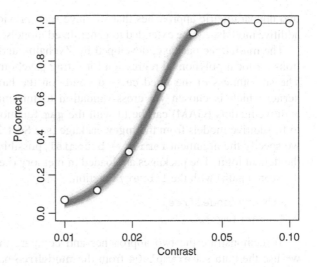

```
Intervals :
Level      Normal                Basic
95%    ( 0.0249,  0.0275 )    ( 0.0249,  0.0275 )

Level      Percentile            BCa
95%    ( 0.0247,  0.0274 )    ( 0.0251,  0.0277 )
Calculations and Intervals on Original Scale
Some BCa intervals may be unstable
```

For estimating quantile points on the tails of the distribution, such as the 95% confidence interval, 1,000 trials will often not be sufficient [55]. At the cost of waiting a few minutes, ten times as many bootstrap replications can be run, and the displayed warning should disappear (and you can have a cup of coffee). It is instructive to compare these confidence intervals with those obtained using the confint method.

The bootstrapped parameter estimates can equally well be used to estimate the precision of the fitted curve. For this, we use the fitted parameters from each bootstrap replication to generate a predicted psychometric function. In this way, we estimate a confidence interval for each point on the psychometric function. An interesting graphic view of the precision is obtained by plotting all of the bootstrap prediction curves using transparency in the color specification as shown in Fig. 5.9.

5.6 Nonparametric Approaches

The choice of link function fixes the shape of the psychometric function and the coefficients of the linear predictor shift and stretch it to fit the data. An alternative approach is to use a nonparametric method to estimate the shape. The two nonparametric approaches that we have seen previously, local regression and additive models, can be extended to generalized models, as well.

The **modelfree** package, developed by Żychaluk and Foster, implements functions for local polynomial regression for fitting psychometric functions [126, 208]. The smoothness of the fitted curve depends on the bandwidth of the smoothing kernel which is chosen via cross-validation to minimize deviance. Generalized additive models (GAM) can be fit with the gam function that we previously used to fit additive models from the **mgcv** package (see Sect. 2.5) It requires simply that we specify the argument family = binomial, possibly with a different link than the default logit. The packages are loaded in memory (i.e., if you recall, attached to the search path) with the library function.

```
> library(modelfree)
> library(mgcv)
```

To demonstrate the two approaches and compare them with a parametric fit, we use the data set example04 from the **modelfree** package. The data are from an experiment by Xie and Griiffin [199] (reported in Żychaluk and Foster [208]) in which observers judged which pair of two pairs of image patches was sampled from the same image. The independent variable is the base 2 logarithm of the patch separation in pixels.

```
> Xie.df <- example04
> Xie.df$p <- with(Xie.df, r/m)
```

The data set contains three components: (1) the image separation, x, (2) the number of correct responses, r, and (3) the number of presentations at each separation, m. We add a column p giving the proportion of correct responses that we will use in plotting the psychometric function.

The data are plotted in Fig. 5.10. Note that the percentage of correct responses is high for small separations and decreases with larger ones resulting in a psychometric function that is monotonic decreasing, for a change. This situation is easily handled by the model fitting functions by finding a negative coefficient for the covariate term.

```
> cxx <- seq(0.01, 0.1, len = 100)
> cc.bt <- weib.boot$t
> l10 <- log(10)
> with(weib.sim, plot(cnt, ny/(ny + nn), type = "n",
+     xlim = c(0.01, 0.1), ylim = c(0, 1), log = "x",
+     xlab = "Contrast", ylab = "P(Correct)"))
> invisible(apply(cc.bt, 1, function(x) {
+     xx <- c(exp(-l10 * x[1]/x[2]), x[2]/l10)
+     lines(cxx, pweibull(cxx, xx[2], xx[1]),
+         col = rgb(0, 0, 0, 0.005) )
+         }))
> points(I(ny/(ny + nn)) ~ cnt, weib.sim, pch = 21,
+     bg = "white", cex = 1.5)
```

Fig. 5.10 Fitted
psychometric functions to the
data of Xie and Griffin [199]
measuring the probability of
correctly identifying the pair
of image samples from the
same image as a function of
image separation. The *curves*
were fit using a GLM (*solid
grey*), a GAM (*solid black*),
and a local polynomial
regression (*dashed*)

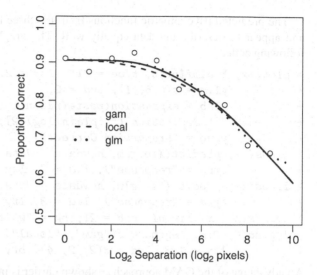

The function `locglmfit` performs the local polynomial fit. It does not take a formula object, but requires vectors of values at which to estimate the psychometric function, the number of correct responses, the total number of trials, and the stimulus values. It also requires a bandwidth estimate. The optimal bandwidth estimate is obtained from a function `bandwidth_cross_validation` that takes similar parameters to the modeling function but also a range of values within which to search for the bandwidth. A list is returned with the optimal bandwidth in a component named `deviance`. The gam function, in contrast, requires a formula object similar to `glm`, the difference being the specification of a smooth term on the right-hand side of the formula with the function s. For fitting a GLM, we note that the experimental paradigm is of type 2AFC and the data seem to asymptote at a value below 1, suggesting the use of the `psyfun.2asym` function from the **psyphy** package introduced in Sect. 5.1.3.

```
> xx <- seq(0, 10, len = 200)
> opt.bw <- with(Xie.df,
+      bandwidth_cross_validation(r, m, x, 10^c(-1, 1)))
> Xie.mf <- with(Xie.df,
+      locglmfit(xx, r, m, x, opt.bw$deviance))
> Xie.gam <- gam(cbind(r, m - r) ~ s(x), binomial,
+      Xie.df)
> Xie.glm <- psyfun.2asym(cbind(r, m - r) ~ x, Xie.df,
+      init.g = 0.5, init.lam = 0.1, mxNumAlt = 100)
lambda =      0.0926       gamma =   0.612
+/-SE(lambda) =     ( 0.0838 0.102 )
+/-SE(gamma) =   ( 0.585 0.639 )
```

Initial tests indicated that `psyfun.2asym` did not converge in the default number of iterations, so we specified starting values for the lower and upper asymptote parameters and doubled the maximum number of iterations.

The predicted psychometric functions from the three fits are shown in Fig. 5.10 and appear to describe the data equally well. The graph was generated with the following code:

```
> plot(xx, Xie.mf$pfit, type = "l", lty = 2,
+          ylim = c(0.5, 1), lwd = 2,
+          xlab = expression(paste(plain(Log)[2],
+             " Separation (", plain(log)[2], " pixels)")),
+          ylab = "Proportion Correct")
> lines(xx, predict(Xie.gam, newdata = data.frame(x = xx),
+          type = "response"), lwd = 2, lty = 1)
> lines(xx, predict(Xie.glm, newdata = data.frame(x = xx),
+          type = "response"), lwd = 3, lty = 3)
> points(p ~ x, Xie.df, pch = 21, bg = "white")
> legend(0, 0.8, legend = c("gam", "local", "glm"),
+          lty = 1:3, lwd = c(2, 2, 4), bty = "n")
```

An advantage of the GAM approach as shown earlier is in the possibility of making model comparisons as with the GLM.

Exercises

5.1. In Sect. 5.1.1, we introduced the data set ecc2 which contains data for two tasks at each of four letter heights. Fit an unconstrained model in which the threshold and steepness of the psychometric function are permitted to vary for each condition (treat Size as a factor). Then, fit a model in which the steepness is constrained to be the same across all conditions. Compare the fits.

For both models, plot the predicted curves with the data points. Which conditions(s) seem to deviate from a constant steepness model most? It may help, also, to examine a plot of the estimated steepness parameter as a function of the variable Size for the unconstrained fit. Create a factor that can be used to allow the steepness to vary just for the aberrant condition(s) and refit the data. Is the fit better?

5.2. For the unconstrained fit to the ecc2 data set, plot the threshold as a function of the Size for each task. Refit the data with Size as a covariate and add the estimated line of threshold vs Size to the plot. Add a quadratic term for Size to the model and compare the two fits.

5.3. In Sect. 5.2 we fit a single psychometric function for the two staircases based on all of the data. Refit the data with a model that estimates separate parameters for each staircase and compare the fit with that obtained in the text using a likelihood ratio test.

5.4. In Sect. 5.3 we described how to estimate thresholds. Use the methods described there to estimate a difference limen that spans the probabilities from 0.25 to 0.75.

Chapter 6
Classification Images

6.1 Ahumada's Experiment

In 1996, Ahumada [2] reported the following experiment. On each trial, the observer was presented with two small horizontal bars separated by a small, lateral gap (of 1 pixel width, corresponding to 1.26 min on his display) on a background. The bars could be either aligned or offset vertically (1 pixel) as shown in the top two images of Fig. 6.1, respectively. The observer's task on each trial was to judge whether an offset was present or not. This is a localization judgment measuring what is called Vernier acuity. Under appropriate conditions, humans are remarkably good at this, detecting offsets that can be an order of magnitude or more finer than the minimum separation that they can resolve between two adjacent bars, hence the term *hyperacuity* [186]. One innovation for this type of experiment in Ahumada's design was that stimuli were presented in "noise," i.e., the luminance of each pixel in the image (128×128 pixels) was increased or decreased by a random amount, as illustrated in the bottom two images of Fig. 6.1. The second was that he analyzed the relation between the spatial distribution of the noise and the observer's responses.

The experiment is a simple Yes–No detection task discussed in Chap. 3 in which the responses are "Offset-Present" and "Offset-Absent" and the stimuli are either "Offset" or "Not Offset." The two responses and the two stimulus conditions give rise to four possible response classification outcomes, indicated in a 2×2 table as follows: In the experiment, the contrast of the bars and the variance of the noise

	Offset	Not offset
Offset-present	Hit	False alarm
Offset-absent	Miss	Correct rejection

were selected so that the observer performed at a value of $d' \approx 1$. The observer makes his decision based on certain cues in the stimulus. Ahumada reasoned that

Offset Absent   Offset Present

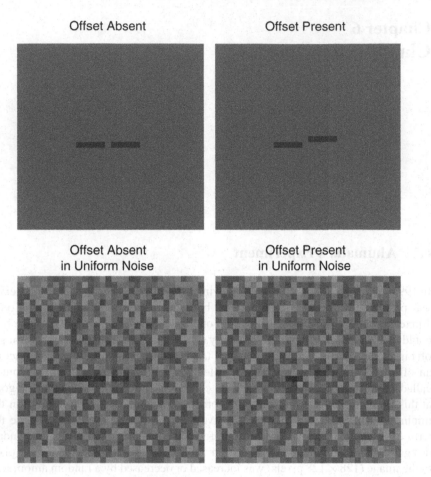

Offset Absent   Offset Present
in Uniform Noise   in Uniform Noise

Fig. 6.1 The *top two images* illustrate the stimulus events offset-absent (*left*) and offset-present (*right*) from Ahumada's experiment. The *bottom two images* show the same two events but with random uniform noise added to each pixel. In the original experiment, the images were 128×128. These are only 32×32

the noise would interfere with the judgments through interactions with these cues. Specifically, when the noise resembled those cues on which the observer based his decision of the offset being present, the observer would be more likely to make the response "offset-present," and conversely with respect to a decision of "offset-absent." Thus, by analyzing the noise conditional on the observer's decisions, it ought to be possible to estimate an image of the cues exploited by the observer in making his decision.

The model assumed by Ahumada is a template-matching model. We suppose that the observer uses a template stored in memory that contrasts the internal responses that he expects when the signal is present with those he expects when it is absent. He compares what he observes with the template and, if the match is close, he responds present, if not, absent. The comparison can be modeled as a scalar product.

Let $T(p)$ be the representation of the internal template in the space of the stimulus, where $p = (x, y)$ indicates the coordinates of a pixel in the image. The stimulus, $\mathscr{S}(p)$, is composed of either noise, $N(p)$,[1] or signal plus noise, $S(p) + N(p)$. The decision variable, Δ, employed by the observer is assumed to be

$$\Delta = T \cdot \mathscr{S} \qquad (6.1)$$

where we have suppressed the stimulus coordinates for simplicity. We assume that the observer responds "present" if $\Delta > c$, where c is a criterion set by the observer, and "absent" otherwise. The stimulus that maximizes the probability of judging the decision to be present has the form of the template [2]. Thus, an analysis of the stimulus on trials in which the observer responds "present" should permit an estimation of the template. False Alarms should occur on signal-absent trials when the spatial pattern of noise accidently resembles the cues critical for detection. It may be less obvious, but the trials on which the observer responds "absent" also provide information on the template, but with the opposite weight. The basic method proposed by Ahumada to estimate the template is very simple. First, one sorts the profiles of noise, $N_i(p)$, from each trial, i, according to the four possible response classifications in the above table. The mean noise profile is then computed separately within each category. Finally, the four mean profiles are combined to generate the classification image, $\mathscr{C}$ with the Hit and False Alarm profiles given a weight of $+1$ and the Miss and Correct Rejection profiles given a weight of -1.

$$\mathscr{C}(p) = (\hat{\mu}_H(p) - \hat{\mu}_M(p)) + (\hat{\mu}_{FA}(p) - \hat{\mu}_{CR}(p)) \qquad (6.2)$$

where $\hat{\mu}_H(x)$ is the mean noise profile for Hits, etc.

We begin by simulating a classification image experiment. Much can be learned from examining a simulation and as R provides excellent means for doing one, we examine a simulated classification image experiment in the next section.

6.1.1 Simulating a Classification Image Experiment

To simplify the analysis, we assume that the stimulus is distributed across one physical dimension such as would occur if the signal is a temporal modulation (see Sect. 6.3.1) or the luminance modulation across an oriented grating [161]. The signal, noise, and stimulus trials are set up as follows:

```
> N <- 32
> w <- 0.65
> Contrast <- 0.7
> x <- seq(-3, 3, len = N)
```

---

[1]That is, pixels of random luminance added to a vernier target with no offset, or zero misalignment signal.

Fig. 6.2 The *solid curve* is the signal added to half of the noise trials and the *dashed line* shows the template used for detection. The *circles* indicate the average over the 5,000 trials that contained a signal and the *triangles* the average over those that did not

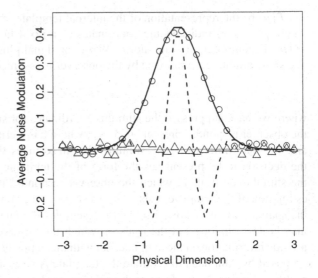

```
> Trials <- 10000
> Signal <- dnorm(x, sd = w)
> Template <- Signal * cos(2 * pi * x * w)
> Stimulus <- matrix(Noise <- rnorm(N * Trials),  N, Trials) +
+    cbind(matrix(rep(Contrast * Signal, Trials/2), N),
+    matrix(0, N, Trials/2))
```

We define the signal as a Gaussian modulation with width parameter (the argument sd) 0.65 indicated in Fig. 6.2 as a solid curve. For the simulation, it is sampled at 32 stimulus values between -3 and 3 and we adjust the height of the signal by a factor of 0.7. To make things interesting, however, we will suppose that the observer uses a template that differs from the signal, a Gabor function obtained by multiplying the signal by a cosine function of frequency 0.65 cycles/stimulus unit and indicated in Fig. 6.2 by the dashed line. In a real experiment, the template is, of course, unknown. The goal of the experiment is to estimate it. Comparing the differences between the estimated template and the signal or a simulation in which the signal, itself, serves as the template can be informative about the information coded by the sensory system and the strategy that the observer uses to make a judgment.

The simulated experiment consists of 10,000 trials, half of which contain just noise and half signal plus noise. The trials are set up in the $32 \times 10{,}000$ matrix Stimulus. Each column corresponds to the stimulus for one trial. Each of the first 5,000 columns contains Gaussian-distributed random numbers to which the signal was added. The second 5,000 columns contain random values, only. In a real experiment, the order of the noise and signal + noise trials would be randomized to minimize the effects of expectation on the judgments. This is unnecessary, however, in the simulation. In Fig. 6.2, the average profile of the signal + noise trials is indicated by the circles and of the noise trials by the triangles. The figure was created with the following code fragment:

```
> plot(x, rowMeans(Stimulus[, 1:(Trials/2)]),
+        type = "p", ylim = c(-0.25, 0.45), cex = 1.5,
+        xlab = "Physical Dimension",
+        ylab = "Average Stimulus Modulation")
> abline(0, 0, col = "grey", lwd = 2)
> points(x, rowMeans(Stimulus[, -(1:(Trials/2))]),
+        cex = 1.5, pch = 2)
> lines(x, Contrast * Signal, lwd = 2)
> lines(x, Contrast * Template, lty = 2, lwd = 2)
```

By comparing the stimulus with the internal template as in (6.1), we decide whether or not the signal is present on each trial.

```
> DecisionVariable <- drop(Template %*% Stimulus > 0)
> Classification <- factor(4 - (rep(c(1, 0), each = Trials/2) +
+        2 * DecisionVariable), levels = 1:4,
+        labels = c("Hit", "FA", "Miss", "CR"))
```

We use a criterion of $\Delta > 0$[2] for a response "present." The drop function transforms the data from a $1 \times 10,000$ instance of class "matrix" to a 10,000 element "vector." We then create a factor that codes the response classification for each trial in terms of the entries in the table on p. 167. To calculate the classification image, we transform the vector Noise into a $32 \times 10,000$ matrix and calculate the average noise vector for each of the four response categories using the function aggregate.

```
> ExpNoise <- matrix(Noise, N, Trials)
> CompImages <- aggregate(t(ExpNoise),
+        list(Classification), mean)[, -1]
> matplot(as.matrix(t(CompImages)), type = "l",
+        lwd = 2, col = "black", ylim = c(-0.5, 0.5),
+        xlab = "Physical Stimulus",
+        ylab = "Average Noise Modulation")
> legend(2.5, 0.5, legend = c("H", "FA", "M", "CR"),
+        lty = 1:4, cex = 1.2, bty = "n")
```

The four-component images have been plotted in Fig. 6.3a, each with a different line type, using the matplot command. Several features are immediately evident. Each of the four images appears to provide a nearly identical estimate but with different weights. For two of the images, the weights would be of opposite sign to those of the other two.

---

[2]This is a biased criterion. If we calculated d' directly, we could establish a criterion for unbiased responding. We leave this as an exercise for the reader.

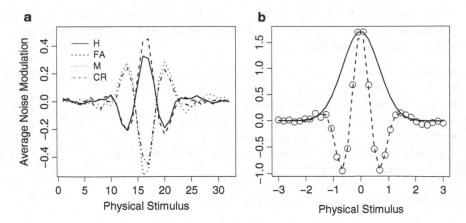

Fig. 6.3 (a) The four-component images obtained by averaging the noise profiles separately from the hits, false alarms, misses, and correct rejections. (b) The *points* indicate the classification image obtained from the weighted sum of the component images in the adjacent figure. The *solid curve* is the scaled value of the signal and the *dashed curve* is the scaled value of the template

Another striking aspect of the result is that the images resemble the template and *not* the signal. Thus, by analogy to the situation in which a human observer makes judgments with respect to a known signal, the difference between the classification image and the signal should inform us about the characteristics of the observer's template. Finally, the images that correspond to a "present" response are weighted with the same sign and vice versa for the two images based on an "absent" response. This explains the assignment of the positive and negative weights in (6.2). Combining the four images in this fashion generates the classification image, plotted as the circles in Fig. 6.3b and calculated as follows:

```
> CIWTS <- c(1, 1, -1, -1)
> SimClassIm <-crossprod(CIWTS, as.matrix(CompImages))
```

Scaled versions of the Signal and Template have been added for comparison as solid and dashed lines, respectively.

Since the characteristics of the noise are known and, in fact, the noise samples are recorded as data, the standard error of each point can be either derived theoretically or estimated. These values are often used to normalize the points so that the ordinate is specified in terms of z-scores.

6.1.2 The Information Used in Offset Detection

Consider again the offset detection task with which we began this chapter. An ideal observer who knows exactly the characteristics of the target and its absolute position would use a template that was based on only the parts of the stimulus that varied

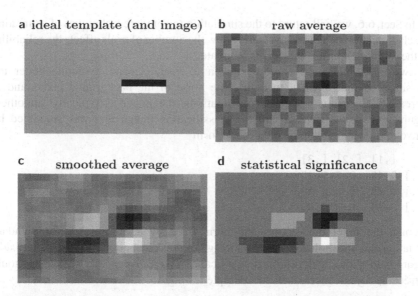

a ideal template (and image)   **b raw average**

c smoothed average   **d statistical significance**

Fig. 6.4 (**a**) Ideal observer's template for offset detection. (**b**) Average classification image for one observer for the offset detection task. The original stimulus presented to the observer was 128×128 pixels. The image here is a 14×24 detail centered on the Vernier targets. (**c**) A smoothed version of the classification image. (**d**) The significantly modulated pixels were detected by dividing each pixel of the smoothed image by twice the standard deviation of the smoothed noise and truncating toward zero

systematically across the experiment [3]. Over the experiment, the only aspect of the target that was systematically varied was the vertical position of the right-hand bar. The left-hand bar was fixed. The ideal classification scheme would construct a template in which the pixels at the upper and lower positions of the right-hand bar were weighted with 1 and −1, respectively, All other weights in the stimulus would be set to 0. This template is obtained by differencing the upper two images of Fig. 6.1 and is illustrated in Fig. 6.4a for a 24×14 pixel region centered on the target stimulus.

The example in Sect. 6.1.1 shows that the classification image would resemble the template. Thus, if the observer behaves in an ideal manner, no pixels along the fixed bar should be modulated significantly in the classification image. The ideal observer weights these pixels with a zero on every trial.

Figure 6.4b shows the classification image for one observer (2,900 trials) obtained for a central 14×24 pixel region of the stimulus. To the right of the central fixation, the image shows the expected bipolar image. However, it also reveals an inverted polarity image to the left, unlike what would be expected from an ideal observer. Thus, the results demonstrate that information at the position of the reference bar to the left influenced the decision. These results might be explained if instead of estimating the absolute position of the offset, the observer judged its position relative to the reference.

In Sect. 6.6, we will return to the simulation to compare techniques of estimating the classification image and evaluating how the number of trials affects the reliability of the estimation of the underlying template.

Two additional steps are often taken to visualize and to define better the classification image. First, to reduce the noise of the background pixels and to increase the sensitivity to gross modulations, the image is typically smoothed. Figure 6.4c shows a version of the classification image that was smoothed by convolution with a low-pass filter of the form

```
      [,1]  [,2]  [,3]
[1,]  0.49  0.7 0.49
[2,]  0.70  1.0 0.70
[3,]  0.49  0.7 0.49
```

Second, the pixel modulations are normalized by a multiple of the standard deviation and thresholded to yield an image of the distribution of significant pixels. Figure 6.4d shows a thresholded significance image. As a final step, one should correct the significance values for multiple testing.

6.2 Some History

Earlier classification image experiments were performed in the auditory domain [4, 5]. Ahumada [2] reintroduced the technique to the vision community in the context of a Vernier acuity experiment which stimulated a flurry of publications and special journal issues on the topic. An excellent review by Murray provides comprehensive coverage of the progress with this technique [130]. From the initial applications in auditory frequency discrimination and visual spatial localization, subsequent studies have ranged over topics from discrimination of luminance contrast [136, 161, 172], of chromatic contrast [18, 75], of retinal disparity [137], of filling-in of subjective contours [71], of brightness [44, 50], and of perception of facial features [102, 125]. The technique has been extended to cover m-alternative forced-choice techniques [1, 130] and rating methods [4, 5, 111]. Ahumada [3] and Murray et al. [131] proposed optimal weights in combining the component response classification images. Alternative techniques to estimate classification images have been considered, including direct maximization of the likelihood [161] and linear regression of observer ratings on a set of smooth basis functions [5, 111]. Abbey and Eckstein [1] mention an approach using generalized linear models. This was developed and extended by Knoblauch and Maloney [95] to generalized additive models (GAM) for providing a principled approach to smoothing (see Sect. 6.5). Ross and Cohen propose using graphical models [152]. Victor [179] considered the classification image technique in the context of related approaches using noise as a probe stimulus, such as in neural receptive field characterization and functional cerebral imaging. He reviews several methods of analyzing the data, some of which will be more efficient than (6.2) when there are correlations or nonlinearities in the data. Finally, although most of the literature on classification images relies on noise to perturb the stimulus, a series of studies have demonstrated that when the stimulus is sufficiently variable and difficult to discriminate without the addition of noise, the underlying approach can still be applied, often with more naturalistic stimuli [89, 90, 118, 187, 203]. These represent only a subset of the many applications of this technique.

6.3 The Classification Image as a Linear Model

Classification image experiments usually require thousands of trials. The data sets used for estimating the image can be very large, especially, for example, if the stimulus is distributed spatially or spatio-temporally. Equation (6.2) offers several efficiencies for calculating classification images. First, the mean and standard deviation can be calculated on the fly using an update method, so that not all of the data need to be kept in memory at once. Second, instead of storing the random numbers used to generate the noise on each trial, a seed can be stored with just the responses of the observer. Then, the random numbers can be regenerated at a later date using the seed. Since R typically performs all of its computations in memory on data that also are in memory, these efficiencies are easily appreciated. Nevertheless, computers with more and more memory are increasingly available, so that it is worthwhile exploring alternative, more memory-intensive techniques, and especially using modeling tools that are basic to R. As noted above, (6.2) can be viewed as just a linear model (LM), i.e., the calculation of the classification image is performed simply by a linear combination of the noise samples. If we could set up the problem to use the lm function, then it would allow us to exploit the strengths of R for modeling data, for example, the use of modeling formulae for specifying models and methods for extracting information from modeling objects. To this end, we will consider a data set from a real observer. Data from real observers are usually based on fewer trials than in a simulation, and human observers are, in general, less reliable than simulated observers. For example, their response criteria may vary over the experiment. Both of these aspects are likely to influence the precision of the obtained images.

6.3.1 Detection of a Gabor Temporal Modulation

The data set Gabor in package **MPDiR** contains a data frame of the results from one observer in one condition of a Yes–No detection experiment reported by Thomas and Knoblauch [172]. On each trial, the peak luminance of a region following a two-dimensional Gaussian spatial profile ($\sigma_s = 2.5°$) was modulated temporally during a 640 ms interval on a calibrated monitor. The modulation consisted of 32

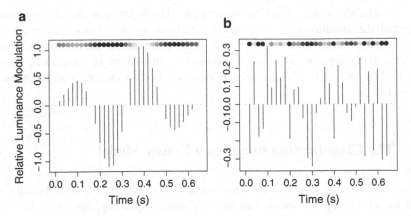

Fig. 6.5 (a) The sampled temporal signal from the experiment from which the Gabor data set was taken. (b) The sampled signal added to random uniform samples. The *grey level circles* at the *top* of each image provide an indication of the appearance of the luminance modulation of the Gaussian disk over time

sequential samples, each of 20 ms duration.[3] On a given trial, the modulations were either set to random luminance values chosen from a uniform distribution centered on the mean luminance of the screen or to random luminance values added to a luminance modulation that followed a Gabor function modulated over time

$$\mathrm{Lum}(t_i) = e^{-0.5\left(\frac{t_i - \mu}{\sigma_t}\right)^2} \sin(\pi t_i / \sigma_t), \qquad (6.3)$$

where t_i is the time of the ith sample, $i = 1, 32$, $\mu = 320$ ms, and $\sigma_t = 160$ ms. The sampled signal is shown in Fig. 6.5a and an example of the signal embedded in uniform noise samples in Fig. 6.5b. The observer's task was to classify each trial as to the presence or absence of the signal. The signal strength and the amplitude of the noise modulation were chosen in pilot experiments so that the observer's performance approximated $d' = 1$.

The data set `Gabor` contains four variables, `resp`, a factor, indicating the response classifications of the observer with four levels ("H," "FA," "M," "CR"), `time`, a 32-level factor indexing the time points of the samples of the stimulus, `Time`, a numeric variable indicating the time of the sample in seconds from the beginning of the trial, and `N`, the instantaneous noise value (in relative luminance difference from the mean level) for the sample. The trial numbers are not indicated but each trial corresponds to a consecutive block of rows over which the factor `time` varies from level "1" to "32." The observer performed 16 sessions, each containing 224 trials with equal numbers of events, signal present and absent. The 114,688 rows of the data frame correspond to 3,584 trials.

---

[3]The screen refresh rate was set to 100 times/s, so that each sample corresponded to two screen sweeps of the electron beam.

Fig. 6.6 The *circles* correspond to the classification image calculated using (6.2) for the data set Gabor. The *solid black line* segments connect points obtained using lm. The *grey curve* is the ideal template

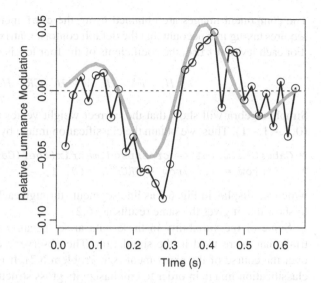

6.3.2   Estimation by a Linear Model

First, we demonstrate how to calculate the classification image with (6.2) when the data are organized in a data frame as here. The images for the four response categories are extracted using the function `tapply` to obtain the mean for each combination of the levels of `time` and `resp`. This generates a 32×4 matrix which is then multiplied by the weight vector to yield the classification image. The steps are shown below:

```
> data(Gabor)
> GaborClsIm.clim <- drop(with(Gabor, tapply(N, list(time,
+                        resp), mean)) %*% c(1, 1, -1, -1))
```

and the estimated image is shown as circles in Fig. 6.6. For reference, the true signal is also plotted as a thick grey curve. The observer's estimated template displays qualitative similarities to the signal while differing quantitatively.

6.3.3   Estimation with lm

We will now show how the function `lm` can be used to perform the same calculations. There are several possibilities, for setting up the model formula. One is to fit separate linear models for each response category. This is performed with the following function call:

```
> Gabor.lm1 <- lm(N ~ time/resp - 1, Gabor)
```

The component images are obtained using the `coef` method but combining them requires taking into account that the default contrast matrix uses treatment contrasts. For each level of `time`, the coefficients of the four levels of `resp` will be

$$H, \quad FA-H \quad M-H \quad CR-H \tag{6.4}$$

Simple algebra will show that the correct weight vector to calculate (6.2) will be $(0, 1, -1, -1)$. Thus, we obtain the classification image by

```
> GaborClsIm.lm1 <- crossprod(matrix(coef(Gabor.lm1),
+     nrow = 4, byrow = TRUE), c(0, 1, -1, -1))
```

which we display in Fig. 6.6 as line segments through each of the estimated points, to show that it gives the same results as (6.2).

Some of the variability in these data arises because they are based on fewer trials than were used in the simulation. The observer's criterion, also, may vary over the course of the experiment (see Problem 6.2). It is common to smooth the classification image in order to emphasize its gross structure. We will demonstrate in Sect. 6.5 how to estimate a smooth classification image using a GAM in which the degree of smoothing is chosen by a model fitting criterion.

It would be interesting (and perhaps less error prone) to parameterize the problem to use the same contrasts as with (6.2). This can be done by specifying the model as

```
> lm(N ~ time:resp - 1, Gabor)
```

This is one of the rare circumstances when it makes sense to use a model that specifies an interaction with no intercept and no main effects. Alternatively, by customizing the contrasts, it is possible to obtain the classification image as a subset of the coefficients. Manipulating the contrasts can be tricky, however, and the procedure is made easier with the `make.contrasts` function in the **gmodels** package [180]. The custom contrasts can be specified directly in the call to `lm`.

```
> library(gmodels)
> lm(N ~ time:resp - 1, Gabor,
+     contrasts = list(resp = make.contrasts(c(1, 1, -1, -1))))
```

In this case, the classification image is found in coefficients 33–64. Verifying that all of these approaches lead to the same classification image is left as an exercise.

6.3.4   Model Evaluation

It is worth noting that within the linear modeling framework, the estimated classification image depends on the interaction terms. A consistent classification strategy by the observer leads to significantly different values of the contrast of the response categories at different times. This suggests that we can evaluate the statistical significance of the obtained image, i.e., the hypothesis that the pattern of modulation in the classification image is different from zero, using the modeling tools in R. The alternative hypothesis would be specified as a model without an interaction term. The `update` function permits the simple removal of the interaction term, and the new model is compared with the former one using the `anova` method to implement a likelihood ratio test of the nested models.

```
> Gabor.lm0 <- update(Gabor.lm1, . ~ . - time:resp)
> anova(Gabor.lm0, Gabor.lm1)


Analysis of Variance Table

Model 1: N ~ time - 1
Model 2: N ~ time/resp - 1
  Res.Df  RSS Df Sum of Sq    F Pr(>F)
1 114656 3439
2 114560 3424 96      15.4 5.37 <2e-16
```

The small p-value supports the significance of the classification image.

In this simple case, in which we are only interested in testing the highest order term, the `drop1` method for `lm` from the **stats** package achieves the same result in one step.

```
> drop1(Gabor.lm1, test = "F")


Single term deletions

Model:
N ~ time/resp - 1
         Df Sum of Sq  RSS      AIC F value Pr(>F)
<none>                 3424 -402476
time:resp 96     15.4 3439 -402154   5.37 <2e-16
```

but note that the test to perform must be specified.

As the dependent variable is composed simply of random numbers, we find that the diagnostic plots that can be generated from objects of class "lm" are of limited use, in this case.

6.4   Classification Image Estimation with `glm`

In the linear model approach described in Sect. 6.3.3, we treat the noise as the dependent variable and the response categories as a term among the independent, predictor variables. Oddly, this is backward with respect to the usual approach of modeling the response as a function of the stimulus.

Instead, consider the problem as one of classification, in which the responses are used to sort the noise profiles from each trial into two groups. The objective is to discover the features of the noise profile that lead the observer to one of the two decisions, present or absent. There are many approaches to solving this problem [150, 178, 179]. One of the simplest is using a GLM. As with psychometric function estimation, we model the expected value of the response by a linear predictor via a link function, η. The linear predictor is defined as a weighted combination of the noise profile from each trial. Thus, each time sample is a covariate taking on the instantaneous value of the noise, and the weights (or coefficients) provide the estimate of the classification image. This conception of the problem corresponds closely to that of the decision rule of (6.1) with the weights in the role of the template

$$\eta(\mathsf{E}[Pr(Y = 1)]) = \beta_1 X_1 + \cdots + \beta_p X_p, \tag{6.5}$$

where the noise profile is composed of p samples. As with the Signal Detection Theory model and the psychometric function, the output of the linear predictor corresponds to the response of the observer in the decision space. An interaction with a factor coding the presence or absence of the signal on each trial completes the model.

To simplify the application of this model, we reshape the data set Gabor to the more usual format of a data frame in which each trial corresponds to one row and the columns contain the responses, factor levels, and covariate values for that trial. We need to recode the factor resp into two factors, corresponding to the decision of the observer and to the status (presence/absence) of the signal. The function unclass returns a copy of the factor resp with its class attribute removed, showing only its levels. We exploit this to extract the responses of the observer, "Present" corresponding to the first two levels, and the signal status of each trial, present corresponding to the first and third levels. Then, noise values, N, must be distributed across columns to make 32 covariates, one for each temporal sample. The following code fragment accomplishes these tasks:

```
> Resp <- factor(unclass(Gabor$resp) < 3, labels = 0:1)[seq(1,
+      nrow(Gabor), 32)]
> Stim <- factor(unclass(Gabor$resp)%%2, labels = 0:1)[seq(1,
+      nrow(Gabor), 32)]
> Gabor.wide <- data.frame(matrix(Gabor$N, ncol = 32,
+      byrow = TRUE))
> names(Gabor.wide) <- paste("t", 1:32, sep = "")
> Gabor.wide <- cbind(Resp, Stim, Gabor.wide)
```

Once the data frame is properly set up, we fit the model by

```
> GaborClsIm.glm1 <- glm(Resp ~ Stim/. - 1,
+          family = binomial(link = "probit"),
+          data = Gabor.wide)
```

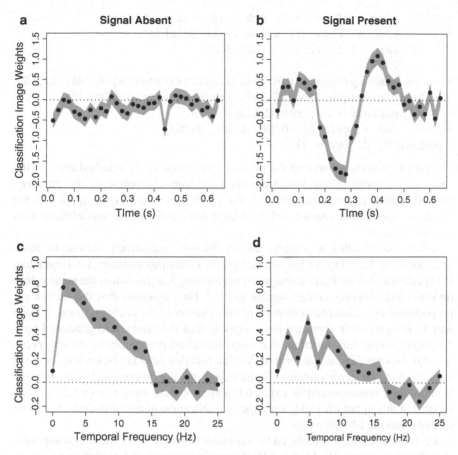

Fig. 6.7 (**a**) The *black points* indicate the classification image for the trials on which the stimulus was absent, estimated using `glm` for the data set Gabor. The *grey zones* in this and the other three figures correspond to standard errors. (**b**) The same as in (**a**), except that the image is calculated for the signal present trials. (**c**) The *black points* indicate the classification image calculated on the modulus of the fast Fourier transform of the noise for the trials on which the signal was absent. (**d**) The same as in (**c**), except that the image is calculated for the signal present trials

The "." is an abbreviation for the sum of all columns in the data frame not already occurring in the formula, i.e., other than `Resp` and `Stim`. This is a tremendous convenience in typing when there are 32 covariates. Note that all stimulus variables are on the right side of the formula, now, and the response variable is on the left.

Classification images for Signal Absent and Signal Present are obtained from the coefficients of the interaction of the covariates, `t1 … t32`, with the terms `Stim0` and `Stim1`, respectively. These are plotted as black points in Fig. 6.7a, b. Here is the code for creating Fig. 6.7b.

```
> ClsIm <- coef(GaborClsIm.glm1)[seq(4, 66, 2)]
> plot(unique(Gabor$Time), ClsIm,
```

```
+          xlab = "TIme (s)", pch = 16, col = "black",
+          ylab = "Classification Image Weights",
+          cex = 1.5, ylim = c(-2, 1.5)
+          )
> se <- summary(GaborClsIm.glm1)$coefficients[seq(4, 66, 2), 2]
> polygon(c(unique(Gabor$Time), rev(unique(Gabor$Time))),
+          c(ClsIm + se, rev(ClsIm - se)),
+          col = rgb(0.25, 0.25, 0.25, 0.25))
> abline(0, 0, lty = 3)
```

The grey region drawn around the points corresponds to ± 1 standard error of the covariate estimates. These values are obtained from the output of the summary method applied to the model object.[4] The coefficients are weights applied to the luminance modulations, so we no longer label them as luminance modulations as in previous figures.

When the stimulus is present, the results are qualitatively similar to those obtained using lm in Fig. 6.6 but appear to yield a smoother estimate. Levi and Klein [111] calculated classification images by performing a regression of the noise values on estimated observer ratings similar to [4].[5] They assumed that the ratings are proportional to the internal responses of the observer in the decision space. A good way to imagine their approach is to suppose that the ratings are generated by a "psychophysical" link function, that maps perceived probability to the observer's internal response. The actual form of this link function need not be known since it is implicit in the observer's ratings. To the extent that the proportionality assumption is valid, their results would be expected to approximate those from a GLM. They noted that this approach yields smoother classification images for a given number of trials than the LM approach.

Each of these images is defined by a contrast between two of the four component classification images (H–M, FA–CR) comparing present and absent responses of the observer. We could combine them as we have done earlier because they often provide similar estimates. However, for this stimulus, they are quite different. Unlike for the carriers of lower temporal frequency reported in [172], the image based on the trials when the stimulus was absent appears flat and not significantly different from zero. The authors conjectured that when the signal was present in the stimulus, it served as a reference so that any noise profiles that tended to match it in frequency and phase would increase the tendency of the observer to respond "Present." The frequency of the signal in this condition was sufficiently elevated, however, that in its absence, the observer would have difficulty in keeping track of its phase. In that case, he might respond only on the basis of frequency information. The classification image calculated in the temporal domain for signal absent trials would then be the sum of images over several phases, which would tend to reduce its amplitude.

---

[4] A bootstrap procedure for obtaining standard error estimates is outlined in Exercise 6.6.

[5] An alternative is to analyze the data with a proportional odds model as in Chap. 3.

To test this hypothesis, they performed a Fourier transform of the noise profiles from each trial and obtained a classification image based on modulus of the transform rather than the temporal noise profiles. The frequency domain analysis is easily performed with the following code fragment.

```
> TmpStim <- with(Gabor, matrix(N, nrow = 32))
> FTmpStim <- apply(TmpStim, 2, fft)
> FTmpStim <- Mod(FTmpStim)
> f.mat <- data.frame(t(FTmpStim))
> names(f.mat) <- paste("f", 1:32, sep = "")
> f.mat <- cbind(Resp, f.mat)
> GaborFreqIm.glm <- glm(Resp ~ Stim/. - 1,
+                 family = binomial(link = "probit"),
+                 data = f.mat)
```

First, the vector of noise profiles from all trials is reshaped into a $32 \times 3,584$ matrix, with trials distributed across columns and temporal samples across rows. The fft function is then applied to each column to generate the fast Fourier transform of each profile. The raw coefficients are of class "complex." Obtaining their modulus, however, leaves the matrix FTmpStim as class "numeric." The moduli are organized in a data frame of similar format to Gabor.wide above, except with the covariates renamed with the prefix 'f' for frequency. By adding the column Resp, we can perform the same analysis as above with glm, but now in the Fourier rather than in the temporal domain. The results and standard errors for the two cases, signal absent and present, are shown in Fig. 6.7c and d, respectively. As they demonstrated, the classification image weights when the stimulus is absent are higher at temporal frequencies around the Gabor carrier frequency than when it is present.

6.5 Fitting a Smooth Model

The techniques presented above for estimating classification images are effective and easy to implement. A criticism that can be raised is that, in each of the techniques, the stimulus dimension variables are treated independently and discretely, as factor levels with lm and as covariates with glm. The noise values at one moment in time are assumed to be uncorrelated with those at other moments; the noise at one pixel is unrelated to the noise at others. Taken over all stimuli in an experiment, this is true, in that the noise values chosen are independent (at least, within the capacity of the random number generator used). However, this will not be the case in the subset of images that the observer classifies similarly. The techniques presented until now do not exploit the dependencies between pixels in these subsets introduced by the observer's choices [179]. One could attempt to approach this by allowing for the possibility of interactions between covariates, but this would not result in parsimonious models.

In practice, classification images are often smoothed by some form of low-pass filtering such as described in Sect. 6.1.2. Implicitly, this procedure recognizes dependencies in the data and explicitly introduces them. The sampling rate of the noise is often finer than the resolution of the judgment made by the observer, so the classification image will tend to be uncorrelated on a fine scale. The smoothing will bring out the large-scale dependencies. Pooling nearby samples will increase the sensitivity of statistical tests at detecting the modulation in the image. The major criticism of such ad hoc smoothing is that the degree of smoothing used is arbitrary. Methods to decide how much smoothing is optimal are typically not employed.[6]

GAMs provide an appealing method for estimating a smooth classification image [77, 198]. They are extensions of the GLM, so the models developed in Sect. 6.4 apply. The difference is that smooth terms that are functions of the stimulus dimensions can be included in the model. Thus, the linear predictor is a function of the continuous stimulus dimension(s). The smooth terms in the model correspond to linear combinations of basis vectors added to the model matrix as linear covariates. More basis terms are chosen than is generally necessary, and smoothness is controlled by adding a penalty for undersmoothing during the process of maximizing the likelihood. The penalty is proportional to the integrated square of the second derivative of the smooth function.[7] In the case of the binomial family for which the scale parameter is known, the degree of smoothing is chosen to minimize a statistic called the un-biased risk estimator (UBRE) [198] that is related to AIC by a linear transformation (see Sect. B.5.2). Like AIC, this measure favors a model that maximizes the predictability of future rather than the actual data and serves to minimize the tendency to overfit the data.

The specification of the GAM model to fit to our example data set would be as follows:

$$\eta(\mathsf{E}[Pr(Y=1)]) = f_0(t)X_\mathrm{P} + f_1(t)X_\mathrm{A}, \tag{6.6}$$

where f_i are smooth terms represented by spline bases in the model matrix and X_P and X_A are the noise covariates for Present and Absent trials, respectively, extracted from the main effects model matrix. The intercept term is absorbed into the specification of the smooth functions. The f_i terms will correspond to the classification images estimates for the present and absent trials.

There are several packages in R that implement GAMs. We will use the **mgcv** package [198] that is one of the recommended packages distributed with R. We first load the package and then create two new columns from the factor resp in the data frame Gabor: one for a factor coding whether the stimulus is present or absent $(0, 1)$, Stim and the second coding the response of the observer, also $(0, 1)$, Resp.

---

[6]But, see [30, 95, 128].

[7]Mineault et al. [128] note that this choice of penalty is equivalent to assuming a Gaussian prior on the coefficients for the basis vectors composing the smooth function to be estimated. They explore an alternative smoothing criterion based on a Laplacian prior that leads to maximizing sparseness in the coefficients.

```
> library(mgcv)
> Gabor$Stim <- factor(with(Gabor, unclass(resp) %% 2))
> Gabor$Resp <- factor(with(Gabor, unclass(resp) < 3),
+        labels = c("0", "1"))
> Imat <- model.matrix(~Stim:N - 1, data = Gabor)
> Gabor$by1 <- Imat[, 1]
> Gabor$by2 <- Imat[, 2]
```

The fourth line creates a model matrix, `Imat`, for the interaction of the factor `Stim` with the noise, `N`, here treated as a function of the covariate `Time`. It contains two columns that take either the value zero or the noise level depending on whether the signal was absent or present, respectively. They are added to the data frame as two columns and will be used in the model formula of the function `gam` used to fit the model.

The function call to fit the model is as follows

```
> Gab.gam <- gam(Resp ~ s(Time, bs = "ts", k = 25, by = by1) +
+        s(Time, bs = "ts", k = 25, by = by2),
+        family = binomial("probit"), data = Gabor)
```

As with the `glm` fit, the observer's response is modeled. New terms, `s`, appear on the right-hand side of the formula, however. These correspond to the smooth terms. Each smooth term is specified as a function of the variable `Time`, the actual time in seconds. The argument `bs` specifies the type of spline basis to use for the smoothing terms. A thin plate regression spline with shrinkage was chosen (see `help(tprs, package = mgcv)`). This basis has some optimality properties even though it is less computationally efficient than some of the other choices for large data sets, as here (see `help(smooth.terms, package = mgcv)`). Also, the shrinkage permits the smoothing to go to zero for some terms, if necessary. The argument k specifies the dimension of the basis used to represent the smooth term. This should be large enough to model the expected variation in the covariate and no larger than number of points to model (here 32). It determines the maximum number of degrees of freedom that will be attributed to the smooth term. Specifying a larger value is computationally less efficient but it can lead to a better solution. The argument by permits the specification of a covariate by which the whole term is multiplied, producing a *variable coefficient model* [77, 198]. This generates the product of the time-dependent smooth term and the noise to estimate the value of the decision variable. The smooth terms will describe the weights of the classification image for present and absent trials. In order to introduce the interaction with the factor `Stim`, individual terms for the two levels need to be introduced. This explains why there are two smooth terms, each obtained from a different column of the model matrix, `Imat`. The `family` and `data` arguments used here are just as those for `glm`. Additional arguments are described in the documentation for `gam` to control the fitting procedure.

The results of the model can be explored to various degrees of detail by invoking either the print, anova, or summary methods to explore the model object. For example, the summary method gives

```
> summary(Gab.gam)
Family: binomial
Link function: probit
Formula:
Resp ~ s(Time, bs = "ts", k = 25, by = by1) + s(Time, bs = "ts",
    k = 25, by = by2)
<environment: 0x12d56ebc0>


Parametric coefficients:
            Estimate Std. Error z value Pr(>|z|)
(Intercept)  0.07296    0.00371    19.7   <2e-16 ***
---
Signif. codes:  0 `***' 0.001 `**' 0.01 `*' 0.05 `.' 0.1 ` ' 1

Approximate significance of smooth terms:
                edf Ref.df Chi.sq p-value
s(Time):by1    1.89   2.07   28.9 6.1e-07 ***
s(Time):by2   10.60  12.90  292.2 < 2e-16 ***
---
Signif. codes:  0 `ï£¡***' 0.001 `**' 0.01 `*' 0.05 `.'
    0.1 ` ' 1


R-sq.(adj) =  0.00274   Deviance explained = 0.214%
UBRE score = 0.3802  Scale est. = 1          n = 114688
```

The main new information is in the smooth terms and the UBRE score. Associated with each smooth term is its estimated degrees of freedom (edf). For the stimulus absent, this is between 1 and 2, suggesting that a linear term provides a reasonable approximation. We see in Fig. 6.8a that this smooth term is approximately flat in accord with the results using glm displayed previously in Fig. 6.7a. The second smooth term requires an edf of about 11. The small p-value indicates a significant modulation over time. The UBRE score is useful in comparing the fits between nested models. The model with a lower UBRE score will be considered to provide a better fit for the number of parameters required. The r^2 and deviance explained are very low, but after all, we are fitting noise.

The plot methods for gam objects allow the estimated smooth components to be visualized as shown in Fig. 6.8a, b for the absent and present images, respectively. As indicated by the summary information, the image for the signal absent trials is nearly flat, as noted above, while the signal present image is more similar to the signal, indicated for comparison as a dashed curve. This image follows the signal more closely than the estimates using GLM or LM methods though it agrees with them in indicating that the observer does not seem to place significant weight on the later part of the stimulus in the decision.

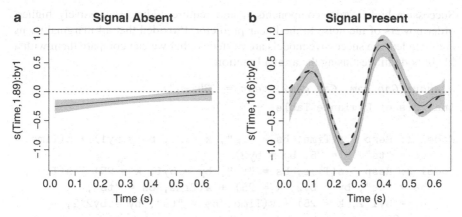

Fig. 6.8 (a) The *solid line* is the estimated smooth classification image for the signal absent trials. The *grey zones* indicate ± twice the standard standard errors. The estimated degrees of freedom are indicated in the ordinate label. (**b**) The estimated smooth classification image for the signal present trials. Other details as in plot (**a**) except that the *dashed curve* indicates the temporal modulation of the signal, scaled to fit the classification image

6.5.1 Higher-Order Classification Images

A more general approach to assessing nonlinearities in the decision process than that described using Fourier analysis in Sect. 6.4 was proposed by Neri and Heeger [134, 136]. They calculated second-order classification images by squaring the noise before applying (6.2). Higher-order classification images could be calculated by using higher powers.[8]

Higher-order classification images can be easily incorporated into the GLM and GAM approaches by including powers of the noise distribution in the linear predictor and estimating their coefficients. For example, the specification of a GAM model with a quadratic component of the classification image would include additional smooth components to weight the square of the noise.

$$\eta\left(\mathsf{E}[Pr(Y=1)]\right) = f_0(t)X_P + f_1(t)X_A + f_2(t)X_P^2 + f_3(t)X_A^2. \qquad (6.7)$$

This simply translates to the following call with the gam function.

```
> Gab.gam2 <- gam(Resp ~ s(Time, bs = "ts", by = by1, k = 25) +
+              s(Time, bs = "ts", by = by2, k = 25) +
+              s(Time, bs = "ts", by = by1^2, k = 25) +
+              s(Time, bs = "ts", by = by2^2, k = 25),
+              family = binomial("probit"), data = Gabor)
```

---

[8]This differs from the definition of higher-order classification images proposed by Nandy and Tjan [133] that is based on correlations of pixel values across regions of an image, thus revealing what they term second-order features.

Successive higher-order components would require adding successively higher-order powers of the noise to the linear predictor. Provided that all marginal terms are included, the successive models are nested so that we can compare them with a likelihood ratio test using the anova function.

```
> anova(Gab.gam, Gab.gam2, test = "Chisq")
Analysis of Deviance Table

Model 1: Resp ~ s(Time, bs = "ts", k = 25, by = by1) + s(Time,
    bs = "ts", k = 25, by = by2)
Model 2: Resp ~ s(Time, bs = "ts", by = by1, k = 25) + s(Time,
    bs = "ts", by = by2, k = 25) + s(Time, bs = "ts",
    by = by1^2, k = 25) + s(Time, bs = "ts", by = by2^2,
    k = 25)
  Resid. Df Resid. Dev   Df Deviance Pr(>Chi)
1    114675     158266
2    114668     149717 6.04     8548   <2e-16 ***
---
Signif. codes:  0 `***' 0.001 `**' 0.01 `*' 0.05 `.' 0.1 ` ' 1
```

As expected from Fig. 6.7, the more complex model significantly improves the fit. However, an extension to include cubic terms is less convincing (left as an exercise for the reader).

The four estimated component classification images are displayed in Fig. 6.9. The top two images correspond to the linear first-order classification images and reproduce the results of Fig. 6.8. The bottom two images show the second-order classification images for stimulus absent (left) and present (right). Here, the pattern is reversed, so that the stimulus present image is nearly unmodulated while the stimulus absent image displays a peak near the center of the trial.

The second-order, stimulus-present image is nearly flat suggesting that the signal present condition can be explained by the linear template-matching model. However, the observer could not know whether the signal was present or not on a trial, and it is improbable that he could modify his strategy in this way. The explanation of the authors is that when "the signal is present, any noise component similar in temporal frequency and phase will increase the effective contrast of the signal..." and thereby bias the observer toward responding "present." Thus, the linearity implied by this behavior is only apparent.

6.6 Comparing Methods

Raw classification images as estimated using linear or generalized linear models are often quite noisy, leading experimenters to introduce some arbitrary amount of smoothing to emphasize the gross features of the image. An important advantage of

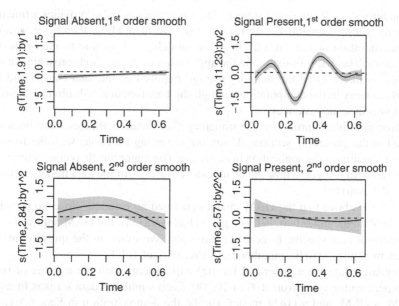

Fig. 6.9 The smooth functions estimated for each of the terms of (6.7) fit to the Gabor data set. The *top two graphs* are the terms for the linear component and the *bottom two* for the quadratic component. The *left two graphs* are fit to the stimulus absent data and the *right two* to stimulus present. The *grey envelopes* correspond to ±2 standard errors for each curve. When using the by argument, as in these analyses, the smooth terms are not automatically centered vertically about an ordinate value of 0 [198]. We have translated the second-order images so that all of the images appear on the same scale

using additive models is that the smoothing is incorporated directly into the fitting process, and the amount of smoothing is optimized with respect to a model selection criterion related to minimizing prediction error rather than data error.

The gam function, also, allows fitting multidimensional smooth terms, so is easily applied to *multidimensional* classification images, such as for spatial, motion, or chromatic signals, as well as their combinations. Analyzing the results from an experiment using such stimuli, however, can quickly consume all of the memory as R usually keeps and analyzes everything in memory and internally often generates multiple copies of objects. To address this issue, the **mgcv** package has recently introduced a new function, bam, to analyze *big* additive models. Its interface and usage resemble that of the gam function, but it exploits hard disk storage (at the cost of speed of computation) to alleviate the memory load. We have successfully fit classification data image sets of 8,000 trials of 64 × 64 images (and we probably could have pushed it further) using bam on a 64-bit machine (and a 64-bit compiled version of R) with 16 Gb of RAM.

The data sets examined in this chapter were for a single observer and a single experimental condition. If a data frame is extended to include several experimental conditions, the model formula is easily modified to include additional terms so that all conditions can be modeled at once. The **mgcv** package also provides a function gamm for fitting generalized additive mixed-effects models, thereby permitting the incorporation of random effects in the models, such as due to inter-observer differences. The major limitation in the application of the methods presented in this chapter[9] is likely to be the memory storage required of the data frames and the model matrices in the computations though the bam function will obviate memory constraints in some instances.

Three possible approaches to estimating classification images have been described in the preceding sections. Using the modeling functions with the formula interface facilitates the application in each case. The major hurdle is conceptualizing the model with respect to how the factors and covariates of the data frame enter into the model matrix.

We noted above that the GAM model generated an estimate closer to the ideal template than the other two. Is this a general advantage of the GAM approach over the others or is it specific to certain conditions? We examine the question in this section by returning to the simulated observer of Sect. 6.1.1.

Simulated data were generated for 100 experiments with the number of trials per experiment varying from 100 to 10,000. Each simulated data set was fit using an LM, a GLM, and a GAM model. Unlike the demonstration in Sect. 6.1.1, the simulated observer used the signal as the template. The estimated classification images were then scaled to fit the template and the residual variance was calculated as a measure of goodness of fit.

We show a code fragment for running the GLM below because it illustrates several uses from the family of apply functions to avoid formal looping constructs. The lapply function applies a function over each member of a list (or, as here, a vector) and returns its results in the list, err. The apply function applies a function over the margins of an array. The sapply function works just like lapply but will return its values in a simpler structure than a list, if possible. Thus, the 100 simulations for an experiment of Trials number of trials are returned as a vector, but the vector for each of the elements of Trials will be stored as an element of the list err. The $N \times$ NumTrials $\times$ NumSims array Stimulus can be viewed as a vector of matrices. The columns of each matrix are the stimuli for the trials of an experiment and each matrix is a different experiment. The decision variable for each trial is given by the dot product of each matrix column and the template in the variable Signal. The crossprod function permits this dot product to be calculated in one step over a whole matrix, returning the decision variable for each trial, and apply iterates the operation over each matrix-slice of the array. A matrix is returned with each column giving the decision variables for the trials of one experiment. Adding LM and GAM models to the simulation is left as an exercise, although beware that the GAM simulations can require several hours to complete when the number of trials is large.

---

[9]Excepting (6.2) for the reasons cited on p. 175.

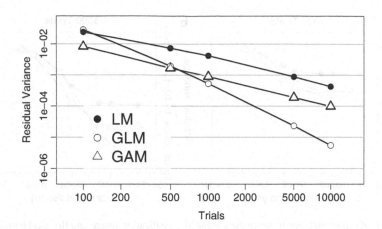

Fig. 6.10 Residual variance between the estimated classification image and the ideal template is plotted as a function of the number of trials in a simulated experiment, using LM, GLM, and GAM methods, as indicated in the legend. Each point is the median of 100 simulations

```
> Trials <- c(100, 500, 1000, 5000, 10000)
> NumSims <- 100
> glc <- glm.control(maxit = 1000)
> err <- lapply(Trials, function(NumTrials){
+    Noise <- array(rnorm(N * NumTrials * NumSims),
+             c(N, NumTrials, NumSims))
+    Stimulus <- Noise + rep(cbind(matrix(rep(Contrast * Signal,
+        NumTrials/2), nrow = N),
+        matrix(0, N, NumTrials/2)), NumSims)
+    resp <- apply(Stimulus, 3, function(x, t) crossprod(x, t),
+            t = Template) > 0
+    Stim <- factor(rep(c(1, 0), each = NumTrials/2))
+    sapply(seq(dim(resp)[2]), function(x, y, z) {
+        Respdf <- data.frame(resp = factor(y[, x],
+            labels = c(0, 1)),
+            N = t(z[, , x]))
+        CI.glm <- glm(resp ~ Stim/. - 1, binomial("probit"),
+            Respdf,  control = glc)
+        CI <- coef(CI.glm)[seq(4, 2 + 2 * N, 2)] +
+            coef(CI.glm)[seq(3, 2 + 2 * N, 2)]
+    res <- var(Template - coef(lm(Template ~ CI - 1)) * CI)
+    res
+        }, y = resp, z = Noise)
+    })
> names(err) <- Trials
```

The results of the simulations are displayed in Fig. 6.10 in which the median residual variance of the fit between the estimated classification image and the template

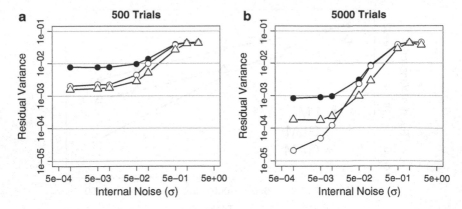

Fig. 6.11 Residual variance between the estimated classification image and the ideal template is plotted as a function of the standard deviation of the random variates added to the template, using LM, GLM, and GAM methods for experiments of (**a**) 500 and (**b**) 5,000 trials (for legend, see Fig. 6.10). Each point is the median value of 100 simulations

is plotted as a function of the number of trials in an experiment. For each of the methods, the residual variance decreases as the number of experimental trials increases. However, the fits using the GLM (white circles) decrease fastest. By 10,000 trials, this method generates a classification image for which the differences from the template are nearly two orders of magnitude less than those given by a linear model. The GAM performs in-between, doing best for small sample sizes but being bettered by the GLM for experiments larger than 500 trials.

In this simulation, only external noise was added on each trial. Since the template was fixed, the observer can be considered noiseless. In testing real observers, it is unlikely that the criterion is perfectly stable from trial to trial. Systematic variations could occur in the guise of fading memory of the template or learning of the template over trials. Random effects could arise from variations in attention and fatigue over the experiment. The effects of random variations on the template were simulated by adding Gaussian random noise to the template on each trial before the decision variable was calculated. The level of the internal noise was parameterized by the standard deviation of the Gaussian random variable and was varied from 0.001 to 2. We tested experiments of 500 and 5,000 trials. The results are shown in Fig. 6.11a, b for the three models. As the internal noise is increased, the difference from the template increases, eventually all three methods reaching the same asymptotic value. For 500 trials, the GAM method does slightly better than the GLM over most of the noise range. For 5,000 trials, the GAM method performs best for an intermediate range of noise levels.

These results are provocative when compared with those obtained from a real observer. The real observer performed 3,584 trials which would put him in a region for which the GLM would perform best, if he used a noiseless template. The results, however, indicated that the GLM and LM classification images were quite similar suggesting the presence of sufficient internal noise to eradicate the difference between the LM and GLM estimates. On the other hand, the GAM method generated a classification image with greater similarity to that of the ideal template, a result consistent with that obtained at the intermediate values of noise tested.

Neri has suggested that human observers are sufficiently noisy that the advantages accrued by the GLM approach cannot be attained in practice [135]. This result may depend on the task performed. Even in the presence of a noisy observer, however, the GAM approach can perform better than GLM in some situations. The simulations above suggest that one never does worse using a generalized approach and one can certainly do much better than the linear model.

Exercises

6.1. For the simulated observer in Sect. 6.1.1, calculate the sensitivity, d', when the decision variable is based on the template used in the example, the vector `Template`. Then, recalculate the distribution of responses and d' using `Signal` for the template. What can you conclude?

6.2. The data set `Gabor` was assembled from 16 sessions of 224 trials each. They are not indexed in the data frame but the order is consecutive. Calculate d' and c for each session. Are they stable? Calculate d' and c separately for the first and the second halves of each session. Are there differences? trends? For which statistic(s)?

6.3. How would you set up the problem to fit separate classification images to the first and second 112 trials of each session? How would you compare the fit of a model with separate classification images for the first and second halves of each session against one with a single image over the whole session, or in other words, a test of whether the observer's template is stable over a session?

6.4. Verify that stimulus present and absent images are similar in the simulated observer of Sect. 6.1.1 for classification images derived using a GLM.

6.5. The residual deviance of the glm fit to the `Gabor` data set is significantly greater than the residual degrees of freedom. This might be suggestive of overdispersion in the data and that a fixed dispersion parameter taken to be 1 does not adequately represent the data. The `quasibinomial` link function can be used to investigate whether the data require a dispersion very different from 1. Analyze the data set with this link. How much does the dispersion change?

6.6. When using `glm` to obtain a classification image, the standard errors can be easily obtained using a bootstrap. The procedure is as follows. (1) Obtain the fitted probabilities, $\hat{p}$, of a response present from the glm object using the `fitted` method. (2) Generate a new vector of responses, r_1^*, by using vector $\hat{p}$ as the probability argument to generate random binomial events for samples of size 1 with `rbinom`.

(3) Substitute the new vector of responses, r_1^*, in place of the column Resp in the data frame in which the noise profiles from the experiment are stored. (4) Generate a bootstrap classification image with this modified data frame. Repeat steps 1–4 a large number of times, storing each bootstrap classification image. The standard deviations of the bootstrap classification images provide an estimate of the standard errors.

Obtain bootstrap standard errors from the data in Gabor.wide. How do they compare with the standard errors provided by the summary method?

6.7. Use the estimated classification images from LM, GLM, and GAM fits for the Gabor data as templates and classify each trial in the Gabor data set as present or absent. Compare the values of d' based on each of these templates with that actually observed in the data set.

6.8. Refit classification images to the modulation terms of the Fourier transforms of the noise profiles from the Gabor data set using a GAM and test the significance of the signal present and signal absent images. Compare the images, as well, with those obtained using a GLM (Fig. 6.7c, d).

6.9. Solomon [161] estimated classification images by performing a direct maximization of the binomial likelihood as described in Appendix B. This approach permits constraining the model in ways that would not be possible using glm. For example, in place of the estimated coefficients, one could specify a parametric function of the stimulus dimensions. Then, the parameters of this function that maximize the likelihood could be estimated using an optimization function like optim. Generally, such a function will have many fewer parameters than the number of samples along the stimulus dimension, yielding a more parsimonious model.

Using the Gabor data set, write a function that calculates the log likelihood but specifies that the template would be a Gabor function with parameters specifying the frequency, phase of the carrier and the position, width and amplitude of the envelope and fit it using optim or an alternative optimizer. In setting up the model, be careful that the parameters are identifiable. Compare the fitted template to the ideal template and that estimated using glm. How do the AIC values compare?

Chapter 7
Maximum Likelihood Difference Scaling

7.1 Introduction

In previous chapters we focused on psychophysical methods such as discrimination and models such as signal detection theory which allow us to measure the ability to discriminate stimuli that are not very different. In this chapter and the following (Chap. 8), we focus on assessing perceptual differences that are readily discriminable, well above the "threshold." The methods we present are, historically, part of the scaling literature. This literature has been closely connected to psychophysics since its inception [59, 165] and the interested reader can read about scaling and its controversies in any of the several excellent references (e.g., [68, 91, 115]).

The key issue in scaling is to devise a method that allows the experimenter to transform psychophysical judgments into a numerical representation that can be used to summarize and predict these judgments. The typical stumbling block is that, while it is easy to have observers *order* stimuli by degree of "loudness," it is not at all easy to elicit interpretable judgments of magnitude of' "loudness" or even differences in magnitude of "loudness." The same holds true for any psychophysical scale. Observers, especially naive observers, will merrily make the judgments we ask them to make even if all of their judgments taken together are incoherent and/or unrepeatable (much like presidential elections).

One achievement of the early mathematical psychology movement [104] was to develop methods to go from ordinal judgments to *interval scales*, scales whose values predict not just that a is "louder" than b but whether the difference in "loudness" between a and b is greater than that between c and d for any choices of stimuli a, b, c, and d. In this chapter we consider physical stimuli that fall along a one-dimensional physical continuum. If the observers' judgments of the order of stimuli agree with the physical order, then the key question concerns observers' perception of differences between stimuli. The experimenter's choice of a physical scale for, for example, "loudness" is arbitrary. We simply wish to learn how physical differences on this scale translate to perceptual differences, the central question of psychophysics since its inception [59].

K. Knoblauch and L.T. Maloney, *Modeling Psychophysical Data in R*, Use R! 32,
DOI 10.1007/978-1-4614-4475-6__7,
© Springer Science+Business Media New York 2012

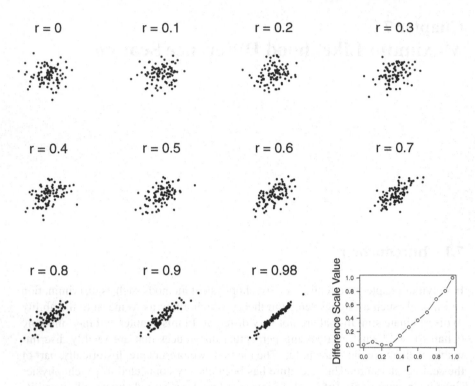

Fig. 7.1 The first 11 graphs are scatterplots of 100 points each drawn from a bivariate Gaussian distribution with the correlation given above each graph. The random values were generated with the function mvrnorm from the **MASS** package. The 12th graph is a perceptual scale for differences of correlation estimated from the judgments of a single observer. Notice that, for values of correlation less than about 0.4, the scatterplots are difficult to discriminate. The corresponding difference scale is nearly constant across this range

In this chapter we present statistical methods and a package [96] suitable for analyzing data from *difference scaling* experiments. Difference scaling is a psychophysical method used to estimate perceptual differences for stimuli distributed along a one-dimensional physical continuum. We begin with an example, illustrating how difference scaling works and what it is intended to do.

Figure 7.1 portrays stimuli drawn from a somewhat unusual "physical" continuum: $p = 11$ scatterplots, each based on a sample of 100 points drawn from a bivariate Gaussian distribution. The continuum corresponds to values of nonnegative correlation $0 \le \rho < 1$, and the stimuli are sampled from that continuum. In presenting them in an experimental context we would recommend resampling stimuli on each presentation so that the observer is unlikely to see the same scatterplot twice over the course of the experiment. The samples were generated with the mvrnorm function from the **MASS** package, one of the packages that come with the basic R installation but is not loaded by default into memory. The function requires the number of samples (here 100, for each image), a vector of means, and a covariance matrix as input. In addition, the argument empirical was set to TRUE so that the mean and covariance of the sample would exactly match those specified as the second and third arguments.

In the figure, the first ten stimuli are equally spaced in ρ from 0.0 to 0.9 while the 11th is 0.98. Each stimulus is a realization controlled by a variable ρ.

The goal of difference scaling is to develop a perceptual scale that predicts the perceived *differences* between stimuli (here scatterplots) on the physical scale. The populations from which the first two stimuli are drawn differ by 0.1 in correlation, yet there is little perceived difference in the resulting scatterplots. In contrast, the population correlations of the last two plots differ by only 0.08, less than 0.1, and they are obviously different. We want to assign numbers to each of the stimuli such that the differences between assigned values reflect the perceived differences between stimuli. The assigned numbers form a scale intended to predict perceived differences, a *difference scale*.

An example of such a scale for one observer is shown as the 12th graph in Fig. 7.1 with physical correlation plotted on the horizontal axis vs difference scale values (intended to capture perceived differences) plotted on the vertical axis. Over the course of this chapter, as part of an extended example, we will describe how to design a difference scaling experiment using these stimuli. We will explain the maximum likelihood difference scaling (MLDS) model, how to fit it to data, and how to interpret and evaluate the resulting difference scales.

In seminal papers, Schneider and colleagues [157–159] applied this procedure to the perception of loudness and proposed a method for estimating the difference scale based on the proportion of times the fitted model reproduced the observer's judgments. This method does not explicitly model stochastic variability in the observer's responses. Subsequently, Maloney and Yang [123] developed a maximum likelihood procedure, *MLDS*, for estimating the parameters of the scale. The method we present here is their method, based on the direct perceptual comparison of pairs of stimuli. We will explain in detail how we estimated these values and how to interpret them in Sect. 7.2.

MLDS has been successfully applied to characterize color differences [114, 123], surface glossiness [56, 139], transparency of thick materials [63], image quality [29], adaptive processes in face distortion [148], and neural encoding of sensory attributes [200].

We first describe the MLDS procedure using direct maximization of the likelihood as initially described by Maloney and Yang [123] and then present an equivalent approach using a GLM [127]. We describe an R package, **MLDS** [96] that contains tools for fitting and analyzing the data from an MLDS experiment. We will demonstrate the package with an extended example in which we show how to use MLDS to evaluate perception of correlation in scatterplots [38].

The mathematical basis for the non-stochastic method, including necessary and sufficient conditions for a difference scale representation in the absence of observer error, can be found in Krantz et al. [104] (see Chap. 4, Definition 1, p. 147). The key difficulty we will encounter in modeling human judgments of perceived difference

Fig. 7.2 An example of a
trial stimulus presentation
from the difference scaling
experiment for estimating
correlation differences in the
scatterplot experiment. The
observer must judge whether
the difference in perceived
correlation is greater in the
lower or upper pair of
scatterplots

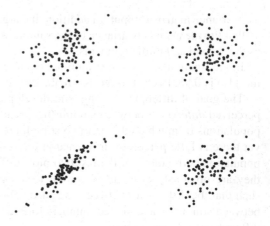

is inconsistency in human judgment, the human propensity to judge one stimulus
greater than the second on one trial and the second greater than the first on a different
trial in the same experiment. We develop a model of the human observer that allows
for such inconsistencies and even predicts that they should occur.

Of course, one possible outcome of an actual experiment is that the data are not
adequately fit by the model. If so, we wish to be in a position to reject the difference
scale model altogether and conclude that judgments of difference recorded for the
stimuli, methods and observers in an experiment are not captured by the model. We
describe methods based on *diagnostic statistics* to evaluate the appropriateness of
the difference scale representation for a set of data.

We first describe difference scaling experiments based on comparison of pairs of
pairs of stimuli, termed *quadruples*. We later describe an alternative method based
on triples of stimuli, termed *triads*, that may prove to be convenient in particular
experimental applications.

On each trial, an observer is presented with a quadruple, (I_a, I_b, I_c, I_d). The
quadruple is drawn from a set of p stimuli, $\{I_1 < I_2 < \cdots < I_p\}$, ordered along
a physical continuum. Each quadruple can be thought of as two pairs, (I_a, I_b)
and (I_c, I_d), each pair defining an *interval* on the physical continuum. On each
trial, the observer judges which pair shows the greater perceptual difference. For
convenience, the two pairs (I_a, I_b) and (I_c, I_d) are often ordered so that $I_a < I_b$ and
$I_c < I_d$ or vice versa on the physical scale but they need not be. Also for convenience,
we will denote quadruples (I_a, I_b, I_c, I_d) by just their indices $(a, b, ; c, d)$ and pairs
such as (I_a, I_b) as (a, b).

Figure 7.2 contains an example of such a quadruple. The observer's task is to
judge whether the upper or lower pair of scatterplots exhibits the larger difference.
The output of the scaling procedure is a vector of scale values $\{\psi_1, \psi_2, \ldots, \psi_p\}$
that best capture the observer's judgments of the perceptual difference between the
stimuli in each pair, (a, b), as we describe in detail below.

7.1.1 Choosing the Quadruples

Over the course of the experiment, the observer sees many different quadruples. The experimenter could choose to present all possible quadruples $(a,b;c,d)$ for p stimuli to the observer or a random sample of all possible quadruples. In past work, experiments have used the set of all possible non-overlapping quadruples $a < b < c < d$ for p stimuli, and the resulting scales have proven to be robust and readily interpretable. Maloney and Yang [123] report extensive evaluations of this subset of all possible quadruples. Consequently we will be primarily concerned with this set of quadruples.

By restricting ourselves to non-overlapping quadruples, we avoid a possible artifact in experimental design. Suppose we included quadruples such as $(a,b;a,c)$ with $a < b < c$ where the same physical scale value appears twice or quadruples of the form $(b,c;a,d)$ with $a < b < c < d$, where one interval is contained in the interior of the other. Now consider an observer who is actually not capable of comparing intervals and whose behavior cannot therefore be captured by a difference scale. If this observer can correctly order the stimuli, then, on a trial of the form $(a,b;a,c)$, he can still get the "right" answer by noting that both intervals have a at one end so that $b < c$ implies that (a,b) must be less than (a,c).

A similar heuristic applied to $(b,c;a,d)$ with $a < b < c < d$ would allow the observer to appear to be ordering intervals consistently when, in fact, he cannot compare intervals. Using only non-overlapping intervals effectively forces the observer to compare intervals—if he or she can do so. Moreover, as noted above, the use of such intervals has proven itself in previous computational and experimental situations.

For p stimuli there are

$$\binom{p}{4} = \frac{p!}{4!\,(p-4)!} \tag{7.1}$$

possible non-overlapping intervals. If $p = 11$, as in the example, there are

$$\binom{11}{4} = \frac{11!}{4!\,7!} = 330 \tag{7.2}$$

different, non-overlapping quadruples. The observer may judge each of the 330 non-overlapping quadruples in randomized order, or the experimenter may choose to have the observer judge each of the quadruples m times, completing $330m$ trials in total. Of course, the number of trials judged by the observer affects the accuracy of the estimated difference scale (see [123]) as illustrated below. The time needed to judge all 330 trials in the scatterplot example was roughly 12–15 min.

The order in which quadruples are presented is randomized. To control for positional effects, on half of the forced-choice trials, chosen at random, the pairs are presented in the order $(a,b;c,d)$ and on the other half, $(c,d;a,b)$.

7.1.2   Data Format

At the end of the experiment, the data can be represented as an $n \times 5$ matrix or data frame in which each row corresponds to a trial: four columns, give the indices, (a, b, c, d) of the stimuli from the ordered set of p, and one column indicates the response of the observer to the quadruple as a 0 or 1, indicating choice of the first or second pair as larger. All columns are of class "integer," although the response can also be of class "logical." As an example,

```
> head(kk3)
  resp S1 S2 S3 S4
1    0  4  8  2  3
2    1  2  3  6 11
3    1  2  6  7 10
4    0  4 11  1  2
5    0  9 11  7  8
6    0  7 10  1  3
```

gives the first 6 rows of the data set kk3, supplied in the **MLDS** package, for the observations that generated the scale shown in the lower right of Fig. 7.1. On the first trial, for example, the observer saw pairs (4,8) and (2,3) and judged the difference between the first pair to be greater.

From these data, the experimenter estimates the perceptual scale values $\psi_1, \psi_2, \ldots, \psi_p$ corresponding to the stimuli, $I_1, \ldots, I_p$, as follows. Given a quadruple, $(a, b; c, d)$ from a single trial, we first assume that a non-stochastic observer would judge a, b to be further apart than c, d precisely when,

$$|\psi_b - \psi_a| > |\psi_d - \psi_c|. \tag{7.3}$$

That is, the difference scale predicts judgment of perceptual difference.

This judgment rule cannot be inconsistent from trial to trial: once the observer has judged that the perceptual difference of (a, b) is greater than that of (c, d) then the observer must *always* judge the perceptual difference of (a, b) to be greater than that of (c, d). It is unlikely that human observers would be so reliable in judgment as to satisfy the criterion just given, particularly if the perceptual differences $|\psi_b - \psi_a|$ and $|\psi_d - \psi_c|$ are small. Moreover, since the stimuli in the scatterplot example are resampled on every trial, the difference between sample realizations will likely contribute to judgment variability.

Maloney and Yang [123] proposed a stochastic model of difference judgment that allows the observer to exhibit some stochastic variation in judgment. Let $L_{ab} = |\psi_b - \psi_a|$ denote the unsigned perceived length of the interval (a, b). The proposed decision model is an equal-variance Gaussian signal detection model [72] where the signal is the difference in the lengths of the intervals,

$$\delta(a, b; c, d) = L_{cd} - L_{ab} = |\psi_d - \psi_c| - |\psi_b - \psi_a| \tag{7.4}$$

The decision variable employed by the observer is δ perturbed by Gaussian error

$$\Delta(a,b;c,d) = \delta(a,b;c,d) + \varepsilon = L_{cd} - L_{ab} + \varepsilon \tag{7.5}$$

where $\varepsilon \sim \mathcal{N}(0,\sigma^2)$: given the quadruple, $(a,b;c,d)$, the observer selects the pair c,d precisely when,

$$\Delta(a,b;c,d) > 0. \tag{7.6}$$

7.1.3 Fitting the Model: Direct Maximization

In each experimental condition the observer completes n trials and we will index trials by $k = 1,n$. The quadruple presented on the kth trial is denoted $q^k = (a^k,b^k;c^k,d^k)$, $k = 1,n$. The observer's response on the kth trial is coded as $R^k = 0$ (the difference of the first pair is judged larger) or $R^k = 1$ (the difference of the second pair is judged larger).

We estimate the parameters $\Psi = (\psi_1, \psi_2, \ldots, \psi_p)$ and σ by maximizing the likelihood,

$$L(\Psi,\sigma) = \prod_{k=1}^{n} \Phi\left(\frac{\delta(q^k)}{\sigma}\right)^{1-R_k} \left(1 - \Phi\left(\frac{\delta(q^k)}{\sigma}\right)\right)^{R_k}, \tag{7.7}$$

where $\Phi(x)$ denotes the cumulative standard normal distribution and $\delta(q^k) = \delta(a^k,b^k;c^k,d^k)$ as defined in (7.5).

Equation (7.7) is the likelihood function for a series of independent Bernoulli variables whose probabilities depend on $\delta(q^k)$. The equation is similar in form to likelihood functions we encountered in fitting psychometric functions in Chap. 4. As we saw there, taking the negative logarithm of likelihood allows the parameters to be estimated simply with a minimization function such as optim in R (e.g., see p. 445 of [178]).

7.1.3.1 Standard Scales

At first glance, it would appear that the stochastic difference scaling model just presented has $p+1$, free parameters: $\psi_1, \ldots, \psi_p$ together with the standard deviation of the error term, σ. However, any linear transformation of the $\psi_1, \ldots, \psi_p$ together with a corresponding scaling by σ^{-1} results in a set of parameters that predicts exactly the same performance as the original parameters. Without any loss of generality, then, we can set $\psi_1 = 0$ and $\psi_p = 1$, leaving us with the $p-1$ free parameters, $\psi_2, \ldots, \psi_{p-1}$ and σ. When scale values are normalized in this way, we refer to them as *standard scales*. Standard scales always run from 0 to 1 and the variable σ has a natural interpretation as a measure of the observer's judgment uncertainty. The value $d' = (2\sigma)^{-1}$ is a measure of the observer's ability to judge differences between intervals expressed in the language of signal detection theory.

7.1.3.2   Unnormalized Scales

There is a second parameterization that is readily interpretable. We refer to this second scale as *unnormalized*. It is obtained by multiplying a standard scale by σ^{-1}. If we denote values on the second scale as ϑ_i then $\vartheta_i = \sigma^{-1}\psi_i$. The values on an unnormalized scale run from 0 to $d'/2 = \sigma^{-1}$ and there are also $p-1$ free parameters. Since each parameterization is a one-to-one transformation of the other, the maximum likelihood estimates $\hat{\psi}_2, \ldots, \hat{\psi}_{p-1}, \hat{\sigma}$ are easily computed from $\hat{\vartheta}_2, \ldots, \hat{\vartheta}_p = \hat{\sigma}^{-1}$ by $\hat{\psi}_k = \hat{\vartheta}_i / \hat{\sigma}$. The justification for changing parameterizations in this way can be found in Sect. B.4.3.

7.1.3.3   Optimization Details

In computing the maximum likelihood, we use a third parameterization. These transformations simply allow us to constrain estimated values to certain intervals (see Sect. B.4.3 for details). We force estimates of ψ_i to be between 0 and 1 on a standardized scale or nonnegative on an unnormalized scale. Similarly, we want estimates of $d' = (2\sigma)^{-1}$ that are nonnegative. We represent the $p-2$ scale values on a standard scale by a logit transformation

$$\eta = \log\left(\frac{\psi}{1-\psi}\right) \tag{7.8}$$

the η values can take on any real values but the corresponding ψ are constrained to be in the interval $(0,1)$. Similarly, we replace σ by its logarithm so that estimates of σ are forced to be positive. Again, these transformations have no theoretical significance. They are used to avoid problems in numerical optimization.

   With these details, a function to compute the negative log likelihood can be written in only a few lines. It takes as arguments a vector containing the current estimates of $\psi_2, \ldots, \psi_{p-1}$ and σ and a data frame containing the observer's responses as the first component and the scale indices for each quadruple as the next four.

```
> ll.mlds <- function(p, d){
+     psi <- c(0, plogis(p[-length(p)]), 1)
+     sig <- exp(p[length(p)])
+     resp <- d[, 1]
+     stim <- as.matrix(d[, -1])
+     del <- matrix(psi[stim], ncol = 4) %*%
+         c(1, -1, -1, 1)/sig
```

```
+     -sum(pnorm(del[resp == 1], log.p = TRUE)) -
+      sum(pnorm(del[resp == 0], lower.tail = FALSE,
+         log.p = TRUE))
+ }
```

The parameter estimates are in a vector in the argument p and the data frame in d. Since at input, the estimates have been transformed, the first two lines of the function undo the transforms. Then, the responses and stimulus indices are obtained from the input data frame and assigned to the names resp and stim, respectively. Subsequently, the decision variable, del, is calculated from the current scale value estimates and scale parameter. The decision variable is then used to calculate the negative log likelihood for the input data. This version has the link function hard-coded as an inverse Gaussian, but it would be simple enough to generalize the function with a third argument to supply the binomial family function with a specified link.

Once defined, ll.mlds can be used with an optimization function such as optim or nlm to obtain a difference scale. For example, with optim, we would write

```
> # transformed init. values by qlogis and log, respectively
> p <- c(qlogis(seq(0.1, 0.9, 0.1)), log(0.2))
> res <- optim(p, ll.mlds, d = kk3, method = "BFGS")
```

The results are stored in a list with the scale values in a component named par. They must be back-transformed with the plogis and exp function to be on the appropriate scale. We use method = "BFGS" because it typically converges faster than the default "Nelder–Mead" method.

7.1.4 GLM Method

Just as with signal detection theory (Chap. 3), the psychometric function (Chap. 4) and classification images (Chap. 6), we can reformulate MLDS as a GLM (2.6). The responses of the observer are modeled as Bernoulli variables. The canonical link function for the binomial case is the logistic transformation from (7.8). However, other choices of link function are possible, and the inverse cumulative Gaussian, or quantile function, corresponds to (7.7). Accordingly, we use it here and have

$$\Phi^{-1}(\mathsf{E}[P(R=1)) = X\beta \tag{7.9}$$

where the link function, Φ^{-1}, is the inverse of the Gaussian cdf, X is the design matrix, and the vector β will contain the estimates of the scale values, ψ_i.

In this section, we assume that we have reordered each quadruple $(a,b;c,d)$ so that $a < b < c < d$ and recode the response, R, as necessary. With this ordering, we can omit the absolute value signs in (7.4) which then becomes

$$\delta = \psi_d - \psi_c - \psi_b + \psi_a$$
$$\Delta = \delta + \varepsilon. \tag{7.10}$$

The observer bases his or her judgment on $\Delta = \delta + \varepsilon$ where $\varepsilon \sim \mathcal{N}(0, \sigma^2)$. The observer therefore selects the second pair (c, d) with probability $\Phi(\delta)$. With the GLM method, it will prove convenient to estimate unnormalized scale values (which we will denote by β_i in keeping with the typical notation for the linear predictor) rather than standard scale values, and accordingly we impose no additional constraints on the scale values.

7.1.4.1 The Design Matrix

The design matrix, X, for the GLM can be constructed by considering the weights of the β_i as covariates. On a given trial, the values in only four columns are nonzero, taking on the values $1, -1, -1, 1$ in that order. This yields an $n \times p$ matrix, X, where n is the number of quadruples tested and p is the number of physical levels evaluated over the experiment. For example, consider a set of seven stimuli distributed along a physical scale and numbered 1–7. The five quadruples

$$
\begin{array}{cccc}
2 & 4 & 5 & 6 \\
1 & 2 & 3 & 7 \\
1 & 5 & 6 & 7 \\
1 & 2 & 4 & 6 \\
3 & 5 & 6 & 7
\end{array}
$$

yield the design matrix

$$
X = \begin{pmatrix}
0 & 1 & 0 & -1 & -1 & 1 & 0 \\
1 & -1 & -1 & 0 & 0 & 0 & 1 \\
1 & 0 & 0 & 0 & -1 & -1 & 1 \\
1 & -1 & 0 & -1 & 0 & 1 & 0 \\
0 & 0 & 1 & 0 & -1 & -1 & 1
\end{pmatrix}.
$$

To render the model identifiable, however, we drop the first column, which has the effect of fixing $\beta_1 = 0$, yielding a model with $p - 1$ parameters to estimate as with the direct method. This yields the model,

$$
\Phi^{-1}(\mathsf{E}[Y]) = \beta_2 X_2 + \beta_3 X_3 + \cdots + \beta_p X_p, \tag{7.11}
$$

where X_j is the jth column of X. If we have a data frame, d, with the response vector, Resp, as the first component and the columns of X excluding the first one as the succeeding components, we obtain the estimated parameters simply with function glm using the following line of code.

```
> glm(Resp ~ . - 1, family = binomial(probit), data = d)
```

The period, ., is a shorthand for an additive model using all of the columns of the data frame except for the ones already specified in the formula, here Resp, and is, thus, conveniently expanded to the model corresponding to (7.11). We suppress the intercept term so that the first estimated coefficient corresponds to β_2.

Unlike the direct method, β_p is unconstrained. Implicitly in the GLM model, $\sigma = 1$. In fact, $\hat{\beta}_p$ from the GLM fit equals $\hat{\sigma}^{-1}$ from the direct method, so that normalizing the GLM coefficients by $\hat{\beta}_p$ should yield the same scale as obtained by the direct method up to numerical precision.

We have compared solutions using direct optimization (optim in R) and GLM fits (glm function) and find them to be in good agreement. This outcome is scarcely surprising since the two methods maximize almost the same equation and differ primarily in details of the numerical maximization procedure used. One difference is that with the direct method, we imposed the constraint that the $\beta \geq 0$ whereas we did not with the glm method. It is sometimes the case that small negative values of β near the bottom of the scale result in better fits.

7.1.5   Representation as a Ratio Scale

While MLDS, as presented in (7.3), is based on a differencing representation, it also permits a ratio representation [104]. In that case, it is supposed that a non-stochastic observer judges stimulus pair (a, b) as further apart than (c, d) precisely when

$$\psi_b/\psi_a > \psi_d/\psi_c. \qquad (7.12)$$

That is, the scale predicts judgment of perceptual ratios. The ratio scale is equivalent to a difference representation, however,

$$\psi_b' - \psi_a' > \psi_d' - \psi_c', \qquad (7.13)$$

where $\psi_i' = \log \psi_i$. Thus, the observer's responses can be fit using the methods described previously, but the estimated scale values are interpreted as the values ψ_i' and the underlying scale values $\psi_i = \exp(\psi_i')$. This generates a ratio scale fixed at 1, instead of 0 with maximum value unconstrained, as with the unnormalized scale. Note that the error term, in this case would be multiplicative, so that an additive error $\sigma = 1$ on the ψ' scale would correspond to a multiplicative term, $\exp(1)$, on the ψ scale.

7.2   Perception of Correlation in Scatterplots

The study of graphical perception has been of interest to statisticians, at least, since the pioneering work of Cleveland and colleagues [37]. Scatterplots have often been the subject of investigation, for example, to determine the plot characteristics that best convey the underlying association in the data to the viewer [36]. Only a few studies have examined how human observers perceive differences in correlation in scatterplots [96, 108, 147]. As noted above, inspection of the plots in Fig. 7.1 hints that perceived correlation is not linear with the correlation parameter ρ. MLDS offers a promising method for assessing human perception of correlation in scatterplots.

The package contains code to run the sample experiment described here. The code

```
> library(MLDS)
> Obs.out <- runQuadExperiment(DisplayTrial = "DisplayOneQuad",
+    DefineStimuli = "DefineMyScale")
```

runs 330 trials of the difference scaling experiment with scatterplot stimuli like those in Fig. 7.2 and records the observer's responses interactively on each trial for the scatterplot example of Fig. 7.1. The stimulus from an example trial is shown in Fig. 7.2. The observer's task is to decide whether the difference in correlation r is greater between the lower pair or the upper and to enter a "1" or "2," respectively, from the keyboard. This function can be readily modified to any difference scaling application by defining the functions DefineStimuli and DisplayTrial that define the stimuli and display the quadruples, respectively, of non-overlapping intervals on each trial. After the observer has completed the experiment, an object of class "mlds.df" is returned which can be used for further analysis. Here, the output is assigned to a variable named Obs.out.

The "mlds.df" object is simply a data frame but it has two additional attributes that some of the methods for its class exploit. The invord attribute is a logical vector that indicates whether the stimuli from the first pair were from higher scale values than for the second pair. The methods for combining data across sessions also combine these attributes. In some circumstances, it is necessary to reorder the trials (see below), and the presence of this attribute assures that the original ordering can be restored. The stimulus attribute is a record of the physical stimulus values and is used by the plot methods for this class. To preserve these attributes, the data should be saved with save or dput. If the data are read into a data frame using read.table, then they may be coerced to the class and format described above with function as.mlds.df.

One of the authors (KK) ran the experiment on himself three times, with the results stored in objects kk1, kk2, and kk3, found in the data sets that come with the **MLDS** package.

After loading the three data sets in memory, we merge them into one object of 990 trials with rbind. In the experiment the quadruple $a < b < c < d$ could have been presented in the order a,b above, c,d below or vice versa. The function SwapOrder revises the order so that a,b is in the first two columns, c,d in the second two, changing the response code as well so that $R = 0$ means a,b judged greater than c,d and $R = 1$ otherwise.

```
> data(kk1, kk2, kk3)
> kk <- SwapOrder(rbind(kk1, kk2, kk3))
```

A second application of SwapOrder restores the original ordering, if desired, using the invord attribute.

7.2.1 Estimating a Difference Scale

We next fit the data using both the direct optimization method and the GLM method. The package **MLDS** contains the function mlds which implements both methods with the choice determined by setting the argument method = "glm" or method = "optim". The glm method is the default. In our experience, the optim method is usually slower and seems to be more sensitive to initial estimates of the parameter σ.

```
> kk.mlds <- mlds(kk)
> summary(kk.mlds)
Method: glm     Link:    probit

Perceptual Scale:
      0      0.1     0.2      0.3     0.4     0.5      0.6
 0.0000 -0.0454  0.0439 -0.0863  0.5682  1.4234   2.0695
    0.7      0.8     0.9     0.98
 2.6661   3.5527  4.4297  5.5739

sigma:   1
logLik: -306
> kkopt.mlds <- mlds(kk, method = "optim",
+    opt.init = c(seq(0, 1, len = 11), 0.2))
> summary(kkopt.mlds)
Method: optim      Link:    probit

Perceptual Scale:
        0       0.1       0.2       0.3       0.4       0.5
0.00e+00  4.70e-05  1.54e-02  1.19e-07  1.10e-01  2.61e-01
      0.6       0.7       0.8       0.9       0.98
3.76e-01  4.83e-01  6.40e-01  7.96e-01  1.00e+00

sigma:   0.175
logLik: -307
```

We needed to provide initial estimates of the parameters for the optim method as an estimate of the 11 scale values opt.init and and an initial value for σ set here to 0.2. For either method, the function returns an object of class "mlds" for which several methods are defined, such as summary above and plot below.

Note the differences in the summaries. The glm method estimates an unnormalized scale and the upper range of scale values is an estimate of the parameter σ^{-1}.

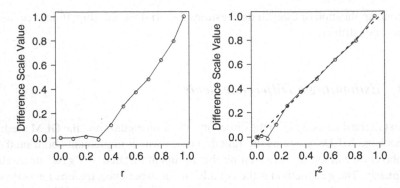

Fig. 7.3 *Left*. Estimated difference scale for observer KK, from 990 judgments, distributed over three sessions for judging differences of correlation between scatterplots. *Right*. The same scale values plotted as a function of the squared correlation or variance accounted for. The *diagonal line* through the origin has slope 1

The reported estimate `sigma` in the summary is always 1. The `optim` method fixes the extreme values of the scale to 0 and 1. The estimate of σ is reported as the parameter `sigma`.

There may be differences in the parameter estimates or the log likelihood These are usually very small, as here, unless `optim` or `glm` has found a local minimum. They are more typically due to the additional constraints of the parameterization employed with the `optim` method. This accounts for the slight discrepancy between the value of σ obtained using `optim` 0.175 and the reciprocal of the maximum scale value using `glm`, 0.179.

We compare the two estimated scales in the right-hand graph of Fig. 7.3 after normalizing the scale estimated by `glm` by its maximum scale value. The second argument of the `plot` method for objects of class "mlds" determines whether the scale is plotted in standard or unnormalized form. To plot the standard scale, set the argument to `standard.scale = TRUE`. The glm and optim scales are shown as points and lines, respectively, and it is clear that the differences between the two are slight.

```
> opar <- par(mfrow = c(1, 2), pty = "s")
> plot(kk.mlds, standard.scale = TRUE,
+        xlab = expression(r), ylab = "Difference Scale Value",
+        cex.lab = 1.5, cex.axis = 1.5, las = 1)
> lines(kkopt.mlds$stimulus, kkopt.mlds$pscale)
> plot(kk.mlds$stimulus^2, kk.mlds$pscale/max(kk.mlds$pscale),
+        xlab = expression(r^2), ylab = "Difference Scale Value",
+        cex.lab = 1.5, cex.axis = 1.5, las = 1, xlim = c(0, 1))
> abline(0, 1, lty = 2, lwd = 2)
> lines(kkopt.mlds$stimulus^2, kkopt.mlds$pscale)
> par(opar)
```

The resulting perceptual scale is fairly flat for correlations up to 0.3 implying that the observer treated them almost interchangeably in judging intervals. If we replot the estimated scale instead as a function of squared correlation, ρ^2, we see that for correlations above 0.4, the observer's judgment effectively matches ρ^2, the variance accounted for. Below this value, the observer seems unable to discriminate scatterplots.

A useful ploy: It is evident that the "physical" scale here is arbitrary. We could as easily index nonnegative correlation by ρ^2 (variance accounted for) as by ρ (correlation). We originally chose stimuli that were close to equal-spaced in ρ but we could as easily have chosen stimuli that were equal-spaced in ρ^2. Had we done so, the resulting difference scale would be close to a straight line and we would have devoted less time to evaluating differences involving the correlations less than 0.4. Indeed, having seen the results just presented, we may want to redo the experiment with equal-spacing in ρ^2 just to check the conjecture that human perception of differences in correlation follows a difference scale that corresponds to ρ^2.

7.2.2 Fitting a Parametric Difference Scale

The results of the difference scaling experiment with scatterplots suggest that the perception of correlation increases quadratically with correlation (Fig. 7.3). We can refit the data but constraining the estimated difference scaling curve to be a parametric curve, such as a power function, using the formula method for mlds. Under the hypothesis that perception of correlation depends on r^2, we would expect the exponent to be approximately 2.

```
> kk.frm <- mlds(~ (sx/0.98)^p[1], p = c(2, 0.2), data = kk)
```

We specify the parametric form for the difference scale with a one-sided formula. The operations take on their mathematical meaning here, as with formulae for nls (Sect. 2.4) but not as for lm and glm without isolation. The stimulus scale is indicated by a variable sx and the parameters to estimate by a vector, p. The fitting procedure adjusts the parameters to best predict the judgments of the observer on the standard scale. The equation is normalized by the highest value tested along the stimulus dimension, $r = 0.98$ so that the fitted curve passes through 1.0 at this value. The optimization is performed by optim and initial values of the parameters are provided by a vector given as an argument in the second position. The last element of this vector is always the initial value for sigma. Finally, a data frame with the experimental results is, also, required.

The estimated parameters are returned in a component par of the model object and the judgment variability, as usual, in the component sigma. Standard errors for each of these estimates can be obtained from the Hessian matrix in the hess component.

Fig. 7.4 Difference scales estimated for the perception of correlation obtained using the default (*points*) and formula (*line*) methods to fit the data. The formula used was a power function

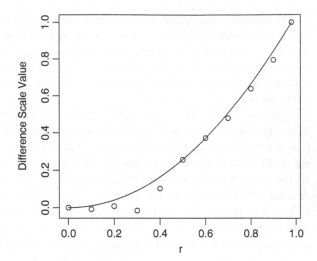

```
> c(p = kk.frm$par, sigma = kk.frm$sigma)
     p sigma
 2.02   0.17
> sqrt(diag(solve(kk.frm$hess)))
[1] 0.0601 0.0104
```

The standard error for the exponent provides no evidence to reject an exponent of 2.

The parametric fit with 2 parameters is nested within the 10 parameter nonparametric fit. This permits us to test the parametric fit with a likelihood ratio test. We extract the log likelihoods from each model object with the `logLik` method. The degrees of freedom of each fit are stored as the "df" attribute of the "logLik" object and obtained with the `attr` function.

```
> ddf <- diff(c(attr(logLik(kk.frm), "df"),
+    attr(logLik(kk.mlds), "df")))
> pchisq(-2 * c(logLik(kk.frm) - logLik(kk.mlds)),
+    ddf, lower.tail = FALSE)
[1] 3.08e-06
```

The results reject the power function description of the data. A predicted curve under the parametric model fit to the data is obtained from the `func` component of the model object, which takes two arguments: the estimated parameters, obtained from the `par` component of the model object and a vector of stimulus values for which to calculate predicted values. The plot of the predicted values against the values obtained by `glm` indicates that the power function fit provides a reasonable description of the data for correlations greater than 0.4 and a lack of fit for lower correlations, for which this observer shows no evidence of discrimination (Fig. 7.4).

```
> plot(kk.mlds, standard.scale = TRUE,
+        xlab = "r",
+        ylab = "Difference Scale Value")
> xx <- seq(0, 0.98, len = 100)
> lines(xx, kk.frm$func(kk.frm$par, xx))
```

7.2.3 Visualizing the Decision Space for Difference Judgments

Additional insight into MLDS can be gained by examining the decision space for this task. To be concrete, we will suppose that the underlying response function is quadratic, similar to the response function for perception of correlation in scatterplots. The approach that we will take, however, can be simply applied to an arbitrary response function, and it would be an excellent exercise to repeat it with other possibilities that might be more appropriate for other tasks.

The decision space is represented by a plane in which each point represents a pair of stimuli. The abscissa value corresponds to the stimulus level of the first element of the pair and the ordinate to the second. For example, if the stimulus levels represent correlations of scatterplots, then the abscissa and ordinate values will indicate the correlations of the first and second stimuli of the pair, respectively, as in the graphs of Fig. 7.5. A trial from a difference scaling experiment is represented by a pair of distinct points in this space. The choice of quadruples is such that the points cannot be on the same horizontal or vertical line (Sect. 7.1.1). To simplify the representation, we will assume that each pair is ordered with the greater stimulus level indicated on the abscissa. Then, all stimuli are represented in the lower triangle equal or below the line $x = y$.

The underlying response determines a set of contours (level sets) for which the difference of response is constant. The following code segment produces a graph with 5 such equal-difference contours for a quadratic response function and was used to make the upper left graph of Fig. 7.5.

```
> RespFun <- function(x) x^2
> x <- seq(0, 1, len = 200)
> xy <- expand.grid(x = x, y = x)
> xy$Diff <- with(xy, ifelse(x - y >=  -0.01,
+   RespFun(x) - RespFun(y), NA))
> contourplot(Diff ~ x * y, xy, cut = 5,
+   aspect = "iso",
+   xlab = "Correlation Stimulus 1",
+   ylab = "Correlation Stimulus 2",
+   main = "Equi-Response Differences")
```

First, the underlying response function is defined, RespFun. We, then, use the function expand.grid to define a 200 × 200 grid of values in the plane of the graph returned as a data frame, xy. A third column is created that gives the difference of

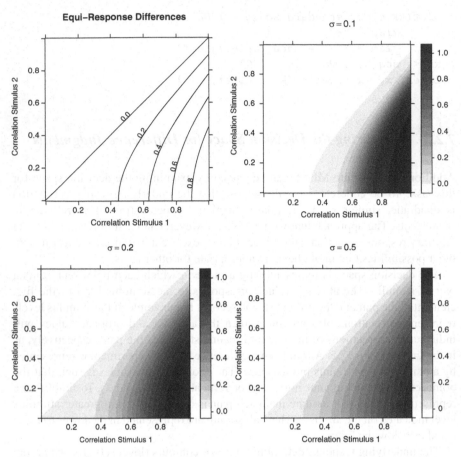

Fig. 7.5 The decision space for a difference scaling experiment. The *upper left graph* displays a set of equi-difference contours for an underlying response that follows a quadratic function of stimulus level (here designated as correlation, as in the example). The remaining three graphs show the distribution of probability for judging a pair to have a greater difference than a pair falling on the equi-difference contour equal to 0.4. The graphs differ in the value of σ, indicated above each figure. A *color bar* beside each of these graphs shows how probability is coded by grey level

response for the nonnegative differences (lower triangle). In fact, we include some negative values to force the `contourplot` function from the **lattice** package to plot the diagonal for which the response difference is 0. The function takes a two-sided formula object and calculates a set of equal-response contours over the grid of values specified by the variables on the right-hand side of the formula. These are plotted and labeled.

When two pairs of stimuli generate points that fall along the same contour, the expected probability that an observer judges the difference between one pair as greater than the other equals 0.5. When one of the pairs corresponds to a point that is on a contour below and to the right of that for the other, then the probability that the observer judges that pair as displaying a greater perceptual difference should be greater than 0.5. Thus, the psychometric function for judging differences in pairs is defined across these contours.

While this description suffices to define the observer's decisions qualitatively, it does not permit a quantitative description of the expected choices. In order to visualize quantitatively how decisions vary as a function of distance in this space, we need to specify a value for σ from (7.7). For a given equal-difference contour, the value of σ controls the rate at which the psychometric function varies across this space; the lower its value, the steeper the slope of the psychometric function and, thus, the more rapidly the probability passes from 0 to 1. Thus, σ determines how sensitive the observer is at discriminating pairs of differences of response.

The remaining three graphs in Fig. 7.5 show how the probability of judging a pair to be greater than a pair with difference equal to 0.4 varies across the plane for each of three values of σ, $(0.1, 0.2, 0.5)$. Probability is specified by grey level with the color bar on the right of each graph indicating the scale. For low values of σ, only a small shift right and downward suffices to generate a large increase in the probability of judging that pair greater; with larger σ, a larger shift is necessary for the same increase of probability of response. The greyscale plots were generated with the levelplot function also from the **lattice** package. The following code fragment shows how the graph for $\sigma = 0.1$ was generated.

```
> xy$Pdiff <-  with(xy, ifelse(x - y >= -0.01,
+    pnorm(RespFun(x) - RespFun(y) - 0.4, sd = 0.1), NA))
> levelplot(Pdiff ~ x * y, data = xy,
+    aspect = "iso",
+    xlab = "Correlation Stimulus 1",
+    ylab = "Correlation Stimulus 2",
+    main = expression(sigma == 0.1),
+    col.regions = rev(grey(seq(0.2, 1, len = 100))))
```

We use the pnorm function to generate probabilities of choosing a pair as greater than a particular difference, here 0.4. The levelplot function takes the same formula as contourplot but includes additional arguments for specifying the color scale for response variable. Note that the distribution of probability across the decision space shown in the figures is specific to the equi-difference contour of 0.4. If a different difference were chosen, the probability gradient would be shifted so that it was centered on the contour corresponding to that difference. The exact distribution of probability would depend on the underlying response function and the value of σ.

In a real experiment, of course, neither the equi-difference contours nor the psychometric functions on the decision space are known. These must be estimated from the set of judgments of the observer on the set of quadruples (or as described in the next section, triads) presented over the course of the experiment.

7.3 The Method of Triads

In the Method of Quadruples, observers compare interval (a,b) and (c,d) and it is not difficult to see how the resulting difference scale would lend itself to predicting such judgments. We can also use a slightly restricted method for presenting stimuli that we refer to as the Method of Triads. On each trial, the observer sees three stimuli (a,b,c) with $a < b < c$ and judges whether or not $(a,b) \succ (b,c)$, i.e., whether the interval between a and b is larger than that between b and c. The observer is still judging the perceived size of intervals but now the intervals judged share a common end point b. Despite this apparent limitation, difference scales estimated using the Method of Triads exhibit about the same variability and lack of bias as difference scales estimated using the Method of Quadruples if the total number of judgments is the same.

The change in method entails a change in the design matrix given that stimulus b participates in the comparison of both intervals. The decision variable for a triad experiment is modeled as

$$(\psi_b - \psi_a) - (\psi_c - \psi_b) = 2\psi_b - \psi_a - \psi_c \qquad (7.14)$$

so that the design matrix has nonzero entries of $(-1, 2, -1)$. The data frame from an experiment has only four instead of five columns, indicating the response and the indices for the three stimuli on each trial. The **MLDS** package includes a function runTriadExperiment that works similarly to runQuadExperiment, but presents triads on each trial and returns an object of class "mlbs.df." The mlds function is generic and methods are supplied for both classes of object. The user need only supply the results of either type of experiment and the function dispatches to the correct method to return the results.

The experimenter is free to choose whichever method proves to be more convenient if, for example, it is easier to fit three stimuli simultaneously on an experimental display than four. For $p > 3$ stimuli, there are

$$\binom{p}{3} = \frac{p!}{3!\,(p-3)!} \qquad (7.15)$$

possible distinct triads. If $p = 11$, as in the example, there are

$$\binom{11}{3} = \frac{11!}{3!\,8!} = 165 \qquad (7.16)$$

different triads. Of course, the number of trials judged by the observer affects the accuracy of the estimated difference scale (see [123]) as we illustrate below. The time needed to judge two repetitions of all 165 trials with triads (330 trials total) is the number of trials for one repetition of all trials with non-overlapping quadruples. In the next section we illustrate the use of simulation to evaluate how number of trials affects the accuracy and bias of parameter estimates and whether the Method of Triads leads to more accurate estimates than the Method of Quadruples when the number of trials is equated.

7.4 The Effect of Number of Trials

The function SimMLDS in the **MLDS** package returns a list of simulations of a difference scaling experiment using either the Method of Triads or of Quadruples. For all repetitions of the simulation, we specify as input the design of the experiment as a $n \times 4$ matrix where n is the number of trials in the experiment. We also specify the true values of a standard scale $\psi_1 = 0, \psi_2, \ldots, \psi_{11} = 1$ and the true value of σ. For this example, we set $\psi_i = \frac{i}{11}^2$ and $\sigma = 0.2$. The output is a list of $n \times 5$ matrices in the format of data for difference scaling. We repeat the simulation $N = 1,000$ times and fit the data from each simulated experiment by the default glm method and record the estimates $\hat{\psi}_1^k, \ldots, \hat{\psi}_{11}^k$ as an $N \times 11$ matrix, psi.quads. We transform the unnormalized scale values from the kth experiment to standard scale values by dividing through by ψ_{11}^k. Of course, we need not record $\hat{\psi}_1 = 0$ or $\hat{\psi}_{11} = 1$ since they never vary but it is convenient to include them.

```
> x <- seq(0,1, len = 11)    # 11 equally-spaced steps
> psi <- x^2                         # true difference scale
> sigma <- 0.2
> n <- length(psi)
> trials.quads <- t(combn(n, 4))    # 330 trials
> N <- 1000
> fit.quads <- lapply(SimMLDS(trials.quads, psi, sigma, N),
+   mlds)
> cc.quads <- sapply(fit.quads, coef)
> psi.quads <- apply(cc.quads, 2,
+   function(x) c(0, x/x[length(x)]))
> mean.quads <- rowMeans(psi.quads)
> sd.quads <- apply(psi.quads, 1, sd)
```

The code fragment above accumulates the fitted scales values as an $11 \times N$ matrix. The last two lines simply compute the mean of each row of the fitted values and the standard deviations. The function rowMeans is one of a suite of functions (e.g., colMeans and rowSums) that perform simple calculations on a matrix and have been optimized for speed. We used the apply function to obtain the standard deviations of the rows from the matrix. More generally, we could estimate the covariance of the scale values. We could also accumulate fitted values of σ and compute their means and standard deviations, but we do not do so for this example.

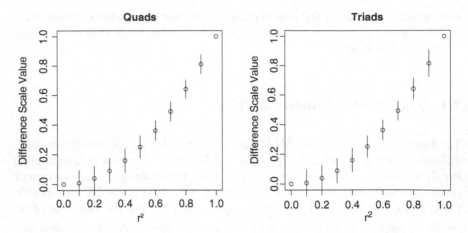

Fig. 7.6 *Left*. Estimated difference scales for 1,000 replications of a difference scaling experiment using 330 trials and the Method of Quadruples. The *error bars* correspond to twice the standard deviation. *Right*. The same but with the method of triads

If the true values were set to the estimates obtained from a subject in an experiment, we would be computing bootstrap confidence intervals [55]. (More about that later.) We plot the mean and 95% confidence intervals for the estimates in Fig. 7.6 together with the true standard scale plotted as a solid line. As can be seen, the mean estimates fall close to the true values. The confidence intervals indicate how reliably we can estimate them with an experiment with this design and number of trials.

We repeat the simulations but now substituting a Method of Triads design, specified as a $n \times 3$ matrix. We need not change the fitting code at all since mlds determines whether the data were collected using the Method of Quadruples or the Method of Triads from the class of the data frame ("mlds.df" or "mlbs.df") or the number of columns if its class is simply "data.frame." For 11 stimulus levels, there are 165 distinct triads, half the number for an experiment with quadruples. To make the results comparable, we double the length of the matrix of trials by repeating each triad twice. We plot these results in the second panel of Fig. 7.6. The outcome for the Methods of Triads experiment with 330 trials is very similar to that for the Method of Quadruples, and, in particular, the error bars are roughly the same in magnitude.

```
> x <- seq(0,1, len =11)   # 11 equally-spaced steps
> psi <- x^2   # true difference scale
> sigma <- 0.2
> n <- length(psi)
> trials.triads <- t(combn(n, 3))    # 165 trials
> trials.triads <- kronecker(matrix(c(1, 1), ncol = 1),
+   trials.triads)   # double number of trials
> N <- 1000
```

```
> fit.triads <- lapply(SimMLDS(trials.triads, psi, sigma, N),
+   mlds)
> cc.triads <- sapply(fit.triads, coef)
> psi.triads <- apply(cc.triads, 2,
+   function(x) c(0, x/x[length(x)]))
> mean.triads <- rowMeans(psi.triads)
> sd.triads <- apply(psi.triads, 1, sd)

> opar <- par(mfrow=c(1,2))
> plot(x, mean.quads, main = "Quads",
+        xlab = expression(r^2),
+        ylab = "Difference Scale Value")
> segments(x, mean.quads + 1.96 * sd.quads,
+        x, mean.quads - 1.96 * sd.quads)
> plot(x, mean.triads, main = "Triads",
+        xlab = expression(r^2),
+        ylab = "Difference Scale Value")
> segments(x, mean.triads + 1.96 * sd.triads,
+        x, mean.triads - 1.96 * sd.triads)
> par(opar)
```

We can use simulations such as these to anticipate how performance of either method is affected by number of trials, the true values of the standard scale and the observer's true value of σ. See [123].

7.5   Bootstrap Standard Errors

Standard errors for the scale values from an MLDS experiment can be estimated using the bootstrap [55] in a fashion similar to that used for psychometric functions. The estimated probabilities of response for each trial are obtained from the model object with the `fitted` method. These are then used with the function `rbinom` repeatedly to generate a set of binary outcomes with the same probabilities as the fitted ones. These serve as bootstrap responses that are analyzed with `mlds` to generate bootstrap estimates of scale values. The means across bootstrap replications of these values compared with the scale values from the experimental data yield bias estimates, and the standard deviations yield standard error estimates.

These steps have been incorporated into a function, `boot.mlds` that takes two arguments, an object of either class "mlds" or "mlbs" and `nsim`, the number of bootstrap simulations to perform.

```
> kk.bt <- boot.mlds(kk.mlds, nsim = 10000)

> str(kk.bt)
```

Fig. 7.7 Estimated scale
values for the perception of
scatterplots from `mlds` using
the `glm` method but plotted on
the standard scale. The *error
bars* correspond to 95%
confidence intervals estimated
using a bootstrap method

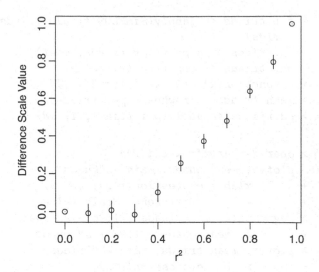

```
List of 4
 $ boot.samp: num [1:11, 1:10000] -0.0029 0.0056 0.0159 0...
 ..- attr(*, "dimnames")=List of 2
 .. ..$ : chr [1:11] "0.1" "0.2" "0.3" "0.4" ...
 .. ..$ : NULL
 $ bt.mean  : Named num [1:11] -0.00879 0.00735 -0.01637 0..
 ..- attr(*, "names")= chr [1:11] "0.1" "0.2" "0.3" "0.4"..
 $ bt.sd    : Named num [1:11] 0.0249 0.0252 0.0285 0.0249..
 ..- attr(*, "names")= chr [1:11] "0.1" "0.2" "0.3" "0.4"..
 $ N        : num 10000
 - attr(*, "class")= chr [1:2] "mlds.bt" "list"
```

As shown above, an object of class "mlds.bt" of four components is returned.
The first component contains a matrix of the bootstrap replications of the scale
estimates, one replication for each column. Each row corresponds to a scale
estimate (in standard scale units) beginning with the second value (the first being by
default 0). The last row contains the bootstrap estimates of σ. The second and third
components, `bt.mean` and `bt.sd`, contain the row means and standard deviations
of `boot.samp`, respectively. The last component, `N`, is the number of bootstrap
replications. We can use the bootstrap standard deviations to assign confidence
intervals to the scale value estimates. A summary method defined for this class
returns a two-column matrix with the first column giving the row means and the
second the standard deviations. The results can be returned on the unnormalized
scale by specifying a second argument `standard.scale` = FALSE. Figure 7.7
shows the code for extracting these values and the figure with the confidence
intervals plotted on the scale values as error bars.

```
> plot(kk.mlds, standard.scale=TRUE,
+        xlab = expression(r^2),
+        ylab = "Difference Scale Value")
> kk.bt.res <- summary(kk.bt)
> kk.mns <- kk.bt.res[, 1]
> kk.95ci <- qnorm(0.975) * kk.bt.res[, 2]
> segments(kk.mlds$stimulus, kk.mns + kk.95ci,
+          kk.mlds$stimulus, kk.mns - kk.95ci)
```

7.6 Robustness and Diagnostic Tests

The model underlying MLDS assumes that human judgment of differences between intervals can be modeled as equal-variance Gaussian signal detection. But what if the underlying distribution is not Gaussian? Would the failure of the distributional assumptions underlying MLDS affect the estimated scale values? This sort of question concerns distributional robustness or, to give it its full title, *robustness to failures of distributional assumptions*.

Maloney and Yang [123] evaluated distributional robustness of the direct optimization method. They varied the distributions of the error term ε in (7.5) and (7.6) while continuing to fit the data with the constant variance Gaussian error assumption. They found that MLDS was remarkably robust to failures of the distributional assumptions. Hence, the GLM approach using the "probit" link is likely to be adequate for most applications of MLDS.

In R, there is a choice between five built-in link functions for the binomial family of glm, including the logit, probit, and cauchit (based on the Cauchy distribution). As of R version 2.4.0, it has become simple for the user to define additional links. In many circumstances, the choice of link is not critical, since over the rising part of these functions, they are quite similar. The difference scaling procedure, however, generates many responses at the tails, i.e., easily discriminable differences and one might think that it could be more sensitive to the choice of link.

7.6.1 Goodness of Fit

As discussed in the introduction to this chapter, we want to be able to evaluate whether the difference scaling model is an appropriate model for human performance in any given experiment. At one extreme, we could imagine an observer who simply cannot carry out a particular difference scaling task and who is simply guessing on every trial. Alternatively, we could imagine an observer who is systematically judging stimuli but whose judgments are not captured by the MLDS model. We would want to detect patterns in the data that suggest that the MLDS model does not capture the observer's performance.

If we use the GLM approach to MLDS, then we can employ any of the several diagnostic tests for generalized linear models. There is an AIC method for extracting this measure of "goodness of fit" from the model object. For example,

```
> AIC(kk.mlds)
[1] 633
```

Alternatively, the "proportion of deviance accounted for" (DAF), suggested by Wood [198], can be easily computed as

```
> with(kk.mlds, (obj$null.deviance - deviance(obj)) /
+    obj$null.deviance)
[1] 0.553
```

This measure indicates the proportional reduction of deviance by the model fit with respect to the deviance of a null model, i.e., a model with only an intercept. Deviance is not variance, however, so some caution must be exercised in the interpretation of this measure.

Some standard GLM diagnostic measures are problematic or difficult to interpret for binary data because the distribution of the deviance (See Sect. B.3.5) cannot be guaranteed to approximate a χ^2 distribution [178, 198]. Here, we implement two diagnostic tests suggested by Wood [198], based on a bootstrap analysis of the distribution and independence of the deviance residuals of the fit. The use of the bootstrap allows us to avoid the problem presented by the lack of convergence to χ^2. The principle on which these tests are based would be applicable to any GLM model. The first involves a comparison of the empirical distribution of the residuals to an envelope of the $1 - \alpha$ proportion of the bootstrap-generated residuals. The second tests the dependence of the residuals by comparing the number of runs of positive and negative values in the sorted deviance residuals with the distribution of runs from the bootstrapped residuals.

We provide a function binom.diagnostics to implement both of these for objects of class "mlds." The function takes two arguments: obj, an "mlds" model object and nsim, the number of bootstrap simulations to run. For example,

```
> kk.diag.prob <- binom.diagnostics(kk.mlds, 10000)
```

An object of class "mlds.diag" is returned that is a list of 5 components, illustrated below

```
> str(kk.diag.prob)
List of 5
 $ NumRuns  : int [1:10000] 199 205 207 209 184 219 208 19..
 $ resid    : num [1:10000, 1:990] -4.67 -4.61 -4.61 -4.54..
 $ Obs.resid: Named num [1:990] 0.458 0.164 0.413 1.155 -1..
  ..- attr(*, "names")= chr [1:990] "1" "2" "3" "4" ...
 $ ObsRuns  : int 173
 $ p        : num 0.0158
 - attr(*, "class")= chr [1:2] "mlds.diag" "list"
```

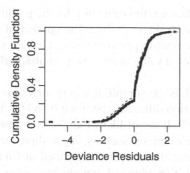 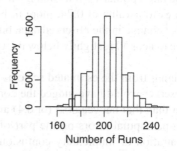

Fig. 7.8 Diagnostic graphs obtained by the `plot` method applied to the binomial diagnostics of the MLDS fit to the `kk` data set. The *left graph* shows the empirical cdf of the deviance residuals (*black points*) compared to the 95% bootstrapped envelope (*dashed lines*). The *right graph* displays a histogram of the number of runs in the sign of the sorted deviance residuals from the bootstrap simulations. The observed value is indicated by a *vertical line*

`NumRuns` is a vector of integer giving the number of runs in the sorted deviance residuals for each simulation. `resid` is a matrix of numeric, each row of which contains the sorted deviance residuals from a simulation. `Obs.resid` is a vector of numeric providing the residuals from the obtained fit. `ObsRuns` is the number of observed runs in the sorted deviance residuals, and finally p gives the proportion of runs from the simulations less than the observed number. A `plot` method is supplied to visualize the results of the two analyses which are shown in Fig. 7.8.

```
> plot(kk.diag.prob, pch = ".", cex = 3)
```

The empirical distribution of the residuals (small black points in left graph of Fig. 7.8) appears to fall within the 95% confidence interval of the envelope (dashed lines). There are two points with evidently extreme residuals on the far left of the graph. The distribution of runs of residuals of the same sign is shown in the right graph of Fig. 7.8. The observed number of runs is indicated by the vertical line and appears to be significantly smaller given that it is greater than only `kk.diag.prob$p` = 0.016 of the residuals.

One possibility is that a different link would provide a better fit to the data and, indeed, the AIC for a logit link is 610 and the proportion of runs is 0.22, not significant (see [96]). The logistic has heavier tails than the Gaussian so that the poorer performance of the latter may reflect its sensitivity to the two outlier points. These two points, as well as a third one are flagged by the diagnostics generated by the glm `plot` method. The three trials to which these points correspond are simply extracted from the data set.

```
> kk[residuals(kk.mlds$obj) < -2.5, ]
    resp S1 S2 S3 S4
295    0  1  2  3 10
857    0  1  2  4 10
939    0  1  2  9 11
```

If these three points are removed from the data set, the value of p for the probit link increases to a value of 0.24, more in line with the logit link. The number of runs does not change in the observed data, but the bootstrapped distribution of runs shifts to a mean near 171, slightly below the observed value of kk.diag.prob$ObsRuns = 173.

Judging from the estimated scale as well as the stimuli, it seems surprising that the observer would have judged the correlation difference between 0 and 0.1 to be greater than that between 0.3 (or 0.4) and 0.9. It seems likely that these correspond to non-perceptual errors on the part of the observer, such as finger-slips, lack of concentration or momentary confusion about the task. As mentioned in Chap. 5, a few such errors nearly always occur in psychophysical experiments, even with practiced and experienced observers. As when fitting psychometric functions to data from detection experiments [188], a nuisance parameter can be introduced to account for these occasional errors.

The error rates are modeled by modifying the lower and upper asymptotes of the inverse link function. This can be performed by adapting the psyfun.2asym function from the **psyphy** package to the MLDS experimental data. The right-hand side of the formula object for this function requires the response to be a two-column matrix of successes and failures. In the case of binary data, this is just one minus the response. The data frame is obtained from the obj component of the "mlds" model object in which the "glm" object from the mlds fit is stored. Finally, we require one of the modified links for this function that specifies both lower and upper asymptotic values of the psychometric function.

```
> library(psyphy)
> kk.mlds2 <- psyfun.2asym(cbind(resp, 1 - resp) ~ . - 1,
+     data = kk.mlds$obj$data, link = probit.2asym)
lambda =       0.0102      gamma =   0.153
+/-SE(lambda) =      ( 0.00632 0.0164 )
+/-SE(gamma) =    ( 0.126 0.184 )
```

The estimated upper asymptote is not much different from 1 but the lower one is considerably elevated, as expected. Interestingly, this error value is similar to the proportion of misclassified trials by the observer with respect to the predictions on the basis of the estimated difference scale. The proportion of misclassified trials is obtained by the pmc function provided in the **MLDS** package.

```
> pmc(kk.mlds)
[1] 0.121
```

Fitting with an elevated asymptote reduces the AIC value by 50 and lowers the number of runs in the distribution of residuals so that the observed value of p is not significant (left as an exercise for the reader).

Fig. 7.9 Six-point condition: Given stimuli $a < b < c$ and $a' < b < c'$ ordered along a scale, if $ab \succ a'b'$ and $bc \succ b'c'$, then $ac \succ a'c'$

7.6.2 Diagnostic Test of the Measurement Model

Even if the data passed an overall goodness-of-fit test, there still may be patterned failures in the data that would allow rejection of the difference scaling model. By analogy, even if a linear regression model accounted for a high proportion of the variance, the pattern of residuals may still allow us to reject the linear model.

In this section, we describe an additional diagnostic test based on the necessary and sufficient conditions that a non-stochastic observer must satisfy if we are to conclude that his judgments can be described by a difference scaling model (see [104, Chap. 4, p. 147, Definition 1]). This test is specific to difference scaling with quadruples.

7.6.2.1 The Six-Point Test

The six-point condition is illustrated in Fig. 7.9. It is referred to as the "weak monotonicity" condition in Definition 1, Axiom 4, of Krantz et al. (Chap. 4, p. 147 and Fig. 1) [104]. It is also known as the "sextuple condition" [15]. We describe the condition with an example. Suppose that there are two groups of three stimuli whose indices are $a < b < c$, and $a' < b' < c'$, respectively. Suppose that a non-stochastic observer considers the quadruple $(a,b;a',b')$ and judges that $ab \succ a'b'$, that the interval ab is larger than the interval $a'b'$. On some other trial, he considers $(b,c;b',c')$ and judges that $bc \succ b'c'$. Now, given the quadruple, $(a,c;a',c')$ there is only one possible response consistent with the difference scaling model: he or she must choose $ac \succ a'c'$. The reasoning behind this constraint is illustrated in Fig. 7.9 and can be demonstrated directly from the model.

For the non-stochastic observer, even one violation of this six-point condition would allow us to conclude that there was no consistent assignment of scale values $\psi_1, \psi_2, \ldots, \psi_p$ in a difference scaling model that could predict his or her judgments in a difference scaling task.

The six-point condition is a slightly disguised test of additivity of contiguous intervals in the difference scale. To see how it might fail, imagine that distances in the scale correspond to chordal distances along a circular segment as shown in the left side of Fig. 7.10. Then the six-point condition in equality form implies that if $ab \sim a'b'$ and $bc \sim b'c'$, then $ac \sim a'c'$ where $\sim$ denotes subjective equality. If the six-point condition and other necessary conditions hold, then the chordal distances on the left side of Fig. 7.10 can be represented along a line as in the previous Fig. 7.9 (See Chap. 4 of Krantz et al. [104] for further discussion). On the right

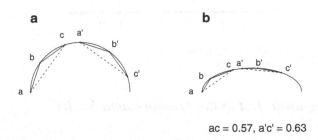

a   **b**

ac = 0.57, a'c' = 0.63

Fig. 7.10 Six-point condition: *Left*. Given stimuli $a < b < c$ and $a' < b < c'$ ordered along a circular (constant curvature) segment, if the chordal distances $ab \approx a'b'$ and $bc \approx b'c'$, then $ac \approx a'c'$. *Right*. The six-point condition fails on a contour with nonconstant curvature

side of Fig. 7.10, in contrast, judgments are based on chordal distances along an ellipse. The six-point condition fails and these judgments cannot be represented by a difference scale.

Human judgments in difference scaling tasks are not deterministic: if we present the same quadruple (or triad) at two different times, the observer's judgments need not be the same. The MLDS model allows for this possibility. In MLDS decisions are based on a decision variable $\Delta(a,b;c,d)$ and, for any given six points a,b,c and a',b',c' there is a nonzero probability that the stochastic observer will violate the six-point condition. In particular, suppose that $\psi_b - \psi_a$ is only slightly greater than $\psi_{b'} - \psi_{a'}$, $\psi_c - \psi_b$ is only slightly greater than $\psi_{c'} - \psi_{b'}$, and $\psi_c - \psi_a$ is only slightly greater than $\psi_{c'} - \psi_{a'}$. Then we might expect that the observer's probability of judging $ab \succ a'b'$ is only slightly greater that 0.5 and similarly with the other two quadruples. Hence, he has an appreciable chance of judging that $ab \succ a'b'$ and $bc \succ b'c'$ but $ac \prec a'c'$ or $ab \prec a'b'$ and $bc \prec b'c'$ but $ac \succ a'c'$, either a violation of the six-point property.

Maloney and Yang [123] proposed a method for testing the six-point property that takes into account the stochastic nature of the observer's judgment and uses a resampling procedure [55] to test the hypothesis that the MLDS model is an appropriate model of the observer's judgments.

Given the experimental design and all of the quadruples used, we can enumerate all six-point conditions present in the experiment, indexing them by $k = 1, n_6$. We count the number of times, V_k, that the observer has violated the kth six-point condition during the course of the experiment and the number of times he has satisfied it, S_k. If we knew the probability that the observer should violate this six-point condition p_k, then we could compute the probability of the observed outcome by the binomial formula,

$$\Lambda_6^k = \binom{V_k + S_k}{V_k} p_k^{V_k} (1 - p_k)^{S_k} \tag{7.17}$$

and we could compute the overall likelihood probability

$$\Lambda_6 = \prod_{k=1}^{N_6} \Lambda_6^k. \tag{7.18}$$

Under the hypothesis that the difference scale model is an accurate model of the observer's judgments, we have the fitted estimates of scale values $\hat{\psi}_1, \ldots, \hat{\psi}_p$ and $\hat{\sigma}$. We can compute estimates of the values $\hat{\Lambda}_6^k$ based on these scale values and compute an estimate of $\hat{\Lambda}_6 = \prod_{k=1}^{N_6} \hat{\Lambda}_6^k$. This is an estimate of the probability of the observed pattern of six-point violations and successes. We next simulate the observer N times with the fitted parameter values $\hat{\psi}_1, \ldots, \hat{\psi}_p$ and $\hat{\sigma}$ of the actual observer used for the simulated observer and perform the analysis above to get N bootstrap estimates $\hat{\Lambda}_6^*$ of $\hat{\Lambda}_6$. Under the hypothesis that MLDS is an accurate model of the observer's judgments, $\hat{\Lambda}_6$ should be similar in value to $\hat{\Lambda}_6^*$ and we employ a resampling procedure to test the hypothesis at the 0.05 level by determining whether $\hat{\Lambda}_6$ falls below the 5th percentile of the bootstrap values $\hat{\Lambda}_6^*$ (see [96,123] for details).

```
> kk.6pt <- simu.6pt(kk.mlds, nsim = 10000, nrep = 3)

> str(kk.6pt)
List of 4
 $ boot.samp: num [1:10000] -1568 -1398 -1439 -1473 -1449 ..
 $ lik6pt   : num -1232
 $ p        : num 0.99
 $ N        : num 10000
```

Using str reveals that the function returns a list of four components. The p-value indicates that the observer did not make a significantly greater number of violations than an ideal observer. Figure 7.11 shows the histogram of the log-likelihoods from the simulation with the observed log-likelihood indicated by a thick vertical line. The results support the appropriateness of the scale as a description of the observer's judgments.

```
> hist(kk.6pt$boot.samp)
> abline(v = kk.6pt$lik6pt, lwd = 2)
```

7.6.2.2 Other Necessary Conditions and Alternative Axiomatizations

Definition 1 of Chap. 4 (p. 147) of Krantz et al. [104] lists six conditions ("axioms") that are jointly necessary and sufficient for a data set to be representable by a difference scale in the non-stochastic case. The six-point diagnostic test was based on one of these conditions (Definition 1, Axiom 4). Of the remaining necessary conditions, two effectively guarantee that the experimental design contains "enough" distinct

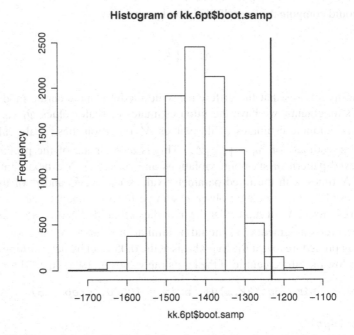

Fig. 7.11 Histogram of the log-likelihood values from the six-point test to data set kk. The *thick vertical line* indicates the observed six-point log-likelihood on the data set

quadruples and that the observer can order intervals transitively (Axioms 1, 2). Axiom 3 is satisfied if the observer can reliably order the items on the physical scale. Axiom 5 is satisfied if the values of the physical scale can be put into correspondence with an interval of the real numbers (evidently true in our example for correlation $-1 \leq r \leq 1$). Axiom 6 precludes the possibility that an interval ab with $a < b$ contains an infinite number of nonoverlapping intervals that are all equal. In the stochastic case, these conditions are either evidently satisfied or are replaced by consideration of the accuracy and stability of the estimated scale. Maloney and Yang [123] have investigated accuracy, stability, and robustness in some detail.

There are alternative sets of axiomatizations of difference scaling such as Suppes [168] and, of course, all sets of necessary and sufficient conditions are mathematically equivalent. We have chosen those of Krantz et al. [104] because they lead to simple psychophysical tasks. Observers are asked only to order intervals. Either they can do so without violating the six-point condition or they cannot and whether they can or cannot is an experimental question. Chapter 4 of Krantz et al. [104] contains useful discussion of the link between the axiomatization that they propose and the task imposed on observers.

The reader may wonder why the observer is asked to compare intervals and not just pairs of stimuli. Chapter 4 of Krantz et al. (pp. 137–142) [104] contains an extended discussion of the advantages of asking observers to directly compare intervals. We note only that pairwise comparison of the stimuli (i.e., given a, b, judge whether $a < b$) does not provide enough information to determine spacing along the scale in the non-stochastic case. Any increasing transformation of a putative difference scale would predict the same ordering of the stimuli. In the stochastic case the observer may judge that $a < b$ on some trials and that $b < a$ on others, and the degree of inconsistency in judgment could potentially be used to estimate scale spacing using methods due to Thurstone [173].

Thurstonian scales, however, have three major drawbacks. First, the scale depends crucially on the assumed distribution of judgment errors (it is not distributionally robust) while MLDS is remarkably robust (see [123]). Second, stimuli must be spaced closely enough so that the observer's judgments will frequently be inconsistent. This typically means that many closely spaced stimuli must be scaled, and the number of trials needed to estimate the scale is much greater than in MLDS. The third drawback is the most serious. It is not obvious what the Thurstonian scale measures, at least not without further assumptions about how "just noticeable differences" add up to produce perceptual differences. The MLDS scale based on quadruples is immediately interpretable in terms of perceived differences in interval lengths since that is exactly what the observer is judging.

Exercises

7.1. Modify the function `11.mlds` to include an argument to specify a link function for the binomial family and the code in the function body to permit testing other link functions than the "probit." Test the function on the kk3 data set for several links and compare the results.

7.2. The data set `AutumnLab` from the **MLDS** package contains the data from one observer for judging an image whose perceived quality depended upon the level of image compression [29]. Analyze this data set to obtain the perceptual scale and bootstrap standard errors for the scale values. Evaluate the goodness of fit by using binomial diagnostics and the six-point test. Repeat the above using the logit and cauchit links and compare the results.

7.3. Rensinck and Baldridge [147] propose the following model for perceived correlation in scatterplots

$$\psi_\rho(r) = \frac{\log(1 - \beta r)}{\log(1 - \beta)}, \tag{7.19}$$

where ψ_ρ is perceived correlation, r is physical correlation, and β a free parameter. Fit this model to the kk data set using the formula method for `mlds` and compare the results with the standard fit.

7.4. The data set `Transparency` in the **MLDS** package contains the data from six observers who judged the transparency of pebble-shaped objects that differed in their simulated index of refraction [63]. Extract the data for each individual observer from the data frame and coerce to an object of class "mlds.df." Then, obtain perceptual scales not only for each individual but also for the whole data set with all observers' data grouped together as a single data set. Compare the mean of the individual perceptual scales with the overall fit on both standard and unnormalized scales. Where are the differences greatest and to what do you think that they are due?

7.5. The Legge–Foley model for contrast perception is described by

$$R(c) = R_{\max} \frac{c^n}{c^n + s^n}, \tag{7.20}$$

where R is the response to contrast c, $R_{\max}$ is the maximum response (see [140]), s the contrast gain and n an exponent, usually greater than 1 that produces an accelerated response to contrast at low contrast levels. Contrast varies between 0 and 1. Choose values for the free parameters and explore the contour plots for equal-response differences in the decision space for a difference scaling experiment involving contrast perception. For a given response difference, use level plots to explore the effects of different values of the judgment uncertainty, σ, on the rate of choosing one pair over another in a difference scaling experiment.

Chapter 8
Maximum Likelihood Conjoint Measurement

8.1 Introduction

Conjoint measurement allows the experimenter to estimate psychophysical scales that capture how two or more physical dimensions contribute to a perceptual judgment. It is most easily explained by an example. Figure 8.1 shows a range of stimuli taken from a study by Ho et al. [80]. Each stimulus was the image of an irregular surface (the original stimuli were rendered for binocular viewing at a higher resolution than those shown in the figure). The experimenters were interested in how the perceived roughness (which they refer to as "bumpiness") and glossiness of a surface were affected by variations in two physical parameters that plausibly affect perceived roughness and glossiness. They reported two experiments that differed only in the judgments observers made. In the first, observers judged perceived roughness, in the second, perceived glossiness.

Physical roughness was defined as the variance of the heights of all the irregular bumps in the surface. It varies from 0 (a smooth surface) to a maximum chosen by the experimenters. *Physical gloss* was defined as the value of a gloss parameter in the rendering package used in preparing the stimuli. The stimuli in each column of Fig. 8.1 vary in physical roughness but not physical glossiness while the stimuli in each row share the same physical roughness, varying in physical glossiness.

Ho et al. sought to test whether human observers could estimate physical roughness or glossiness without "cross-talk" or "contamination" from the other physical dimension. They sought to measure and model any contamination of perceived roughness by physical gloss or perceived gloss by physical roughness.

At one extreme we might imagine that each combination of physical gloss and physical roughness evokes an idiosyncratic perception of gloss and roughness. If so, then increasing physical gloss might sometimes lead to a perception of increased roughness and sometimes lead to a perception of decreased roughness, depending on the initial values of physical gloss and physical roughness. At the other, we might find that each perceptual dimension is unaffected by the other physical dimension.

K. Knoblauch and L.T. Maloney, *Modeling Psychophysical Data in R*, Use R! 32, DOI 10.1007/978-1-4614-4475-6__8,
© Springer Science+Business Media New York 2012

Fig. 8.1 Surface material perception

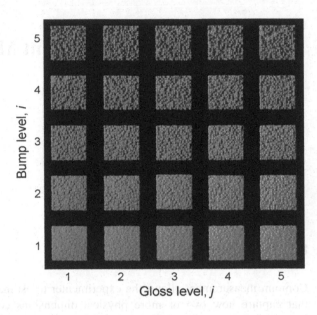

Ho et al. [80] developed and tested models of gloss and roughness perception (described below) that spanned these two extremes.

We could imagine approaching this problem using difference scaling as presented in the immediately preceding chapter. If we used difference scaling to estimate psychophysical functions for perceived glossiness for each row we would obtain possibly different difference scales, one for each row. We would need to decide how to test whether they are the same. Moreover, even if we conclude that all the rows share a common difference scale, we still need to estimate it by combining data across rows. If the difference scales for different rows proved to be different, then we would still need to model how roughness differences (between rows) affect perceived glossiness. We would have similar problems in modeling perceived roughness. These problems are not insurmountable, but application of a different method, *additive conjoint measurement*, allows us to obviate them all.

Additive conjoint measurement permits us *simultaneously* to measure and model the contributions of both physical dimensions to perceived roughness and perceived glossiness. The results, if all goes well, are estimates of two functions that we refer to as *perceived roughness* and *perceived gloss* that capture these contributions. If all does not go well, we reject the model as inappropriate to model how physical gloss and roughness determine perceived gloss and roughness.

The psychophysical task that drives the method is remarkably simple. In Ho et al.'s first experiment, observers viewed all possible pairs of the 25 stimuli in Fig. 8.1 and judged which of each pair was "rougher." In a second experiment, a different set of observers viewed the same stimuli and judged which of each pair was "glossier." We emphasize that the stimuli in the two experiments were identical but that observers were instructed to extract a perceptual analogue of each physical dimension in turn while "squelching" any effect of the other.

Let φ^r denote values on the physical roughness scale and let φ^g denote values on the physical gloss scale. The value of roughness is constant across the ith row in Fig. 8.1 and is denoted φ^r_i and the value of gloss is similarly constant in the jth column and is denoted φ^g_j. Consider now a comparison between the surface in the ith column and jth row and the surface in the kth column and lth row. The first surface has physical roughness and physical gloss that can be represented as an ordered pair, $(\varphi^r_i, \varphi^g_j)$ and the second as an ordered pair $(\varphi^r_k, \varphi^g_l)$. Ho et al. first assumed that the visual system computes an estimate of perceived roughness for the first stimulus that could potentially depend on both its physical roughness φ_{r_i} and its physical glossiness, φ_{g_j}. We write this estimate as $\psi^R(\varphi^r_i, \varphi^g_j)$. We will consistently use uppercase letters ("R") for perceptual measures and lowercase letters for physical variables ("r").

When the model observer judged roughness, Ho et al. assumed that the resulting estimates of perceived roughness were compared by computing the difference between perceived estimates of the relevant perceptual dimension,

$$\Delta(i,j,k,l) = \psi^R(\varphi^r_k, \varphi^g_l) - \psi^R(\varphi^r_i, \varphi^g_j) + \varepsilon \tag{8.1}$$

where the random variable $\varepsilon \sim \mathcal{N}(0, \sigma^2_R)$ is normally distributed *judgment error*. The model observer for glossiness was computed in an analogous fashion but based on the perceived glossiness $\psi^G(\varphi^r_i, \varphi^g_j)$ and $\psi^G(\varphi^r_k, \varphi^g_l)$ of the two stimuli compared and with a possibly different magnitude of judgment error for glossiness comparisons σ^2_G.

Ho et al. considered three model observers, *the saturated observer, the additive observer* and *the independent observer*. The first and last of these correspond to the two extreme models discussed above. These three models form a nested series with the saturated model as least constrained, the independent as most constrained. They all refer to the judgment process just described and differ only in how the estimates of roughness (alternatively glossiness) to each of the stimuli on a trial $(\varphi^r_i, \varphi^g_j)$ and $(\varphi^r_k, \varphi^g_l)$ are affected by the physical dimensions of roughness and gloss.

The additive conjoint model replaces $\psi^R(\varphi^r_i, \varphi^g_j)$ by $\psi^{R:r}_i + \psi^{R:g}_j$ where $\psi^{R:r}_i$ is an additive contribution of physical roughness to perceived roughness that is constant across the ith row of Fig. 8.1 and $\psi^{R:g}_j$ is a similar contribution but now that of physical gloss to perceived roughness that is constant across the jth column.

The values $\psi^{R:r}_i$ and $\psi^{R:g}_i$ are parameters of the additive model that we wish to estimate from data. The values $\psi^{R:r}_i$ form a psychophysical scale mapping physical to perceived roughness. There is an evident comparison between such a scale and its counterpart obtained from difference scaling. Estimating the difference scale involves ordering of differences while in conjoint measurement we simply order the two stimuli in a single pair. Moreover, if the conjoint model is valid, then the same values $\psi^{R:r}_i$ apply *even as we vary the second dimension of gloss*. It is an open question as to whether scales derived with difference scaling would agree with scales derived from additive conjoint measurement.

The additive model is based on the assumption that physical gloss and physical roughness interact in determining perceived roughness but that the contribution of a particular level of gloss, $\psi_j^{R:g}$, to perceived roughness is independent of the physical roughness of the surface. The second term, $\psi_j^{R:g}$, can be thought of as a "contamination" of perceived roughness due to changes in physical glossiness. The additive conjoint model is based on the assumption that the contamination is additive, and one of the goals of Ho et al. was to test that assumption. Moreover, they also wished to test whether changes in physical glossiness affect perceived roughness at all, testing the hypothesis that $\psi_j^{R:g} = 0$ for all j.

For the additive model observer, (8.1) can be rewritten as

$$\Delta(i,j,k,l) = \psi_k^{R:r} + \psi_l^{R:g} - \psi_i^{R:r} - \psi_j^{R:g} + \varepsilon \qquad (8.2)$$

which can be regrouped as

$$\Delta(i,j,k,l) = [\psi_k^{R:r} - \psi_i^{R:r}] + [\psi_l^{R:g} - \psi_j^{R:g}] + \varepsilon. \qquad (8.3)$$

The additive model observer for roughness bases its judgment of a comparison of the perceived roughness levels of the two stimuli with an additive contamination from the difference in perceived glossiness.

The *independent model observer* of roughness is identical with the additive model observer except that there is *no* contamination of perceived roughness by gloss and vice versa. The decision variable is then just

$$\Delta(i,j,k,l) = [\psi_k^{R:r} - \psi_i^{R:r}] + \varepsilon. \qquad (8.4)$$

The perceived difference in roughness depends only on the physical roughness levels of the stimuli compared. Analogous equations for saturated, additive and independent model observers for glossiness are easily derived.

The goals of the experimenters in [80] were, first of all, to estimate the two functions $\psi_i^{R:r}$, $\psi_j^{R:g}$, as well as σ_R from each observer's data in the condition where observers judged roughness and the two functions $\psi_j^{G:g}$, $\psi_i^{G:r}$ from each observer's data in the condition where observers judged glossiness. Second, they wished to test the additive conjoint model against the independent model. This test is in effect a test of whether physical gloss has a significant additive effect on perceived roughness and vice versa. Ho et al. rejected the independent model. Third, they wished to test the saturated model against the additive to test whether the effect of physical gloss on perceived roughness and/or physical roughness against perceived roughness could be adequately captured by the additive model. The additive model they use is an example of an *additive conjoint measurement* model and, before going further, we state the general model which applies to two or more dimensions.

8.2 The Model

Conjoint measurement was developed independently with slightly different assumptions by several researchers (see [104] for review). In its more recent manifestations it included the assumption that observers' judgments are deterministic: presented with the same stimuli, the observer would make the same judgment. The axiomatic theory of conjoint measurement is treated in [117] with textbook level presentations in [104] and in [151]. It focuses on the conditions on data that preclude or permit its representation by an additive conjoint model. What we present here is a version of conjoint measurement that allows for the possibility that the observers judgments may be in part stochastic. The changes entailed by this assumption are analogous to those encountered in developing Maximum Likelihood Difference Scaling in the preceding chapter.

In the example just presented, we modeled the contributions of two dimensions, physical gloss and physical roughness to observers' judgments of perceived glossiness and perceived roughness of surfaces. In the general case, we consider stimuli that can vary along two or more physical dimensions. The number of dimensions is $n \geq 2$. If, for example, Ho et al. had also varied the size of the elements, then n would have been 3. In the general case, we assume that each stimulus is represented by magnitudes of contribution of $n \geq 2$ distinct dimensions to each stimulus. There is no restriction on the choice or units of such dimensions. We could consider stimuli consisting of varying amounts of "shoes and ships and sealing wax, and cabbages and kings" [27]. In fact, the earliest uses of conjoint measurement were concerned with preferences between bundles of goods that might include varying numbers of cabbages and amounts of sealing wax. Of course, whether the observer can meaningfully order such stimuli is a separate question. We return to the issue of testing the additive conjoint model below.

We denote the physical dimensions of a stimulus by $\varphi^1, \ldots, \varphi^N$. For simplicity we assume that each physical dimension is continuous. Just as in the example, the observer is presented with two stimuli on each trial, the first denoted by $\varphi^1, \ldots, \varphi^N$. and the second by $\varphi'^1, \ldots, \varphi'^N$. We will typically refer to any stimulus by listing its physical values on all dimensions. The observer is asked to order these two stimuli by some criterion ("roughness," "glossiness," preference, etc.) and we assume the following model of the observer's judgment. There are functions $\psi^i(\varphi^i), i = 1, \ldots, n$ such that the value of the decision variable

$$\Delta = \sum_{i=1}^{n} \psi^i(\varphi^i) - \sum_{i=1}^{n} \psi^i(\varphi'^i) + \varepsilon. \qquad (8.5)$$

Once again, $\varepsilon \sim \mathcal{N}(0, \sigma^2)$, a Gaussian random variable with mean 0 and variance σ^2, intended to capture observers' uncertainties in judgment. If $\Delta > 0$, the observer selects the stimulus corresponding to $\varphi^1, \ldots, \varphi^N$ and otherwise the stimulus corresponding to $\varphi'^1, \ldots, \varphi'^N$.

We can rearrange the terms in (8.5) to emphasize that, in the additive model, the observer effectively computes the differences on each scale (dimension) separately (after transformation by the ψ^k) and then adds together all of the differences across scales to arrive at the decision variable

$$\Delta = \sum_{i=1}^{N} \left[\psi^i(\varphi^i) - \psi^i(\varphi'^i) \right] + \varepsilon. \tag{8.6}$$

The goal of maximum likelihood conjoint measurement (MLCM) is to recover estimates of the functions $\psi^i(\varphi^i)$, $i = 1, \ldots, N$ and, as the name suggests, we will use the method of maximum likelihood estimation (Sect. B.4.2) to do so. A key decision for the experimenter is to decide whether to restrict attention to functions $\psi^i(\varphi^i)$, $i = 1, \ldots, n$ drawn from a parametric family (e.g., power functions $\psi(\varphi) = a\varphi^b$ with parameters a and b) or to consider a model that requires no such assumptions.

Somewhat imprecisely, we refer to the latter approach as *nonparametric* and the results for the example drawn from [80] (which we present below) are examples of nonparametric fits. The example of Ho et al. above is an example of a nonparametric approach with $N = 2$ and $n_1 = n_2 = 5$. The nonparametric approach is only nonparametric in the sense that we have not constrained the estimated scale values (ψ^i) by any parametric function of the stimulus dimensions. Even with the nonparametric approach, we are still engaged in estimating parameters. Moreover, the experimenter may choose to fit both the nonparametric and one or more parametric models and use model selection methods (Sect. B.5) to decide among them. The nonparametric model is the natural alternative model to use in testing the goodness of fit of any parametric model nested within it (Sect. B.5).

We first consider the nonparametric approach, where the experimenter has selected n_k discrete levels of φ^k for each of the k scales (dimensions). We denote the values for the kth scale by φ_j^k, $j = 1, \ldots, n_k$. The experimenter generates $M = n_1 n_2 \cdots n_N$ stimuli corresponding to every possible combination of physical levels. The experimenter presents pairs of the M stimuli and records the observer's response.

An important aspect of conjoint measurement in psychophysical applications is that, while we are trying to determine how each physical dimension contributes to each perceptual judgment, we recognize that the particular scales chosen for each physical dimension are arbitrary. Ho et al., for example, spaced their stimuli according to the variance of roughness height but could have instead used the standard deviation. Similarly, the physical scale for gloss was inherited from a gloss parameter used in rendering the stimuli. There is no reason to accept that observers' perception of gloss of stimuli is proportional to (or even in the same order) as the values of this gloss parameter.

Moreover, all the data we have are orderings of pairs of stimuli. We have no way to decide if the perceived difference between two stimuli judged in isolation is large or small. Given the arbitrariness of the scales and the ordinal nature of the judgments demanded of the observers, it is not obvious that we can recover the functions $\psi^i()$, $i = 1, \ldots, n$. As in MLDS, we seem to be playing " ...on a

Fig. 8.2 The two *solid curves* are indifference contours. Two points that are on the same indifference contour are judged to be equivalent. We can use these indifference contours to compare intervals on the vertical axis to intervals on the horizontal axis. The marked intervals on the *vertical axis* represent differences that effectively cancel the difference on the *horizontal axis* and must therefore be equal to each other

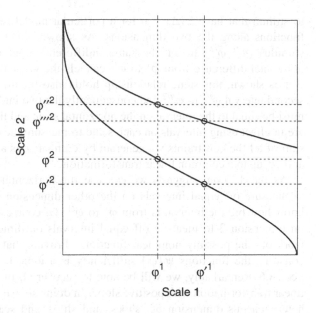

cloth untrue, with a twisted cue, and elliptical billiard balls." (*The Mikado*). In fact, we will be able to recover considerable information about the $\psi^i(),i=1,\ldots,n$ as described next. We will end this section with an intuitive description of how MLCM is possible and what underlies it.

The key idea is that when we compare two stimuli that differ in two dimensions, we can rewrite the decision variable,

$$\Delta = \psi^1(\varphi^1) + \psi^2(\varphi^2) - \psi^1(\varphi'^1) - \psi^2(\varphi'^2) + \varepsilon \tag{8.7}$$

as

$$\Delta = (\psi^1(\varphi^1) - \psi^1(\varphi'^1)) + (\psi^2(\varphi^2) - \psi^2(\varphi'^2)) + \varepsilon, \tag{8.8}$$

which is the sum of contributions of two differences (intervals) on the first and second dimension. Suppose now that we fix the interval on dimension 1 and set

$$C = \psi^1(\varphi^1) - \psi^1(\varphi'^1) \tag{8.9}$$

so that

$$\Delta = C + (\psi^2(\varphi^2) - \psi^2(\varphi'^2)) + \varepsilon. \tag{8.10}$$

For any choice of φ^2 we attempt to adjust φ'^2 until the observer is indifferent and picks the first and second stimulus equally often. This corresponds to the point where $\mathsf{E}(\Delta) = 0$. Thus, although we do not know the magnitude of $\psi^1(\varphi^1) - \psi^1(\varphi'^1)$ we now treat it as equal in absolute magnitude to $\psi^2(\varphi^2) - \psi^2(\varphi'^2)$ since they cancel. This situation is shown in Fig. 8.2 where the solid curves correspond to stimuli that have $E(\Delta) = 0$ for a particular model of the underlying response functions along the two dimensions. As shown, the stimulus (φ^1, φ^2) and the stimulus (φ'^1, φ'^2) lie on the same indifference contour. This implies that the horizontal difference from φ^1 to φ'^1 cancels the vertical distance from φ^2 to φ'^2. But as shown, this same relationship holds also for the interval φ^1 to φ'^1 and a second interval φ''^2 to φ'''^2 and consequently the two intervals on the vertical scale must be equal to the interval on the horizontal scale and therefore to each other. We are in effect using intervals on each scale to measure the other, and remarkably the sum of all the constraints we generate by comparisons allows us to determine ψ^1 and ψ^2 up to a common linear transformation.

As the diagram suggests we can use the horizontal interval from φ^1 to φ'^1 to measure off equal intervals on the other dimension—even though we do not know how big the interval is from φ^1 to φ'^1. Of course, we can also use intervals on dimension 2 to measure off equal intervals on dimension 1 and in this way work out the possibly nonlinear functions. Proving that all of these comparisons constrain the functions $\psi_i(\varphi_i)$ sufficiently is a topic to which we will return in Sect. 8.6. Remarkably, we will be able to recover all of the $\psi_i(\varphi_i)$ up to a single linear transformation with positive slope. In doing so, we have reduced the possible heterogeneous dimension of "shoes, and ships, and sealing wax" to a common perceptual scale. Any failure in the model indicates that there exists no such common scale.

8.3   The Nonparametric Model

We develop the fitting procedure for the condition in [80] in which observers judged roughness. We first reexpress (8.1) in terms of the additive decomposition in (8.2) to get

$$\Delta(i, j, i', j') = \beta_R(r^i) + \beta_g(g^j) - \beta_R(r^{i'}) - \beta_g(g^{j'}) + \varepsilon. \tag{8.11}$$

We can simplify our notation by writing $\beta_R(i) = \beta_R(r_i)$ and $\beta_g(j) = \beta_g(b_j)$. Then (8.11) can be rewritten as

$$\Delta(i, j, i', j') = \beta_R(i) + \beta_g(j) - \beta_R(i') - \beta_g(j') + \varepsilon \tag{8.12}$$

and the two functions $\beta_R(r)$ and $\beta_g(g)$ affect the outcome only through the discrete points $\beta_R(1), \ldots, \beta_R(5)$ and $\beta_g(1), \ldots, \beta_g(5)$. These ten numbers together with the variance σ^2 determine the probability of any response the subject may make. Consequently, we can set as our goal recovering these 11 parameters. However, a moment's consideration indicates that we cannot recover all of these parameters. If, for example, we add a constant c to all of the $\beta_R(1), \ldots, \beta_R(5)$, then $\Delta(i, j, i', j')$ in (8.13) is unaffected since the constant c cancels itself in the difference $\beta_R(i) - \beta_R(i')$. Consequently, given any estimates $\beta_R(1), \ldots, \beta_R(5)$, the estimates $\beta_R(1) + c, \ldots, \beta_R(5) + c$ will account for the data equally well. Similarly, any estimates $\beta_g(1), \ldots, \beta_g(5)$ can be replaced by $\beta_g(1) + d, \ldots, \beta_g(5) + d$ without affecting any of the experimental judgments. We take advantage of this indeterminacy to arbitrarily constrain $\beta_R(1)$ and $\beta_g(1)$ to both be 0 leaving us with nine free parameters including σ. But there is a further indeterminacy. We can multiply (8.13) by any positive constant without affecting the outcome of the test $\Delta > 0$ which determines the model observer's response. We can remove this indeterminacy by constraining that $\sigma = 1$ leaving us with eight free parameters $\beta_R(2), \ldots, \beta_R(5)$ and $\beta_g(2), \ldots, \beta_g(5)$. We will estimate these parameters by the method of maximum likelihood (ML). We will also show that because the decision rule is linear, the ML fitting procedure is a special case of a GLM.

The data from two observers for the experiment of [80] can be found in two data sets in package **MLCM** [97]. The data sets are named BumpyGlossy for an observer who judged "roughness" and GlossyBumpy for an observer who judged "glossiness." Note that, following the terminology of Ho et al., we retain the terms "bumpy" and "bumpiness" in the package but use the terms "rough" and "roughness," respectively, in this chapter. Each data set is a data frame with class "mlcm.df." We use the BumpyGlossy data set in the examples that follow and leave analysis of the second data set as an exercise.

Each trial is specified as a quadruple $[i, j, i', j']$ and a response designating either the first stimulus b_i, b_j or the second, $b_{i'}, b_{j'}$. For convenience, we represent the selection of the first interval by 0, the selection of the second by 1. The following code loads the BumpyGlossy data set from the package and displays the first six lines.

```
> data(BumpyGlossy, package = "MLCM")
> head(BumpyGlossy)
  Resp G1 G2 B1 B2
1    1  3  4  4  3
2    1  2  2  3  3
3    1  3  5  4  2
4    0  1  1  1  4
5    0  2  3  1  2
6    1  1  1  3  3
```

The first column is a two-level factor indicating the response, the next two columns specify the indices of the physical gloss indices i for the first and second stimulus (in that order) and the last two columns those for the physical roughness index j for the two stimuli. Thus, the first line corresponds to a trial in which the stimulus $(3,4)$ was compared to the stimulus $(4,3)$ and the observer judged the second stimulus as rougher. On the second trial, the stimuli are $(2,3)$ and $(2,3)$ and the observer judged the second as rougher. Note that the physical scale values are equal on both scales for both images presented on this trial. This does not mean that the stimuli were identical, however. In Fig. 8.1, the positions of the roughness in each image are random, but their density from image to image is approximately equal. In the actual experiment, different exemplars, i.e., different instances of the distribution of

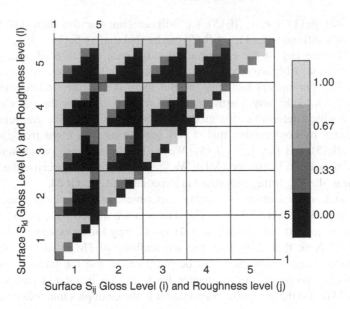

Fig. 8.3 A conjoint proportions plot for the data set `BumpyGlossy`

roughness, were presented on each trial, so that when both images were of equal value on both physical scales, the images were not identical. The observer should choose the left or right stimulus with equal probability for these trials. While they make no contribution to the estimation of the perceptual scale values, they can be used to test if the observer displayed a positional bias in making his (or her) judgments (see Problem 8.4).

Subjects in the experiment saw every possible pairing of the 25 stimuli in Fig. 8.1 three times in randomized order, a total of $975 = 3 \times ((24 \times 25)/2 + 25)$ trials. Consequently, the data frame for one observer has 975 rows and 5 columns. A first overview of the pattern of the observer's responses can be obtained from a grey level image indicating the proportion of "Rougher" (or "Bumpier") responses emitted by the observer for each combination of stimuli, i.e., the proportion of times that stimulus S_{kl} was judged rougher as a function of stimulus S_{ij}. Conveniently, the `plot` method for objects of class "mlcm.df" was designed to produce such a graph. We have added a color bar beside the plot to indicate the values indicated by each grey level. In order to do so, we first adjusted the margins of the region to make space for the color bar. This is done with the "mar" argument of the `par` function. It requires a 4 element vector specifying the number of lines for each margin of the plot. A second invocation of `par` is used with the argument "xpd" to turn off clipping outside of the plot region, so that the bar will appear when plotted. The actual color bar was created with the function `gradient.rect` from the **plotrix** package [110]. Examination of the pattern of grey levels in Fig. 8.3 shows that as roughness increases for stimulus S_{kl} or glossiness decreases for stimulus S_{ij}, the proportion of judgments that S_{kl} is rougher increases.

```
> library(MLCM)
> library(plotrix)
> mr <- par("mar")
> op <- par(mar = mr + c(0, 0, 0, 4.5), pty = "s")
> plot(BumpyGlossy,
+    xlab = expression(paste("Surface ", S[ij],
+    " Gloss Level (i) and Roughness level (j)")),
+        ylab = expression(paste("Surface ", S[kl],
+    " Gloss Level (k) and Roughness level (l)"))
+ )
> par(xpd = NA)
> gradient.rect(5.9, 1, 6.3, 5,
+        col = grey.colors(4, 0, 0.83), gradient = "y")
> text(6.375, seq(1.5, 4.5, 1),
+    formatC(round(seq(0, 1, len = 4), 2), 2, format = "f"),
+    adj = 0)
> par(op)
```

Given any such data set we want to compute the likelihood of the observer's response and we begin by computing the likelihood of a single row, $[r,i,j,i',j']$ in terms of the free parameters $\beta_R(2),\ldots,\beta_R(5)$ and $\beta_g(2),\ldots,\beta_g(5)$. For convenience, we include $\beta_R(1)$ and $\beta_g(1)$ in the notation although both of these values are fixed at 0. The likelihood of the row $[r,i,j,i',j']$ is just the probability of the response given the values $\beta_R(1),\ldots,\beta_R(5)$ and $\beta_g(1),\ldots,\beta_g(5)$. We first compute the probability that the response is 1 (the observer chooses the second interval) as,

$$P[\Delta(i,j,i',j') > 0] = P[\beta_R(i) + \beta_g(j) - \beta_R(i') - \beta_g(j') + \varepsilon > 0]$$
$$= P[\varepsilon > \beta_R(i') + \beta_g(j') - \beta_R(i) - \beta_g(j)]$$
$$= 1 - \Phi(\beta_R(i') + \beta_g(j') - \beta_R(i) - \beta_g(j)), \qquad (8.13)$$

where $\Phi(z)$ is the cumulative distribution function of the Gaussian with mean 0 and variance $\sigma^2 = 1$. Since the Gaussian has the symmetry property $1 - \Phi(z) = \Phi(-z)$ (Sect. B.2.5) we have probability of a response $r = 1$ is

$$P[r = 1] = \Phi(\beta_R(i) + \beta_g(j) - \beta_R(i') - \beta_g(j')) \qquad (8.14)$$

and, of course, $P[r = 0] = 1 - P[r = 1]$. The likelihood of the actual response for the row $[r,i,j,i',j']$ is then

$$\mathscr{L}(r|i,j,i',j') = P[r = 1]^r P[r = 0])^{1-r} = P[r = 1]^r (1 - P[r = 1])^{1-r} \qquad (8.15)$$

where $P[r = 1]$ is computed from i,j,i',j'.

If we let $\mathscr{L}_k$ denote the likelihood for the kth row, then the likelihood for the data set, given any choice of the free parameters $\beta_R(2),\ldots,\beta_R(5)$ and $\beta_g(2),\ldots,\beta_g(5)$ is just

$$\mathscr{L}(\text{Data}|\beta_R(2),\ldots,\beta_R(5),\beta_g(2),\ldots,\beta_g(5)] = \prod_{k=1}^{M}\mathscr{L}_k \qquad (8.16)$$

when there are M trials indexed $k = 1,\ldots,M$. It is usually convenient to work with the logarithm of the likelihood as discussed in Sect. B.4.2. Then

$$\log\mathscr{L}(\text{Data}|\beta_R(2),\ldots,\beta_R(5),\beta_g(2),\ldots,\beta_g(5)] = \sum_{k=1}^{M}\log\mathscr{L}_k. \qquad (8.17)$$

The ML estimates of the parameters $\beta_R(2),\ldots,\beta_R(5),\beta_g(2),\ldots,\beta_g(5)$, denoted as $\hat{\beta}_R(2),\ldots,\hat{\beta}_R(5),\hat{\beta}_g(2),\ldots,\hat{\beta}_g(5)$, can be estimated numerically by using a numerical optimizer, such as optim.

8.3.1 Fitting the Nonparametric Model Directly by Maximum Likelihood

We first define a function, lmlcm to calculate (8.17) for the data set. It takes two arguments, a vector giving the current estimates of the eight parameters and the data frame; it returns the negative of the log likelihood. Then, we specify initial estimates of the parameters in a vector p. The initial parameter estimates and the function to minimize are the first two formal arguments of optim. It is possible to specify a function to calculate the gradient as a third formal argument, though we do not do so here.[1] The additional argument to lmlcm is specified as a named argument after these three but before the argument specifying the method. We use the "BFGS" method because we find it to be more rapid and to converge more reliably than the default here.

```
> lmlcm <- function(p, d){
+       n <- max(d[, 2:3]) - 1
+       pcur <- c(0, p[1:n], 0, p[-(1:n)])
+       rv <- pcur[d[, 2]] - pcur[d[, 3]] +
+           pcur[d[, 4] + 5] - pcur[d[, 5] + 5]
+       -sum(ifelse(d[, 1] == "1", pnorm(rv, log.p = TRUE),
+           pnorm(rv, lower.tail = FALSE, log.p = TRUE)))
+       }
> p <- c(seq(0.1, 1, len = 4), seq(1, 10, len = 4))
> bg.opt <- optim(p, lmlcm, d = BumpyGlossy,
+       method = "BFGS", hessian = TRUE)
```

---

[1] See the example in [178] on p. 445.

Fig. 8.4 Estimated perceptual scales obtained by fitting the Additive MLCM model using optim to the BumpyGlossy data set. The *circles* and *solid line* correspond to the contributions of roughness and the *triangles* and *dashed line* to the contributions of glossiness to the perceived roughness

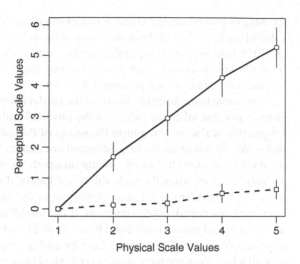

The results are returned in a list that includes, among other details, the final estimates of the parameters and the value of the negative log likelihood at the final evaluation.

```
> bg.opt$par
[1] 0.132 0.185 0.504 0.630 1.693 2.947 4.282 5.275
> bg.opt$value
[1] 238
```

The first four values give the estimates of perceived roughness for gloss scale values 2–5 and the last four for perceived roughness for roughness scale values 2–5. Recall that scale value 1 is constrained to be 0 for both scales. By specifying hessian = TRUE, the matrix of second partial derivatives of the log likelihood at the minimum is returned as a component as well. This can be used to obtain standard errors of the parameter estimates as follows:

```
> sqrt(diag(solve(bg.opt$hessian)))
[1] 0.153 0.152 0.154 0.156 0.243 0.284 0.320 0.337
```

The resulting coefficients are plotted in Fig. 8.4. We also include 95% confidence intervals based on the standard errors. We describe an alternate method to estimate confidence intervals based on a bootstrap approach in Sect. 8.5 below.

8.3.2 Fitting the Nonparametric Model with glm

The decision rule for the MLCM model presented in Sect. 8.3 is based on a linear combination of the internal responses. Thus, as we have repeatedly seen in the preceding chapters, we may conceptualize and fit the model as a GLM. As with the MLDS procedure, we create a model matrix of signed indicator variables, the value of each, $\{-1, 0, 1\}$, being determined by its coefficient in the decision variable (8.11). When there are two physical scales, as with the Ho et al. example, there will be four nonzero terms in each row, corresponding to the two physical scale values for each of the two stimuli presented in a trial.

First, some housekeeping. To create the model matrix, we write a helper function, make.wide that takes the indices of the physical scale values from the two stimuli for one physical scale and returns the subset of the columns of the model matrix for that scale. To constrain the first estimated coefficient for each scale to 0, we return the model matrix with the first column dropped. We need to include a condition to identify the cases when the scale values are identical on one scale, i.e., image pairs from Fig. 8.1 that are in the same column or in the same row. We then use lapply to create a list containing the columns of the model matrix for each physical scale. The final model matrix is obtained by a call to cbind using do.call to paste them together column-wise. By using lapply and do.call, the procedure works just as well when there are more than two physical dimensions along which the stimuli vary. Finally, we give names to the columns of the matrix for later identification.

```
> make.wide <- function(d){
+       nr <- nrow(d)
+       wts <- rep(c(1, -1), each = nr)
+       ix.mat <- matrix(0, ncol = max(d), nrow = nr)
+       ix.mat[matrix(c(rep(1:nr, 2), as.vector(unlist(d))),
+           ncol = 2)] <- wts
+       ix.mat <- t(apply(ix.mat, 1, function(x)
+           if (sum(x) == 0) x else
+               rep(0, max(d))))
+       ix.mat[, -1]
+ }
> bg.lst <- lapply(seq(2, length(BumpyGlossy) - 1, 2),
+       function(x, d){ make.wide(d[, x:(x+1)]) },
+       d = BumpyGlossy)
> X <- do.call(cbind, bg.lst)
> colnames(X) <- c(paste("G", 2:5, sep = ""),
+       paste("B", 2:5, sep = ""))
```

Compare the first six lines of the model matrix with the first six lines of the data frame shown in the last section.

```
> head(X)
     G2 G3 G4 G5 B2 B3 B4 B5
[1,]  0  1 -1  0  0 -1  1  0
[2,]  0  0  0  0  0  0  0  0
[3,]  0  1  0 -1 -1  0  1  0
[4,]  0  0  0  0  0  0 -1  0
[5,]  1 -1  0  0 -1  0  0  0
[6,]  0  0  0  0  0  0  0  0
```

Fig. 8.5 Estimated perceptual scales obtained by fitting a MLCM model using glm (*points*) and optim (*lines*) to the BumpyGlossy data set. The *circles* and *solid line* correspond to the roughness scale and the *squares* and *dashed line* to the glossy scale

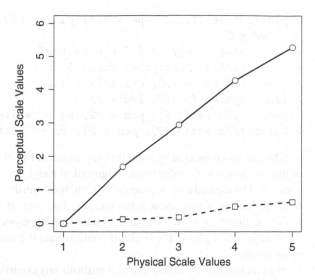

Row 1 of BumpyGlossy indicates that the first trial consisted of stimuli (3,4) and (4,3). Thus, the first row of the model matrix contains 1s in columns 2 (level 3 of dimension 1) and 7 (level 4 of dimension 2) and −1s in columns 3 (level 4 of dimension 1) and 6 (level 3 of dimension 2). Recall, that for the second trial of this data set, the images were equal on both physical scales. Thus, the 1 and −1 fall in the same column for each dimension and cancel out, leaving a row of zeros.

The model fit to the data is the standard GLM

$$E[P(R = 1)] = \boldsymbol{X}\beta, \tag{8.18}$$

where $\boldsymbol{X}$ is the matrix defined above and the vector β contains the coefficients of the estimated perceptual scale.

After joining the vector of responses to the index matrix in a data frame, we can fit the model using glm. The columns of the model matrix are treated as covariates and the coefficients will give the estimated values for the two perceptual scales. We eliminate the intercept term as we are only interested in the coefficients of the covariates.

```
> bg.df <- data.frame(resp = BumpyGlossy$Resp, X)
> bg.glm <- glm(resp ~ . -1, binomial(probit), bg.df)
```

A comparison of the coefficients obtained by glm and optim shows them to be identical (to numerical precision). This is illustrated in Fig. 8.5 in which the lines are connected to the coefficients estimated with optim and the points are the coefficients estimated with glm.

```
> opt.est <- c(0, bg.opt$par[1:4], 0, bg.opt$par[-(1:4)])
> glm.est <- c(0, bg.glm$coef[1:4], 0, bg.glm$coef[-(1:4)])
```

```
> plot(opt.est[1:5], type = "l", ylim = c(0, 6), lty = 2,
    lwd = 2,
+        xlab = "Physical Scale Values",
+        ylab = "Perceptual Scale Values",
+        cex.lab = 1.5, cex.axis = 1.3)
> lines(opt.est[6:10], lwd = 2)
> points(glm.est[1:5], pch = 22, bg = "white", cex = 1.5)
> points(glm.est[6:10], pch = 21, bg = "white", cex = 1.5)
```

The circles connected by solid line segments in Fig. 8.5 indicate the contribution of the roughness scale value to the judgment of roughness in the images. The squares connected by dashed lines segments indicate the contribution of the gloss scale value to the judgment of roughness in the images. The contribution of physical roughness to the judgments is 8–16 times greater than that of physical gloss. Nevertheless, the data suggest that gloss may begin to contaminate the roughness judgments at high gloss levels.

We can test this by fitting a model without any contribution of gloss. We use the update method to remove the terms involving gloss from the model to arrive at the independent model. The independent model is nested within the additive model, so we can compare them using the anova method.

```
> bg.Ind <- update(bg.glm, . ~ . - G2 - G3 - G4 - G5)
> anova(bg.Ind, bg.glm, test = "Chisq")
Analysis of Deviance Table

Model 1: resp ~ B2 + B3 + B4 + B5 - 1
Model 2: resp ~ (G2 + G3 + G4 + G5 + B2 + B3 + B4 + B5) - 1
  Resid. Df Resid. Dev Df Deviance Pr(>Chi)
1       971        500
2       967        476  4     23.6  9.5e-05 ***
---
Signif. codes:  0 `***' 0.001 `**' 0.01 `*' 0.05 `.' 0.1 ` ' 1
```

The results reject the independent model and indicate that the additive contribution of gloss to the roughness judgments, though small, is significant.

Ho et al. also consider a saturated model containing 24 terms, one for each pairing of the levels of roughness with each of the levels of gloss, except for the first level on each dimension. For model identifiability, this term is fixed at 0. We refer to this model as the saturated model to conform to typical terminology in model testing.

To fit this model, we must create a model matrix in which each combination of levels along the two dimensions has its own column, i.e., each stimulus will have its own column instead of having a column for each of its attributes. For each trial, the comparison will be between two stimuli (columns) and not 4 columns (2 attributes for each stimulus). The helper function make.wide.full indicated below takes as input the data frame of indices of the physical scale values and returns a 24-column matrix for the saturated model.

```
> make.wide.full <- function(d){
+    cn <- function(y, mx) (y[, 2] - 1) * mxd[2] + y[, 1]
+    nr <- nrow(d)
+    mxd <- c(max(d[, 1:2]), max(d[, 3:4]))
+    nc <- prod(mxd)
+    nms <- sapply(seq(1, ncol(d), 2), function(x)
+        substring(names(d)[x], 1, nchar((names(d)[x])) - 1))
+    fnm <- mapply(paste, nms,
+        list(seq_len(mxd[1]), seq_len(mxd[2])),
+        sep = "", SIMPLIFY = FALSE)
+      nms.f <- interaction(do.call(expand.grid, fnm),
+      sep = ":")
+    ix.mat <- matrix(0, ncol = nc, nrow = nr)
+    ix.mat[cbind(seq_len(nr), cn(d[, c(1, 3)], mxd))] <- 1
+    ix.mat[cbind(seq_len(nr), cn(d[, c(2, 4)], mxd))] <- -1
+    ix.mat <- t(apply(ix.mat, 1, function(x) if (sum(x) == 0)
+        x else rep(0, nc)))
+    colnames(ix.mat) <- levels(nms.f)
+    ix.mat[, -1]
+ }
```

For example, we use make.wide.full to create the matrix required to fit the saturated model to the BumpyGlossy data and add it to a data frame with the observer's responses as a column named resp. The data frame, bg.dff is a bit unwieldy, with 25 columns so we just compare its first row with the first row of BumpyGlossy

```
> Xf <- make.wide.full(BumpyGlossy[, -1])
> bg.dff <- data.frame(resp = BumpyGlossy[, 1], Xf)
> bg.dff[1, ]
  resp G2.B1 G3.B1 G4.B1 G5.B1 G1.B2 G2.B2 G3.B2 G4.B2 G5.B2
1    1     0     0     0     0     0     0     0     0     0
  G1.B3 G2.B3 G3.B3 G4.B3 G5.B3 G1.B4 G2.B4 G3.B4 G4.B4 G5.B4
1     0     0     0    -1     0     0     0     1     0     0
  G1.B5 G2.B5 G3.B5 G4.B5 G5.B5
1     0     0     0     0     0
> BumpyGlossy[1, ]
  Resp G1 G2 B1 B2
1    1  3  4  4  3
```

The two representations are consistent, indicating that the first stimulus is a comparison of levels G4.B3 with G3.B4 and the observer responded that the first stimulus appeared rougher than the second. The model can be fit with glm, but a simpler strategy is to use the update method and change the data argument.

```
> bg.saturated <- update(bg.glm, data = bg.dff)
> anova(bg.glm, bg.saturated, test = "Chisq")
```

```
Analysis of Deviance Table
```

```
Model 1: resp ~ (G2 + G3 + G4 + G5 + B2 + B3 + B4 + B5) - 1
Model 2: resp ~ (G2.B1 + G3.B1 + G4.B1 + G5.B1 + G1.B2 +
    G2.B2 + G3.B2 + G4.B2 + G5.B2 + G1.B3 + G2.B3 +
    G3.B3 + G4.B3 + G5.B3 + G1.B4 + G2.B4 + G3.B4 +
    G4.B4 + G5.B4 + G1.B5 + G2.B5 + G3.B5 + G4.B5 +
    G5.B5) - 1
  Resid. Df Resid. Dev Df Deviance P(>|Chi|)
1       967        476
2       951        462 16     14.9      0.53
```

The difference between the two models does not attain statistical significance so we retain the simpler additive model.

Conveniently, the operations to fit the three models illustrated above have been wrapped into a function mlcm in the R package **MLCM** [97]. Fitting the additive model to the data is as simple as

```
> library(MLCM)
> bg.add <- mlcm(BumpyGlossy)
```

which returns an object of class "mlcm" for which several method functions are available.

```
> methods(class = "mlcm")
 [1] anova.mlcm*    coef.mlcm*     fitted.mlcm*   formula.mlcm*
 [5] lines.mlcm*    logLik.mlcm*   plot.mlcm*     points.mlcm*
 [9] predict.mlcm* print.mlcm*    summary.mlcm* vcov.mlcm*

   Non-visible functions are asterisked
```

The other models are fit by specifying the model argument.

```
> bg.ind <- mlcm(BumpyGlossy, model = "ind", whichdim = 2)
> bg.saturated <- mlcm(BumpyGlossy, model = "full")
```

For the independent model, the argument whichdim = 2 specifies that the coefficients will be estimated with respect to the physical scale values of the second dimension in the data set, here roughness. The estimated coefficients of the perceptual scale are stored in a component of the object named pscale.

```
> bg.add$pscale
          G     B
Lev1 0.000 0.00
```

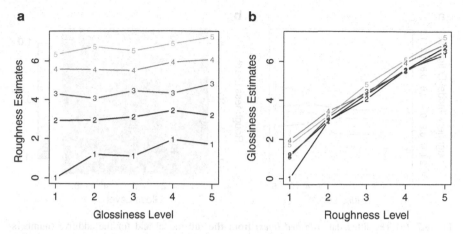

Fig. 8.6 (a) Roughness estimates as a function of the glossiness level for each level of roughness obtained by the `plot` method from the model object for the fit of the saturated model. (b) Glossiness estimates as a function of the roughness level for each level of glossiness obtained as in the adjacent figure except with the argument `transpose = TRUE`

```
Lev2 0.132 1.69
Lev3 0.185 2.95
Lev4 0.504 4.28
Lev5 0.630 5.28
```

For the (default) additive model object, the `plot` method will generate Fig. 8.5. We can also plot the saturated model from its returned model object, `bg.saturated`, but we have a choice of plotting the values as a function of the scale values on either dimension with the argument `transpose`. Figure 8.6a displays the estimates at fixed values of roughness while in Fig. 8.6b they are replotted for fixed levels of glossiness. The near independence of the effect of the gloss levels on the roughness judgments is revealed in both plots.

```
> opar <- par(mfrow = c(1,2), cex.lab = 1.5, cex.axis = 1.3)
> plot(bg.saturated, type = "b", lty = 1, lwd = 2,
+         col = c("black", paste("grey", 10 + 15 * (1:4))),
+         xlab = "Glossiness Level",
+         ylab = "Roughness Estimates")
> mtext("a", 3, adj = 0, cex = 2, line = 1)
> plot(bg.saturated, transpose = TRUE, type = "b", lty = 1,
    lwd = 2,
+         col = c("black", paste("grey", 10 + 15 * (1:4))),
+         xlab = "Roughness Level",
+         ylab = "Glossiness Estimates")
> mtext("b", 3, adj = 0, cex = 2, line = 1)
> par(opar)
```

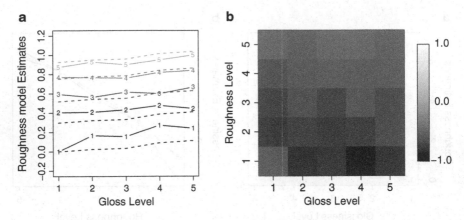

Fig. 8.7 (**a**) The fitted data (*dashed lines*) from the full model and for the additive (numbers connected by *solid line* segments). There is no obvious pattern in the differences between the models. (**b**) The differences for each level of gloss and roughness plotted as *grey* levels

```
> mr <- par("mar")
> opar <- par(mfrow = c(1, 2), mar = mr + c(0, 0, 0, 3))
> bg.add.pred <- with(bg.add, outer(pscale[, 1],
    pscale[, 2], "+"))
> plot(bg.saturated, standard.scale = TRUE,
+            type = "b", lty = 1, lwd = 2,
+            col = c("black", paste("grey", 10 + 15 * (1:4))),
+            ylim = c(-0.2, 1.2),
+            xlab = "Gloss Level",
+            ylab = "Roughness model Estimates")
> cf <- coef(lm(as.vector(bg.saturated$pscale/bg.saturated$
    pscale[5, 5]) ~
+            as.vector(bg.add.pred) + 0))
> matplot(cf * bg.add.pred, type = "l", lty = 2, lwd = 2,
+            col = c("black", paste("grey", 10 + 15 * (1:4))),
+            add = TRUE)
> bg.saturated.sc <- bg.saturated$pscale/bg.saturated$
    pscale[5, 5]
> bg.add.adj <- cf * bg.add.pred
> bg.res <- (bg.add.adj - bg.saturated.sc) + 0.5
> mtext("a", 3, adj = 0, cex = 2, line = 1)
> par(xpd = NA)
> gcl <- grey.colors(100, min(bg.res), max(bg.res))
> image(1:5, 1:5, bg.res,
+            col = gcl,
+            xlab = "Gloss Level", ylab = "Roughness Level"
+            )
```

```
> dgcl <- grey.colors(100, 0, 1)
> gradient.rect(5.8, 1, 6.2, 5, col = dgcl,
+            gradient = "y")
> ylab <- -1:1
> ypos <- seq(1, 5, len = 3)
> text(6.7, ypos, formatC(ylab, 1, format = "f"), adj = 1)
> mtext("b", 3, adj = 0, cex = 2, line = 1)
> par(opar)
```

We showed above that the saturated model does not fit the data significantly better than the additive model. We can compare the two fits more easily using the format of Fig. 8.6a. We compute the additive model predictions for each level using the outer function to add each estimated glossiness coefficient to each roughness level. Since the judgments are unchanged if the coefficients are multiplied by a scale value, we normalize the saturated model coefficients by specifying standard.scale = TRUE in the plot method. Then, we estimate the value by which to scale the additive model predictions using lm. Finally, we add the scaled additive model predictions to the plot using the matplot function as dashed lines. The comparison of these two models is displayed in Fig. 8.7a. The residuals between the two models are shown as a grey-level map in Fig. 8.7b.

8.4 Fitting Parametric Models

In the preceding sections, we fit an additive model that included an estimated scale value at each level along each dimension. No specific parametric form of the variation of the estimates with intensity along the dimensions was imposed upon the data.

The simple form of the variation of the scale estimates with physical glossiness and roughness suggests that we might predict the observer's judgments almost as well by specifying simple curves with only a few parameters for each dimension to describe the data. Examination of the data in Fig. 8.5 suggests that power functions might provide reasonable descriptions along both dimensions.

To illustrate how to fit this model, we continue the example drawn from Ho et al. but now assuming that we can capture the transformation from physical parameters to perceptual by power transformations ($y = x^\alpha$) of the corresponding physical dimensions. That is, perceived roughness is now controlled by the sum of a power transformation of physical roughness and any contamination of perceived roughness, by a second power transformation of physical gloss. Thus, for any stimulus whose physical roughness and gloss are represented by the ordered pair (φ^r, φ^g), the mean perceived roughness is

$$\psi^R(\varphi^r, \varphi^g) = a_r(\varphi^r)^{b_r} + a_g(\varphi^g)^{b_g}. \tag{8.19}$$

The two exponents b_r and b_g and the scale factors a_r and a_g are the free parameters that replace all of the parameters $\psi_i^{R;r}$ and $\psi_j^{R;g}$ in the nonparametric model. There is one more free parameter, the judgment uncertainty σ_R. However, the five parameters are redundant since the perceived roughness scale ψ^R is determined only up to a linear transformation with positive slope. For convenience, we arbitrarily set σ_R to be 1 and fit the four remaining parameters, b_r, b_g, a_r, and a_g. We have formulated the additive power function model so that the corresponding independent power function model is evidently nested within it with $a_g = 0$.

We cannot formulate the power function model as a GLM (it is an example of a generalized nonlinear model) so we must use optim (or another optimizer) in place of glm. The ML fit of the additive model (and the independent model within it) is essentially unchanged except that optim now searches on the restricted parameter set consisting of b_r, b_g, a_r, and a_g. The additive power function model is nested within the nonparametric additive model (Sect. B.5) and so we can also test the additive power model against the additive nonparametric model, in effect testing whether power functions are appropriate.

The **MLCM** package contains a formula method for mlcm that is a wrapper for optim. It allows the parametric response function to be specified as a formula object. It can fit either the additive or independent model and we illustrate its use to fit additive and independent power function models.

```
> bg.frm <- mlcm(~ p[1] * (x - 1)^p[2] + p[3] * (y - 1)^p[4],
+    p = c(0.1, 1, 1.5, 0.8),
+    data = BumpyGlossy)
```

Subtracting 1 from the physical scale indices fixes the first scale value at 0 for both scales. The two following arguments are, p, the initial estimates for the parameters, and data, the data frame. The fitted parameters are in the component par of the returned object, from which we see that glossiness makes a contribution of about 5% to perceived bumpiness (ratio of first and third coefficients).

```
> bg.frm$par
[1] 0.0976 1.3582 1.6971 0.8174
```

In addition, perceived roughness is a slightly compressive function of the stimulus level (fourth coefficient < 1). The standard errors returned by the summary method suggest that the exponent for glossiness does not differ significantly from unity, however.

The fitted curves are shown in Fig. 8.8 as solid lines. The plot is drawn with the following code fragment.

```
> xx <- seq(1, 5, len = 100)
> plot(bg.add, xlab = "Physical Scale Values",
+    ylab = "Perceptual Scale Values")
> lines(xx, predict(bg.frm, newdata = xx)[seq_along(xx)])
> lines(xx, predict(bg.frm, newdata = xx)[-seq_along(xx)])
```

Fig. 8.8 Estimated
perceptual scales obtained by
fitting a nonparametric
MLCM model (*plotted*
numbers) and a parametric
power-law model (*curves*)

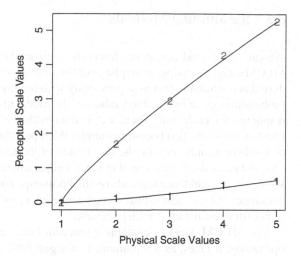

The curves are obtained by using the `predict` method applied to the model object.
The "newdata" argument allows predictions to be obtained at values intermediate
to the scale values used in the experiment. Atypically, this argument takes a vector
rather than a data frame. The method returns a prediction for each of the scales in
one long vector. The `seq_along` function is used to extract the values for the first
curve because the values in the variable xx are not integers. Negative indexing is
conveniently used to extract the values from the second curve.

The parametric curves follow the nonparametric estimates very closely. We can
evaluate this statistically using the `anova` method which performs a likelihood ratio
test. As noted above, such a test is permissible because the parametric model can be
considered to be nested within the nonparametric additive model. The high p-value
shows that the 4-parameter power functions predict the observer's responses as well
as the 8-parameter model.

```
> anova(bg.add, bg.frm)
Analysis of Deviance Test

Model 1 :   Resp ~ (G12 + G13 + G14 + G15 + B12 + B13 +
        B14 + B15) -      1
Model 2 :   ~p[1] * (x - 1)^p[2] + p[3] * (y - 1)^p[4]
  Df Deviance      p
1  4     1.97 0.741
```

Superficially, the resulting combination of nonparametric and parametric fits
resembles the fits of psychometric functions to data in Chap. 4 but, of course, the
nonparametric coefficients plotted as points are fitted summaries of data, not the
data themselves in raw form.

8.5   Resampling Methods

We can readily add confidence intervals to either the nonparametric or parametric MLCM estimates using resampling (or "bootstrap") methods. We follow a procedure that is similar to that used previously with binomial GLMs. First, we obtain the probabilities given by the fitted values of the model to predict new random binomial responses for each trial. Then, we fit data with the new responses to obtain the coefficients for the first bootstrap sample. We repeat this process many times (10,000 as a rule of thumb) and use the covariance of the resulting parameter estimates as an estimate of the covariance of the original parameters, based on the original data. We can base confidence intervals on the bootstrap standard error estimated for each parameter. We can also estimate confidence intervals from the histogram of the set of bootstrap estimates of each parameter.

The **MLCM** package contains a function boot.mlcm that performs the above operations. It takes, as a minimum, two arguments. The first is an object of class "mlcm," typically returned by fitting the results of an experiment with the mlcm function. The second argument indicates the number of bootstrap repetitions to perform. For example, to obtain bootstrap estimates of the coefficients for the additive model fit, one would execute the following function.

```
> bg.add.boot <- boot.mlcm(bg.add, nsim = 10000)
```

The object returned is a list containing four components. The first is a matrix named boot.samp. Each column contains the coefficients from a bootstrap sample. The second component is a vector of the means of the coefficients, i.e., the mean across columns for each row of boot.samp. The third is the estimate of the bootstrap standard error for each coefficient, obtained by the standard deviation across columns boot.samp. The last component is simply the number of bootstrap simulations that were performed.

We will not demonstrate this function here as it works in exactly the same manner as the equivalent functions seen in earlier sections. We leave this as an exercise (Problem 8.2).

8.5.1   The Choice of Stimuli

Ho et al. asked each observer to judge each possible comparison of stimuli three times [80]. This exhaustive approach was possible since there were only two dimensions (physical roughness and physical glossiness). With five levels on each dimension, there is a total of 25 stimuli, as illustrated in Fig. 8.1, thus yielding a total of $25 \times 24/2 = 300$ different comparisons. If we were to attempt to carry out

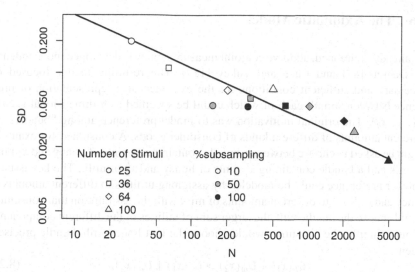

Fig. 8.9 Mean standard deviations of perceptual estimates as a function of number of trials. The *plotted line* (in log–log coordinates) assumes that the confidence intervals decrease with the square root of sample size

an experiment with three scales (e.g., gloss, roughness and albedo) each with five levels, then the total number of stimuli would be 125 and the number of trials needed to compare each pair of possible stimuli would be 7,750, which would be rather impractical. Similarly, if we were to restrict attention to only two dimensions but increase the number of levels sampled on each dimension to 10, the total number of trials required to compare each of 100 stimuli to all others would be 4,950, still prohibitive.

An alternative to exhaustive sampling is to sample as many trials as possible, chosen at random from all possible comparisons. To evaluate this approach, we simulated observers whose parameters were similar to those estimated in the first experiment of Ho et al. [80] but varying the number of simulated trials. We simulated stimulus sets with 25, 36, 64, and 100 stimuli, i.e., 5, 6, 8, and 10 levels per dimension. For each stimulus set, we ran simulations for random subsamples of 10, 50, and 100% of the stimulus set. Each simulation was repeated 200 times. We calculated the standard deviation of the coefficient estimates for each condition. In Fig. 8.9 we plot the average standard deviation of estimated parameters as a function of the total number of trials, N, on log–log coordinates. The estimates fall roughly on a straight line with slope -0.5, indicating that SDs decrease as the $\sqrt{n}$ with n the number of comparisons, even when the sampling is not exhaustive. This preliminary result suggests that MLCM can be carried out successfully with a large number of stimuli where exhaustive comparisons of all possible stimuli are prohibitive.

8.6   The Axiomatic Model

As already mentioned, additive conjoint measurement was developed independently
by Debreu [47] and Luce and Tukey [117]. The resulting theory focused on
necessary and sufficient conditions for the existence of a representation of preference between stimuli each of which could be specified as n-dimensional vectors
$(x_1,\ldots,x_n)$. The original motivation was to model preference among "bundles" of
different amounts of different kinds of consumer goods. A consumer, for example,
might be asked to choose between a bundle containing x_1 pints of honey and x_2 pints
of milk and a bundle containing x_1' pints of honey and x_2' of milk. The key issue is
whether preference could be modeled by assigning utilities to different amounts of
honey and utilities to different amounts of milk with the assumption that consumers
would choose the bundle with the larger sum of utilities of the different components.
In our example, the consumer would choose the first honey-milk bundle precisely
when

$$U_h(x_1) + U_m(x_2) > U_h(x_1') + U_m(x_2'). \qquad (8.20)$$

where $U_h(x)$ and $U_m(x)$ are the subjective utility functions for different amounts of
honey and milk, respectively.

   The key theoretical insight into the development of additive conjoint measurement is that, if such a representation is possible, then the consumer's preference
also constrains the subjective utility functions so that they are determined up to a
common linear transformation with positive slope. Remarkably, if we only asked
the consumer to order amounts of honey or only order amounts of milk, we could
not recover either subjective utility function. In the preceding chapter, we showed
how difference scaling and MLDS could be used to recover such functions by
asking observers to judge the order of pairs of pairs of stimuli, effectively judging
the relative magnitude of differences. In the additive conjoint model, with two or
more dimensions, we can recover these functions with only judgments of preference
between stimuli.

   We can represent the problem graphically in terms of the indifference contours
of Fig. 8.2. Given the plots of all possible indifference contours, are there transformations of the two axes $f_1(x_2)$ and $f_2(x_2)$ that turn all of the indifference contours
into lines with slope -1? If so, then these transformations are unique except for an
overall linear transformation with positive slope. The equations of these lines are, in
our example, just

$$f_1(x_1) + f_2(x_2) = C \qquad (8.21)$$

for any choice of the constant C and two transformations f_1 and f_2. That is, we can
magnify or displace the transformed contours while they all remain lines with slope
-1 but that is the limit of freedom we have in the choice of scale.

   The key issue in the axiomatic approach to conjoint measurement is to specify
conditions on preference that permit or disallow a conjoint measurement representation of preference and conditions that guarantee that the presentation is unique up

to a linear transformation with positive slope [47, 117]. Some of these conditions are mathematical (e.g., that the f_i are continuous or increasing). Others apply to the physical conditions of application of the theory, for example, that there is an amount of honey corresponding to any possible number. The details of available proofs of existence and uniqueness of a conjoint measurement representation are beyond our scope. The interested reader can consult Krantz et al. [104] for details and further references.

A typical example of a necessary condition is that, if a bundle (x_1, x_2) is preferred to (x'_1, x_2) then, for any choice of x'_2, the bundle (x_1, x'_2) is preferred to (x'_1, x'_2). This condition is easily derived by writing out the conditions for preference in (8.20)

$$U_h(x_1) + U_m(x_2) > U_h(x'_1) + U_m(x_2) \tag{8.22}$$

then we can subtract $U_m(x_2)$ from both sides and add $U_m(x'_2)$ to both sides to get

$$U_h(x_1) + U_m(x'_2) > U_h(x'_1) + U_m(x'_2). \tag{8.23}$$

We emphasize, however, that previous work for the most part has considered performance by observers whose judgments never vary. That is, observers, having expressed a preference for one stimulus over a second, cannot later reverse their preference (except in the very narrow case that the two stimuli fall exactly on the same indifference contour). As discussed in the preceding chapter on MLDS, human observers typically have graded responses to stimuli and at best we can expect to estimate the probabilities of preferences where these preferences are never precisely equal to 0 or to 1. The models developed for MLCM in this chapter and MLDS in the previous use a simple signal detection model of preference that allows for stochastic reversals. The resulting theory differs from the axiomatic theory in one crucial respect, a consequence of assuming that judgment is contaminated by Gaussian error (or, equivalently, by error drawn from any distribution with support on the entire real line). In any experiment, *any* pattern of response is consistent with any conjoint measurement representation or the absence of any representation. Even if the probability of choosing A over B is 0.999, there is a probability 0.001 of choosing B over A. Any pattern of responses to any series of trials is possible and we seek the explanation with greatest likelihood, not the only possible representation.

The shift to a stochastic model is in a sense trivial and necessary if we attempt to give a realistic account of human performance. One intriguing consequence, however, is that violations of necessary conditions for a representation as a difference scale or an additive conjoint representation are not only possible but, under certain circumstances, probable. See Sect. 7.6.2.1 for example where we expect violation as a necessary condition. Indeed, the persistent absence of violations predicted by the fitted model of human performance is evidence against the underlying model.

Exercises

8.1. Repeat the analyses of this chapter on the GlossyBumpy data set from the **MPDiR** package. Consider how the contribution of physical roughness contributes to perceived glossiness and how this compares to the contribution of physical glossiness to perceived roughness that we analyzed in the BumpyGlossy data set.

8.2. Use the function boot.mlcm in the **MLCM** package to obtain bootstrap standard errors for the estimated coefficients from the MLCM fit to the BumpyGlossy and GlossyBumpy data sets for an additive model. Set nsim = 10000 which will take a bit of time to complete. Compare the standard errors estimated from the bootstrap approach with the Wald estimates obtained by apply the summary method to the obj component of the "mlcm" model objects.

8.3. Evaluate the goodness of fit of the additive MLCM models to the same two data sets used above with the function binom.diagnostics from the **MLCM** package. Set nsim = 10000 which, as in the above example, will take a while to run.

8.4. We noted above that Ho et al. included trials where observers selected between stimuli that had the same physical roughness and gloss values. Since stimuli were randomly generated on each trial, the actual stimuli presented were almost certainly different despite sharing the same expected roughness and gloss. The observers' judgments on these stimuli were not used in fitting perceptual scales of roughness and glossiness. They could, however, be used to test whether observers favored the first or second stimulus presented on each trial. Using the data frame BumpyGlossy check whether observers tended to favor the first or second interval in responding. Repeat for the data frame GlossyBumpy. Yeshurun et al. (2008) [202] analyzed previous data and report an experiment documenting biases in temporal forced-choice.

Chapter 9
Mixed-Effects Models

With one exception the models that we have treated before this chapter contain a single source of variability. In the linear models with Gaussian error, the variance of the population was estimated from the residuals. In the GLM models used to describe the judgments of observers, the variability of observer's choices is given by that of the binomial distribution and depends on the estimated probability of a particular choice, $\hat{p}$, and the number of trials, n.

The only model considered so far that included multiple sources of variance was the linear mixed-effects model briefly presented in Chap. 2. In this chapter, we introduce the extension of the GLM to such models. These are called generalized linear mixed-effects models or GLMMs. The need for these in psychophysics arises most frequently when data are collected from a sample of different observers, and we want to generalize the results beyond the specific observers tested to the population of observers [49, 154]. A less frequent, but equally important, use would be to model the variability due to the choice of stimuli used in an experiment, when these stimuli can be considered a subset from a larger population. Examples could include natural images, faces, words, odors, melodies, etc. We cannot possibly test all the possible face stimuli in an experiment and we instead use a limited, but hopefully, random sample from the population. In order to generalize the results to the population of such stimuli, we should treat the stimulus items as random effects [7, 35, 154]. One of the advantages of mixed-effects models is that we reduce the number of parameters to model by estimating the variance of the effects due to the sample rather than the effect of each member of the sample.

As a second example, we return to the case of collecting data from observers to estimate their probability of making a particular choice, for example, that a particular stimulus was louder, of higher pitch, etc. Good estimates of the experimental parameters require many trials: as n increases, the variance decreases. Typically, the trials are distributed across several sessions, in order to avoid changes in p due to fatigue. If we simply combine trials across sessions, however, we act as if p were constant. We assume that the variability observed across sessions can be attributed simply to the binomial variance.

K. Knoblauch and L.T. Maloney, *Modeling Psychophysical Data in R*, Use R! 32,   257
DOI 10.1007/978-1-4614-4475-6_9,
© Springer Science+Business Media New York 2012

An alternative is that p can change across sessions. An interesting model of two possible sources of such variability can be constructed by considering the linear predictor of the GLM as the decision variable in a signal detection model. As we previously saw, the probability of a particular response is related to the decision variable through the psychometric function. In terms of the GLM, we would say that the expected value of the response is related to the linear predictor through a link function.

$$\eta(\mathsf{E}[Y]) = \eta(\mathsf{E}[Pr(Y = \text{``Correct''})]) = \beta_0 + \beta_1 X, \qquad (9.1)$$

where Y, the observer's response, is distributed as $Binomial(p;N)$ and X is the stimulus intensity level. We demonstrated in Chap. 3 that the two parameters, β_0 and β_1, were related to the criterion and sensitivity of the observer, respectively. Suppose that each of these may fluctuate over time, specifically across sessions, and that each follows a Gaussian distribution. Then, we can rewrite the model as

$$\eta(Pr(Y = \text{``Correct''} \,|\, b_0, b_1)) = (\beta_0 + b_0) + (\beta_1 + b_1)X, \qquad (9.2)$$

where $b_0 \sim N(0, \sigma_0^2)$ and $b_1 \sim N(0, \sigma_1^2)$. This is an example of a GLMM because the model contains both fixed and random effects to be estimated. Note that the response is specified as conditional on the random effects. A matrix form of this equation is obtained by separating the fixed and random components and is typically notated as

$$\eta(Y \,|\, b) = X\beta + Zb, \qquad (9.3)$$

where X is the design matrix for the fixed effects components and Z for the random effects, and β is a vector of fixed effects coefficients and b for the random effects. It is important to stress that it is not the random coefficients of the vector b that are estimated in fitting the model but their variances. This model then includes three sources of variance, the variance associated with the binomial response and the variances associated with the criterion and the sensitivity of the observer.

The logic that we are applying to repetitions of the experiment across sessions could equally well be applied to repetitions across observers. Different observers are likely to use different criteria in making their judgments, and they may also have different sensitivities to the stimulus. Such random variations across observers can be modeled, as well, by the introduction of random effects in the model. In this way, the fixed effects estimated can be extrapolated from the specific observers tested to the population from which the observers were sampled. Note that this is a different strategy than that typically employed in modeling psychometric functions. That strategy is to fit individual curves to each observer's data and then perform statistical analyses on the extracted parameters. *Linear* mixed-effects (LMM) models could be applied to the extracted parameters and in some cases, for example, the large data sets obtained in cerebral imaging experiments, this is the most practical path to take [107]. When feasible, however, the approach through GLMM is more comprehensive because it permits modeling all of the data and all of the parameters in one step rather than several.

The implementation of GLMMs is much more challenging than LMM or GLM models because the calculation of marginal effects involves an integral that does not have a closed form solution; it must be reevaluated repeatedly during any optimization process. In general, fitting such models is harder than fitting GLMs and LMMs, also, because the properties of such models are less well understood than for simpler types of models. Advances in computing speed and optimal coding, such as with sparse matrix representations, have permitted significant progress in the application of such models even if significant progress remains to be made. R includes several packages that are useful for analyzing mixed-effects models. The recommended package **nlme** has extensive methods for analyzing linear and nonlinear mixed-effects models, but does not contain any functions for GLMMs.

There are several approaches to performing GLMMs and packages to implement them. To mention just a few, there are **lme4** [9], **MCMCglmm** [74], **glmmML** [23], and **glmmAK** [101] among others, and the reader is encouraged to explore them when in need of solving a particular modeling problem. Here, we demonstrate the **lme4.0** package [10] which includes the function `glmer` for modeling GLMMs.

9.1 Simple Detection

9.1.1 Random Criterion Effects

Before attempting to fit psychometric functions, we treat the simpler cases of detection and rating scale data in a signal detection paradigm. Recall that the main stimulus variable in these models is a two-level factor indicating the presence/absence of the signal. However, other explanatory variables related to the experimental conditions may enter the model as well. We begin with a simple experiment that involves detection of a somewhat unusual signal.

In Chap. 3, we examined the Faces data set in which observers rated the relatedness of pairs of images of children that represented either siblings or not [122]. In the same article, the experimenters performed a study on a separate group of 32 observers in which they simply classified the images as siblings or not. Thus, the task of the observers was simply to detect whether the pairs of images were of related children or not. The data from this experiment are in the data set Faces2 in our package **MPDiR**.

The data set has 960 observations with four factor components

```
> library(MPDiR)
> str(Faces2)
'data.frame':   960 obs. of  4 variables:
 $ Resp : Factor w/ 2 levels "0","1": 2 2 2 1 2 2 2 2 2 2 ..
 $ Stim : Factor w/ 2 levels "A","P": 2 2 2 2 2 2 2 2 2 2 ..
```

```
$ Obs  : Factor w/ 32 levels "S1","S2","S3",..: 1 1 1 1 1..
$ Image: Factor w/ 30 levels "Im1","Im2","Im3",..: 1 2 3 ..
```

indicating the responses of the observer as a 2-level factor with (0)1 for (un)related siblings, a 2-level factor for whether or not the signal (siblings) was absent or present (A/P), a 32-level factor identifying the observer and a 30-level factor identifying the particular image pair. The data frame is arranged so that the first 15 images for each observer were siblings and the second 15 were not. In the actual study, the order in which the images were presented was randomized.

The four different combinations of the levels of the factors Resp and Stim correspond to the different type of classifications in a Yes-No signal detection experiment. As we saw in Chap. 3, the criterion and sensitivity can be calculated from this information using glm. The number of responses per observer is rather small for an accurate calculation of d', and it is often the case that responses are combined across individuals for an overall measure of sensitivity.

```
> coef(summary(glm(Resp ~ Stim, binomial(probit), Faces2)))
            Estimate Std. Error z value Pr(>|z|)
(Intercept)   -0.357     0.0586    -6.1 1.04e-09
StimP          0.999     0.0851    11.7 7.15e-32
```

The value of d' is given by the StimP term and is about equal to that obtained independently from the rating scale data in the Faces data set.

Combining responses across observers, however, confounds potential observer differences in criterion and sensitivity. A second independent source of variability could be attributed to varying difficulty of the pairs of images, some being harder to identify than others. The grouping factors for the two effects, Obs and Image, are completely crossed, i.e., each observer was tested on every image, leading to what are called crossed random effects. This situation contrasts with that of nested random effects, an example of which was given in Sect. 2.3, using a linear mixed-effects model. We can incorporate such multiple sources of variability into the model with mixed-effects models.

In the following code fragment, we load the **lme4.0** package and fit a detection model similar to the one above, but we also include terms for random intercepts for both Obs and Image grouping factors. No specialization of the formula object is needed to indicate that these are crossed random effects. The software can identify this during the construction of the design matrices. Notice that the positions of the formal arguments data and family are reversed in glmer with respect to glm.

```
> library(lme4.0)
> F2.glmm1 <- glmer(Resp ~ Stim + (1 | Obs) + (1 | Image),
+   data = Faces2, family = binomial(probit))
> F2.glmm1


Generalized linear mixed model fit by the Laplace approximation
Formula: Resp ~ Stim + (1 | Obs) + (1 | Image)
```

```
   Data: Faces2
   AIC  BIC logLik deviance
   1016 1035   -504     1008
Random effects:
 Groups Name          Variance Std.Dev.
 Obs    (Intercept) 0.0136   0.117
 Image  (Intercept) 0.5573   0.746
Number of obs: 960, groups: Obs, 32; Image, 30

Fixed effects:
            Estimate Std. Error z value Pr(>|z|)
(Intercept)   -0.479     0.206   -2.33     0.02 *
StimP          1.260     0.289    4.35  1.3e-05 ***
---

Correlation of Fixed Effects:
       (Intr)
StimP -0.703
```

Recall that the intercept in this model corresponds to a measure of the observer's criterion, $-\Phi^{-1}(P(FA))$. The standard deviations indicate that variability associated with the random factor Image on the linear predictor scale is over six times greater than that for the random factor Obs. When the variance attributed to these two random factors is modeled, the sensitivity to sibling differences, given by the term StimP increases by 26%.

There is a fairly high correlation between the two fixed effect parameters, (Intercept) and StimP, the criterion and sensitivity, respectively. It is possible to reparameterize the model so that the terms are orthogonal. Recall that the criterion is sometimes defined as

$$c = -0.5(\Phi^{-1}(P(H)) + \Phi^{-1}(P(FA))) \qquad (9.4)$$

instead of, as here, $-\Phi^{-1}(P(FA))$. This parameterization defines a criterion of zero as occurring at the abscissa value of the intersection of the Noise and Signal + Noise distributions. In particular, the coefficients of c are orthogonal to those for the calculation of d'

$$d' = \Phi^{-1}(P(H)) - \Phi^{-1}(P(FA)), \qquad (9.5)$$

In order to fit this parameterization, we must create the model matrix for the intercept and factor Stim to replace the one generated by the default "treatment contrasts."

The procedure, defined in [178], is to define the desired contrasts as rows of a matrix. Then, the corresponding values for the model matrix will be found in the columns of the inverse (or pseudo-inverse, using the ginv function from **MASS**, if the contrast matrix is not square). In our case, we would like the contrasts to correspond to the rows of the matrix

```
> ( m <- matrix(c(-0.5, -0.5,
+                   -1, 1), 2, 2, byrow = TRUE) )
      [,1] [,2]
[1,] -0.5 -0.5
[2,] -1.0  1.0
```

with inverse given by

```
> solve(m)
      [,1] [,2]
[1,]    -1 -0.5
[2,]    -1  0.5
```

Thus, we assign -1 to the first column and the presence and absence of the signal are coded by 0.5 and -0.5, respectively. We create the fixed effects model matrix with the following code

```
> Stm <- ifelse(Faces2$Stim == "P", 0.5, -0.5)
> Stm <- cbind(-1, Stm)
> colnames(Stm) <- c("_Criterion", "_d'")
```

Then, we substitute this matrix for the fixed effect Stim in the formula object specified in the call to glmer and remove the, now redundant, intercept from the model.

```
>  glmer(Resp ~ Stm + (1 | Obs) + (1 | Image) - 1,
+    Faces2,binomial(probit))


Generalized linear mixed model fit by the Laplace approximation
Formula: Resp ~ Stm + (1 | Obs) + (1 | Image) - 1
   Data: Faces2
  AIC  BIC logLik deviance
 1016 1035   -504     1008
Random effects:
 Groups Name         Variance Std.Dev.
 Obs    (Intercept) 0.0136   0.117
 Image  (Intercept) 0.5573   0.746
Number of obs: 960, groups: Obs, 32; Image, 30

Fixed effects:
              Estimate Std. Error z value Pr(>|z|)
Stm_Criterion   -0.151      0.146   -1.03      0.3
Stm_d'           1.260      0.289    4.35  1.3e-05 ***
---


Correlation of Fixed Effects:
        Stm_Cr
Stm_d' -0.001
```

The correlation between the fixed effects is now eliminated and all other parameter estimates are unchanged except for the intercept, now specified with respect to the intersection of the two underlying response distributions.

9.1.2 Random Sensitivity Effects

If we want to model the variability of the sensitivity associated with these grouping factors, then we can use the factor Stim on the left side of a random term as if we were modeling a random slope. For example,

```
> (Stim | Obs)
```

for including a random effect of criterion and sensitivity with respect to Obs. Stim is a two-level factor, so each such term will add two columns to the random effects design matrix. It is not clear in advance, however, whether the data would be modeled best by including random effects for either or both grouping factors. Therefore, we consider a series of models

```
> F2.glmm2 <- glmer(Resp ~ Stim + (Stim | Obs) + (1 | Image),
+    Faces2, binomial(probit))
> F2.glmm3 <- glmer(Resp ~ Stim + (1 | Obs) + (Stim | Image),
+    Faces2, binomial(probit))
> F2.glmm4 <- glmer(Resp ~ Stim + (Stim | Obs) + (Stim | Image),
+    Faces2, binomial(probit))
```

and perform likelihood ratio tests on the nested sequences, two different ones in this case.

```
> anova(F2.glmm1, F2.glmm2, F2.glmm4)
Data: Faces2
Models:
F2.glmm1: Resp ~ Stim + (1 | Obs) + (1 | Image)
F2.glmm2: Resp ~ Stim + (Stim | Obs) + (1 | Image)
F2.glmm4: Resp ~ Stim + (Stim | Obs) + (Stim | Image)
          Df  AIC  BIC logLik Chisq Chi Df Pr(>Chisq)
F2.glmm1   4 1016 1035   -504
F2.glmm2   6 1016 1045   -502  4.13      2       0.13
F2.glmm4   8 1019 1058   -501  0.91      2       0.63
> anova(F2.glmm1, F2.glmm3, F2.glmm4)
Data: Faces2
Models:
F2.glmm1: Resp ~ Stim + (1 | Obs) + (1 | Image)
F2.glmm3: Resp ~ Stim + (1 | Obs) + (Stim | Image)
F2.glmm4: Resp ~ Stim + (Stim | Obs) + (Stim | Image)
          Df  AIC  BIC logLik Chisq Chi Df Pr(>Chisq)
```

```
F2.glmm1  4 1016 1035    -504
F2.glmm3  6 1019 1048    -503  1.30      2      0.52
F2.glmm4  8 1019 1058    -501  3.74      2      0.15
```

The results for both nested sequences suggest that a model without random sensitivity effects describes the data reasonably well. However, these tests, based on a χ^2 distribution, are only approximate and the p-values can be conservative (i.e., too high) [11, 141]. This is because variance components for the smaller models are forced to 0; they are at the boundary of the parameter space. The χ^2-based p-value can be up to twice as high [11].

Taking this on board, F2.glmm2, the simplest model incorporating a random effect for sensitivity across observers, could approach significance ($p = 0.065$). The AIC values give no evidence to distinguish the models though the more stringent BIC supports the simpler model.

9.2 Rating Scale Data

Random factors can also be incorporated into models of rating scale data. Recall that such models estimate multiple intercepts associated with cut-points of cumulative proportions of the ratings. The intercepts or cut-points correspond to multiple criteria in the decision space employed by the observer. To examine such models, we return to the Faces data set presented in Chap. 3 and modeled there with the clm function from the **ordinal** package. As mentioned in the preceding section, the data were obtained with the same stimulus set of 30 images, 15 sibling and 15 non-sibling pairs, as used with the Faces2 data set, but with a different set of 32 observers and a different task, giving a confidence rating that the image pairs were siblings on an 11-point scale (0–10).

As with the Faces2 data set, we would like to include the possibility of random observer and item effects when modeling performance on this task. The function clmm in the **ordinal** package provides for fitting ordinal response data with the possibility of incorporating random effects. For example, to model a random effect of observer, we would load the package in memory and specify the model as

```
> library(ordinal)
> clmm(factor(SimRating) ~ sibs + (1 | Obs), data = Faces,
+    link = "probit", Hess = TRUE)
Cumulative Link Mixed Model fitted with the Laplace
    approximation

formula: factor(SimRating) ~ sibs + (1 | Obs)
data:    Faces

 link   threshold nobs logLik   AIC     niter     max.grad
 probit flexible  960  -2124.22 4272.44 30(2628)  6.60e-05
```

```
Random effects:
      Var Std.Dev
Obs 0.163   0.403
Number of groups:  Obs 32

Coefficients:
sibs1
 1.16

Thresholds:
    0|1      1|2      2|3      3|4      4|5      5|6      6|7
-0.7130 -0.3842 -0.0624  0.2420  0.4945  0.7425  0.9779
    7|8      8|9     9|10
 1.2970  1.7333  2.2170
```

The response variable must be specified as a factor. We model it as a function of the 2-level stimulus factor sibs and specify the grouping factor for the random effect with a separate term. Notice that we use the same notation for random factors as that used in the **lme4.0** package. The argument Hess = TRUE specifies that the Hessian should be computed, which is necessary for the calculation of standard errors with the summary and vcov methods.

The variance and standard deviation of the random factor of Obs are the first items displayed after the call, followed by the sensitivity estimate (sibs1) and finally the estimates of the ten criterion points (Threshold estimates) for an average observer. The random factor is associated with the intercepts and corresponds to a rigid shift of the cut-points as a random effect for each observer. The function does not currently permit estimating random effects associated with the sensitivity. This package is under active development, however, and such functionality will be eventually incorporated. We could similarly run this model with a random intercept associated with the factor Image or both at once.

The glmer function permits random effects associated with the sensitivity but does not currently interpret ordinal response variables. Recall that the cumulative ordinal response model conceptualizes the ordinal levels as a set of binary responses on the cumulative ordinal response.

$$g(P(Y \leq k \,|\, x)) = \alpha - X\beta, \tag{9.6}$$

where g is the link function, the cumulative responses, Y, are conditional on the explanatory variables, x, and the linear predictor is subtracted from a vector, α parameterizing one intercept per ordinal cut-point. We could reformat the ordinal responses to a binary format for cumulative responses, for example, for each ordinal level, a binary response is assigned for that level or lower. Then, in the model matrix an intercept column assigned to each of these binary responses with the negation of the model matrix for the linear predictor terms adjoined would be suitable for fitting such a model with glmer.

We illustrate the mechanics of this process for the Faces data set in the following code fragment.

```
> SimRating <- ordered(Faces$SimRating)
> Cuts <- seq(0, length(levels(SimRating)) - 2)
> cumRat <- as.vector(sapply(Cuts,
+    function(x) Faces$SimRating <= x))
> fRat <- gl(10, nrow(Faces), nrow(Faces) * length(Cuts))
> X <- model.matrix(~ fRat - 1)
> X <- cbind(X, -(Faces$sibs == "1"))
> ColNames <- c("Resp", sapply(Cuts, function(x)
+    paste(x, '|', x + 1, sep = "")), "sibs")
> Rat.df <- data.frame(Resp = cumRat, X = X)
> names(Rat.df) <- ColNames
> Obs <- rep(Faces$Obs, length(Cuts))
> Image <- rep(Faces$Image, length(Cuts))
> g1 <- glmer(Resp ~ . - 1 + (sibs - 1 | Obs),
+    Rat.df, binomial("probit"))
```

We have wrapped similar code into a function polmer in our **MPDiR** package that calls glmer so that we can fit ordinal response models with random sensitivity effects. If no random factors are specified in the formula object, then glm is called instead. To fit a model, a formula object similar in format to those used in the **lme4.0** package is used. The response should be a numeric or integer vector with *n* distinct values for the ratings. It will be coerced to an ordered factor as in the code fragment above. There is no implicit intercept in the model and none should be added. In addition, the intercept should be removed in the specification of all random factors. Finally, the random factors should be specified with respect to the second level of the factor as shown below. The levels of the factor sibs are 0/1 so sibs1 is the second level.

```
> Faces.glmm <- polmer(SimRating ~ sibs + (sibs1 - 1 | Obs) +
+    (sibs1 - 1| Image), data = Faces, lnk = "probit")
> print(Faces.glmm, cor = FALSE)


Generalized linear mixed model fit by the Laplace approximation
Formula: cumRat ~ . + (sibs1 - 1 | Obs) +
    (sibs1 - 1 | Image) - 1
   Data: p.df
  AIC  BIC  logLik deviance
 8443 8536  -4209     8417
Random effects:
 Groups Name  Variance Std.Dev.
 Obs    sibs1 0.305    0.553
 Image  sibs1 0.455    0.675
Number of obs: 9600, groups: Obs, 32; Image, 30
```

```
Fixed effects:
        Estimate Std. Error z value Pr(>|z|)
`0|1`   -0.7428     0.0551  -13.48  < 2e-16 ***
`1|2`   -0.4259     0.0506   -8.41  < 2e-16 ***
`2|3`   -0.1151     0.0481   -2.39    0.017 *
`3|4`    0.1969     0.0471    4.18  2.9e-05 ***
`4|5`    0.4635     0.0472    9.81  < 2e-16 ***
`5|6`    0.7281     0.0483   15.07  < 2e-16 ***
`6|7`    0.9862     0.0503   19.62  < 2e-16 ***
`7|8`    1.3431     0.0544   24.69  < 2e-16 ***
`8|9`    1.8473     0.0633   29.18  < 2e-16 ***
`9|10`   2.4279     0.0784   30.97  < 2e-16 ***
sibs1    1.2576     0.2027    6.20  5.5e-10 ***
---
```

The function returns an object of class "mer" (or "glm" when no random factors have been specified) so all of the methods for this class of object are available. We assign the object to a variable Faces.glmm and print it out explicitly so that we can suppress the large correlation matrix for the fixed-effects in order to simplify the display. As observed with the Faces2 data set, the variability associated with Image is larger than that associated with Obs though the two estimates are much closer in this data set. Including these random factors also increases the estimated sensitivity of observers to the signal, d' increasing by about 26%. The predicted ROC curve is obtained by taking the cut-points as normal quantiles and plotting the estimated probabilities or cdfs with means set to 0 and to sibs1, respectively, against each other using pnorm.

9.3 Psychometric Functions: Random Intercepts and Slopes

When the main explanatory variable in (9.2) is a covariate representing a stimulus dimension, then we are fitting a psychometric function. On the scale of the linear predictor, we can estimate random effects associated with either the intercept term as above and/or for the slope of the covariate. To appreciate the influence of such random explanatory variables on the structure of the obtained data and the fitted models, we begin this section with a simulation of an observer who is subject to cross-session variation of his criterion and sensitivity.

To be concrete, we suppose that the observer is tested on 7 intensity levels of a stimulus, each is presented 50 times within a session and that he is run in 10 sessions. While the number of intensity levels and the within session repetitions correspond to typical values that one might observe in a set of psychophysical data, the number of runs is probably larger than most experimenters would perform. If instead we named this variable "Observer," then 10 would not be such an atypical value. In any case, we assign these values to variables so that it would be easy to modify and investigate their influence on the results. If we choose a Gaussian psychometric function, then its location and steepness are determined by its mean and standard deviation. These parameters are set by the following code. Again, these are representative values and the reader is encouraged to experiment further with the simulation.

```
> NLevs <- 7
> Intensity <- 10^seq(-2, -0.5, len = NLevs)
> Ntrials <- 50
> NRun <- 10
> mu <- 0.1
> sigma <- 0.05
```

From the mean and standard deviation of the Gaussian, we calculate the coefficients of the linear predictor and the underlying psychometric function.

```
> beta0 <- -mu/sigma
> beta1 <- 1/sigma
> LinPred <- function(Intensity, mu, sigma)
+    -mu/sigma + (1/sigma) * Intensity
> xx <- 10^seq(-2, -0.5, len = 200)
> TrueFunc <- pnorm(LinPred(xx, mu, sigma))
```

Note that the value chosen for the intercept corresponds to a False Alarm rate of

```
> pnorm(-beta0, lower.tail = FALSE)
[1] 0.0228
```

Now, we can simulate the decision variable possibly with additional sources of variance, such as in (9.2). First, let's set up the decision variable in the case that there are no additional sources of variability, similar to how we have simulated psychometric functions previously.

```
> DV0 <- rep(beta0, NLevs * NRun * Ntrials) +
+    rep(beta1, NLevs * NRun * Ntrials) *
+    rep(Intensity, NRun * Ntrials)
```

We put each term contributing to the decision variable on a separate line to facilitate understanding its construction. The first line of code creates the intercept term of the linear predictor, composed of the fixed coefficient, β_0 repeated for every trial across intensity levels and runs. The second line generates the slope term of the linear predictor that is a fixed coefficient, β_1 repeated for all trials across the experiment. This coefficient multiplies the intensity level, included on the third line. There are 7 intensity levels, repeated 50 times during each of the 10 runs, yielding 3,500 trials.

As we did previously, we use the pnorm function to convert the values of the decision variable to probabilities and use these probabilities to generate random Bernoulli events $(0, 1)$, with rbinom. We store these values in a data frame, Sim0.df along with the intensity values and a factor that codes the run. It is then straightforward to extract the proportion of correct responses for each intensity and

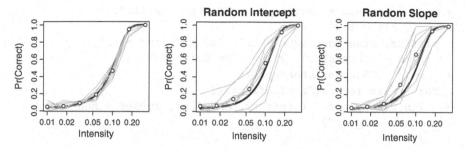

Fig. 9.1 Simulated psychometric functions for 10 runs of an experiment with no additional random effects (*left*), a random intercept in the linear predictor (*center*) or a random slope in the linear predictor (*right*). The *grey lines* connect the proportion correct from individual simulated runs and the *black line* is the true underlying psychometric function. The points are the proportion correct aggregated over all runs

run using `table` from which we construct a second data frame, `Prop0.df` with these derived data also coded by intensity and run.

```
> SimResp0 <- rbinom(NLevs * NRun * Ntrials, 1,
+    pnorm(DV0))
> Sim0.df <- data.frame(Intensity = Intensity,
+    Resp = SimResp0,
+    Run = rep(paste("R", seq(NRun), sep = ""),
+       each = NLevs * Ntrials))
> Prop0.mat <- with(Sim0.df,
+    table(Resp, round(Intensity, 3), Run))[2, , ]/Ntrials
> Prop0.df <- stack(as.data.frame(Prop0.mat))
> Prop0.df$Intensity <- Intensity
> names(Prop0.df) <- c("PCorrect", "Run", "Intensity")
```

In the left panel of Fig. 9.1, we plot the psychometric functions for each run as grey lines with the true underlying function from which all trials were generated as a thick black line. The overall proportion correct, combining responses from all runs is indicated by the points. The variability in the individual curves is that due to the intrinsic variability of a binomial variable. The points fall close to the true curve, indicating little bias in the estimates.

```
> opar <- par(mfrow = c(1, 3))
> plot(xx, TrueFunc, type = "l", lwd = 3, log = "x",
+    xlab = "Intensity", ylab = "Pr(Correct)")
> for (ix in levels(Prop0.df$Run)){
+    lines(PCorrect ~ Intensity, Prop0.df, subset = Run == ix,
        col = "grey")
+    }
> points(Intensity, t(with(Sim0.df,
+    table(Resp, round(Intensity, 3))))[, 2] /
```

```
+        (Ntrials * NRun), pch = 21, bg = "white",
+        cex = 1.2)
> plot(xx, TrueFunc, type = "l", lwd = 3, log = "x",
+    xlab = "Intensity", ylab = "Pr(Correct)",
+    main = "Random Intercept")
> for (ix in levels(Prop1.df$Run)){
+    lines(PCorrect ~ Intensity, Prop1.df, subset = Run == ix,
+        col = "grey")
+  }
> points(Intensity, t(with(Sim1.df,
+    table(Resp, round(Intensity, 3))))[, 2] /
+        (Ntrials * NRun), pch = 21, bg = "white",
+        cex = 1.2)
> plot(xx, TrueFunc, type = "l", lwd = 3, log = "x",
+    xlab = "Intensity", ylab = "Pr(Correct)",
+    main = "Random Slope")
> for (ix in levels(Prop2.df$Run)){
+    lines(PCorrect ~ Intensity, Prop2.df, subset = Run == ix,
+        col = "grey")
+  }
> points(Intensity, t(with(Sim2.df,
+    table(Resp, round(Intensity, 3))))[, 2] /
+        (Ntrials * NRun), pch = 21, bg = "white",
+        cex = 1.2)
> par(opar)
```

Now, let's consider what happens if an additional random component that varies across sessions is added to the criterion. The sensitivity is still considered as only a fixed effect.

```
> SD1 <- 0.5
> b0 <- rnorm(NRun, sd = SD1)
> DV1 <- rep(beta0, NLevs * NRun * Ntrials) +
+    rep(b0, each = NLevs * Ntrials) +
+    rep(beta1, NLevs * NRun * Ntrials) *
+    rep(Intensity, NRun * Ntrials)
```

As before, we put each term on a separate line to facilitate understanding its construction. The intercept term is now composed of the sum of two terms, the fixed coefficient, β_0 repeated for every trial across intensity levels and runs, added to the random effect of run, b_0, drawn from a Gaussian distribution with mean equal to 0 and standard deviation SD1. Plus or minus this amount of variability in the linear predictor corresponds to

```
> pnorm(-beta0 + c(-SD1, SD1), lower.tail = FALSE)
[1] 0.06681 0.00621
```

variation in the False Alarm rate. There is one random value drawn per run so each value is repeated within a run for each intensity level and trial. The third and fourth lines, for generating the sensitivity term, are as before.

We use the pnorm function again to convert the values of the decision variable to probabilities and use these probabilities to generate random Bernoulli events $(0, 1)$, with rbinom, again, and store the responses by simulated run in a data frame, Sim1.df.

```
> SimResp1 <- rbinom(NLevs * NRun * Ntrials, 1, pnorm(DV1))
> Sim1.df <- data.frame(Intensity = Intensity,
+    Resp = SimResp1,
+    Run = rep(paste("R", seq(NRun), sep = ""),
+        each = NLevs * Ntrials))
```

Again, we summarize the proportion of correct responses across intensity by run in order to visualize the psychometric functions obtained for each run.

```
> Prop1.mat <- with(Sim1.df,
+    table(Resp, round(Intensity, 3), Run))[2,, ]/Ntrials
> Prop1.df <- stack(as.data.frame(Prop1.mat))
> Prop1.df$Intensity <- Intensity
> names(Prop1.df) <- c("PCorrect", "Run", "Intensity")
```

The psychometric functions from the individual runs are shown as thin lines in the center panel of Fig. 9.1. The plotting conventions are the same as those for the graph to its left. The curves from the individual runs are considerably more variable than those in the graph to the left. The points at the bottom of the curve deviate systematically from the true underlying function. The random variability that we introduced in the intercept of the linear predictor is more evident in the lower part of the curves and has less effect at the higher ends. Thus, it may affect the rate at which the curves rise toward the asymptotic value as well.

Finally, we consider what the effect of a random variation of sensitivity would be. We follow the same steps as above, but now add a random component that varies across runs to the slope coefficient and not to that of the intercept. The random slope component in the code fragment below is in line 3 of the definition of the decision variable, DV2 and is added to the slope, β_1 prior to multiplying the stimulus term, Intensity.

```
> SD2 <- 10
> b1 <- rnorm(NRun, sd = SD2)
> DV2 <- rep(beta0, NLevs * NRun * Ntrials) +
+    ( rep(beta1, NLevs * NRun * Ntrials) +
+        rep(b1, each = NLevs * Ntrials) ) *
+    rep(Intensity, NRun * Ntrials)
```

As before we organize the information in data frames that we will use to extract average proportion correct and for plotting.

```
> SimResp2 <- rbinom(NLevs * NRun * Ntrials, 1, pnorm(DV2))
> Sim2.df <- data.frame(Intensity = Intensity,
+    Resp = SimResp2,
+    Run = rep(paste("R", seq(NRun), sep = ""),
+        each = NLevs * Ntrials))
> Prop2.mat <- with(Sim2.df,
+    table(Resp, round(Intensity, 3), Run))[2,, ]/Ntrials
> Prop2.df <- stack(as.data.frame(Prop2.mat))
> Prop2.df$Intensity <- Intensity
> names(Prop2.df) <- c("PCorrect", "Run", "Intensity")
```

The right-hand graph of Fig. 9.1 shows the results of simulating a random slope. The variability is now distributed around the rate at which the individual functions approach asymptote and much less so at the extremes. Bias is evident in the middle range of points, where the proportion correct is changing most rapidly and would lead, in this case, to underestimates of the true threshold.

Another way to visualize the influence of the random intercept is to fit the psychometric function from each session individually and compare the parameter estimates. This can be easily done using the function lmList in the **lme4.0** package that fits a linear or generalized linear model to a data set for each level of a grouping factor.

```
> Sim0.lst <- lmList(Resp ~ Intensity | Run,
+    Sim0.df, binomial(probit))
> Sim1.lst <- lmList(Resp ~ Intensity | Run,
+    Sim1.df, binomial(probit))
> Sim2.lst <- lmList(Resp ~ Intensity | Run,
+    Sim2.df, binomial(probit))
```

The function takes a formula object that includes a conditioning variable specified to the right of the "|," as used in **lattice** plotting commands and that indicates the grouping factor in the data, Run. By specifying the family argument, a glm is fit to the data. for each level of the grouping factor using the formula to the left of it. The model objects from each fit are returned as components of an object of class "lmList." A confint method has also been defined for this class of objects for which a plot method permits visualizing the estimates and their confidence intervals. The left two plots of Fig. 9.2 show the parameter estimates for the intercept and covariate coefficients for the data set Sim0.df. The segments indicate the 95% confidence intervals. In this data set, there were no random effects added to either the intercept or the slope terms. There is individual variation in the values of both parameters but all of the confidence intervals overlap. This is usually taken as evidence that fixed effects would be sufficient to model both of these parameters.

```
> print( plot(confint(Sim0.lst)),
+    more = TRUE,
+    split = c(1, 1, 3, 1)
+    )
```

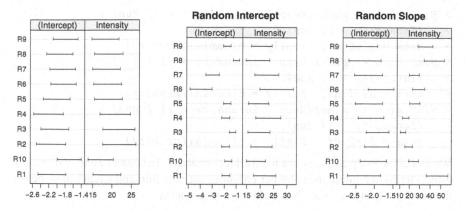

Fig. 9.2 Parameter estimates and 95% confidence intervals for each Run of the simulated data sets Sim0.df (*left*), Sim1.df (*center*) and Sim2.df (*right*). The underlying model of the *center plot* includes a random effect of the intercept term and the one on the *right* a random intercept of the slope

```
> print( plot(confint(Sim1.lst), main = "Random Intercept"),
+   more = TRUE,
+   split = c(2, 1, 3, 1)
+   )
> print( plot(confint(Sim2.lst), main = "Random Slope"),
+   more = FALSE,
+   split = c(3, 1, 3, 1)
+   )
```

The middle pair of graphs illustrate the results when a random component is added to the intercept term. Here, the slope estimates of the covariate Intensity overlap but not the estimates for the intercept term. This is typically taken as diagnostic of the need to include a random effect when modeling a term. The right-hand pair of graphs give a complementary picture, indicating non-overlap of confidence intervals for the slope term but not the intercept. Thus, we would be led to model the intercept as a fixed effect but would include a random effect for the slope.

The above examples suggest that not taking into account random effects in a model would lead to biased estimates of the psychometric function. In addition, without modeling the random effects, we are likely to overestimate the significance of differences since we would be underestimating the true variability in the data. To incorporate such sources of variability in our models, we need to consider modeling functions for mixed-effects models.

We use again the glmer function to model the simulated data sets. We model each of the data sets including terms for the random effects in the intercept and the slope with the following code.

```
> Sim0.glmm <- glmer(Resp ~ Intensity + (1 | Run) +
+    (Intensity - 1| Run),
+     data = Sim0.df,  family = binomial(probit))
> Sim1.glmm <- glmer(Resp ~ Intensity + (1 | Run) +
+    (Intensity - 1| Run),
+     data = Sim1.df,  family = binomial(probit))
> Sim2.glmm <- glmer(Resp ~ Intensity + (1 | Run) +
+    (Intensity - 1| Run),
+     data = Sim2.df,  family = binomial(probit))
```

We model the Resp as a sum of a fixed effect covariate, Intensity and two random effects terms, a random intercept per grouping variable Run and a random slope per the same grouping variable. We could use a single term

```
> (Intensity | Run)
```

that employs the usual convention in R of an implicit intercept. By including separate terms for the two random effects, we assume that they are independent. Now, we examine the summary of the fit to the data simulated without random effects.

```
> summary(Sim0.glmm)
Generalized linear mixed model fit by the Laplace approximation
Formula: Resp ~ Intensity + (1 | Run) + (Intensity - 1 | Run)
   Data: Sim0.df
  AIC  BIC logLik deviance
 2142 2167  -1067     2134
Random effects:
 Groups Name        Variance Std.Dev.
 Run    (Intercept) 0.000162 0.0127
 Run    Intensity   0.000000 0.0000
Number of obs: 3500, groups: Run, 10

Fixed effects:
            Estimate Std. Error z value Pr(>|z|)
(Intercept)  -1.9390     0.0528   -36.7   <2e-16 ***
Intensity    19.0138     0.5855    32.5   <2e-16 ***
---
Signif. codes:  0 `***' 0.001 `**' 0.01 `*' 0.05 `.' 0.1 ` ' 1

Correlation of Fixed Effects:
          (Intr)
Intensity -0.814
```

Unsurprisingly, the random effect of the intercept is quite small. That for the slope is sufficiently small to yield an estimate of 0. The fixed effects give close approximations to the parameters used to simulate the data. We can test the significance of the random effects by running the model with one or the other removed, but we cannot test a model in `glmer` without at least one random effect specified in the model. We could fit a model without random effects using `glm` but we cannot directly compare the models with the `anova` method as they do not share any inheritance of classes. If the data arose from an experiment in which the terms would be expected to be random, such as observer differences, then we might just want to include them anyway, even if they are small.

We could look at the summary of each model, but to simplify and continue to demonstrate how to obtain just the information needed from a model object, we extract the standard deviations for the random effects from each of the three simulations.

```
> sapply(Sim.lst <- list("Sim0" = Sim0.glmm,
+     "Sim1" = Sim1.glmm, "Sim2" = Sim2.glmm),
+     function(x){
+         VC <- VarCorr(x)
+         sapply(VC, attr, which = "stddev")
+     }
+ )
                 Sim0   Sim1  Sim2
Run.(Intercept) 0.0127 0.569  0.00
Run.Intensity   0.0000 0.000  8.54
```

The `VarCorr` method is used to extract the variances and correlations of the random effects from the "mer" model object. The standard deviations are stored as the attribute "stddev" extracted with the `attr` function. We put the three model objects into a list, `Sim.lst` for easy indexing and for ease of extraction of other quantities later. The estimated random effects approximate those that we programmed into the simulation for both simulations in which we included random effects.

Conveniently, we can extract the fixed effect coefficients using the list of the three model objects that we set up above and note that fixed effects also provide reasonable estimates of the underlying parameters, although the slope for `Sim2` looks to be slightly overestimated.

```
> sapply(Sim.lst, fixef)
             Sim0   Sim1   Sim2
(Intercept) -1.94  -1.92  -2.06
Intensity   19.01  20.36  26.57
```

Finally, what if we had ignored the random effects in fitting the data? We can see what happens by fitting each simulated data set with a `glm` ignoring the term `Runs`. The code below performs the three fits and displays just the summary tables.

```
> lapply(list(Sim0 = Sim0.df,
+     Sim1 = Sim1.df, Sim2 = Sim2.df),
+     function(x) summary(
```

```
+              glm(Resp ~ Intensity, binomial(probit), x) )$coef
+     )
$Sim0
              Estimate Std. Error z value  Pr(>|z|)
(Intercept)     -1.94      0.0526   -36.8  3.61e-297
Intensity       19.01      0.5855    32.5  2.41e-231

$Sim1
              Estimate Std. Error z value  Pr(>|z|)
(Intercept)     -1.68      0.0476   -35.3  2.87e-272
Intensity       17.88      0.5665    31.6  1.55e-218

$Sim2
              Estimate Std. Error z value  Pr(>|z|)
(Intercept)     -1.77      0.0501   -35.3  1.12e-272
Intensity       20.20      0.6414    31.5  9.29e-218
```

Ignoring the random effects seems to bias the parameter estimates. The intercept terms underestimate the true value for Sim1 and Sim2. This is also true for the slope estimate in Sim1. Comparisons of these tables with the random effects models indicate that the errors associated with these estimates are underestimated as well. With real data both the intercept and slope might vary across conditions which would complicate things even further. Of course, we do not know what the true underlying parameters are in that case. This leads us to consider a real data set.

9.3.1 Context Effects on Detection

The data set Context from our package **MPDiR** contains the results of the first experiment reported by Zhaoping and Jingling [205], who studied the effect of context on detection threshold. In a Yes-No paradigm, observers were required to report the presence of a short vertical target bar that appeared as the center element in an array of vertically aligned flanker bars (see Fig. 9.3 for example). Four contrast levels of the flanker bars were tested including a contrast of zero corresponding to the absence of flankers. The data set contains five components

```
> str(Context)
'data.frame':   120 obs. of  5 variables:
 $ Obs     : Factor w/ 6 levels "A","B","C","D",..: 1 1 1 ..
 $ ContCntr: num  0 0 0 0 0 0.01 0.01 0.01 0.01 0.01 ...
 $ TargCntr: num  0 0.002 0.004 0.006 0.008 0 0.002 0.004 ..
 $ NumYes  : int  0 0 5 18 23 24 22 22 22 24 ...
 $ NumNo   : int  24 24 19 6 1 0 2 2 2 0 ...
```

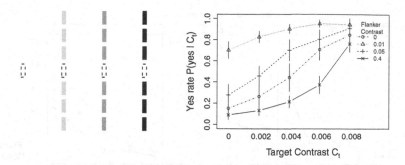

Fig. 9.3 *Left*. Four examples of the stimulus configuration used by Zhaoping and Jingling [205]. The location of the target stimulus is indicated by the *dashed rectangle*. The *flanker bars* are shown schematically at four contrast levels. The actual levels are indicated in the legend of the figure to the *right*. *Right*. Average proportion of reports that the target stimulus was present as a function of its contrast for each of four contrast levels of the flankers. The *error bars* indicate standard errors of the mean

coding for observer, Obs, the contrast of the contextual bars, ContCntr, the contrast of the target bar, TargCntr and the number of "Yes" and "No" responses, NumYes and NumNo, respectively.

The right hand graph of Fig. 9.3 shows the average for the six observers of the proportion of times on which the target was reported as present as a function of its contrast for each of the four levels of flanker contrast tested. An intermediate contrast results in a greater tendency for the observer to respond that the target was present than a lower or higher contrasts. This graph was created with the following code.

```
> Context <- within(Context, Pc <- NumYes/(NumYes +
+     NumNo))
> Pc.sd <- with(Context, tapply(Pc, list(TargCntr,
+     ContCntr), sd))
> Pc.sem <- Pc.sd/sqrt(6)
> Pc.mean <- with(Context, tapply(Pc, list(TargCntr,
+     ContCntr), mean))
> with(Context, interaction.plot(TargCntr, ContCntr,
+     Pc, mean, pch = 1:4, bg = "white", cex = 1,
+     type = "b", ylim = c(0, 1), fixed = TRUE,
+     trace.label = " n   Flanker n   Contrast",
+     xlab = expression(paste("Target Contrast ",
+         C[t])), ylab = expression(paste("Yes rate P(yes | ",
+         C[t], ")")), ))
> for (ix in 1:4) segments(1:5, (Pc.mean + Pc.sem)[,
+     ix], 1:5, (Pc.mean - Pc.sem[, ix])[, ix])
```

The average data of Fig. 9.3 show a clear effect of flanker contrast on the observers' responses. The data from each observer, however, were collected with

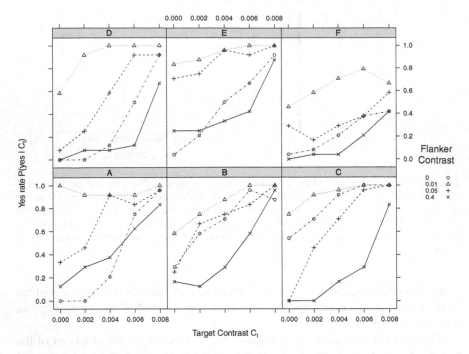

Fig. 9.4 Psychometric functions indicating performance for detecting target bar as a function of its contrast. The *four curves* indicate different contrasts of the contextual, *flanker bars* as indicated in the legend to the *right*. Each panel corresponds to a different observer

only 24 observations per point, not a large number of observations, and, thus, we might expect a fair amount of variability in the individual curves. This raises the question as to how consistent the effect is across observers. Figure 9.4 shows the four conditions plotted by observer with a lattice plot. We use the same coding of symbols and line types to facilitate comparison with the previous figure. While the effect of an increased bias to respond present is evident for all observers for the flanker contrast of 0.05, there is considerable variation among the curves and in their ordering. We would like to see if we can account for this variation within the framework of a GLMM.

```
> print(
+ xyplot(Pc ~ TargCntr | Obs, data = Context,
+   groups = ContCntr,  type = c("l", "p"),
+   xlab = expression(paste("Target Contrast ", C[t])),
+   ylab = expression(paste("Yes rate P(yes | ", C[t], ")")),
+   auto.key =  list(space = "right",
+        title = "Flanker nContrast ", cex = 0.75),
+   par.settings =
+        list(strip.background = list(col = "lightgrey"),
```

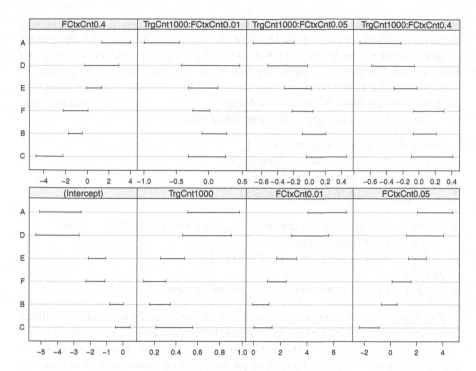

Fig. 9.5 Parameter estimates and 95% confidence intervals for each observer fit individually, for each of the parameters estimated by fitting a GLM to each observer's data set

```
+           superpose.symbol = list(col = "black", pch = 1:4),
+           superpose.line = list(col = "black", lty = 4:1)))  )
```

In attempting to construct a model, we use lmList to examine which components might be best treated as random effects. Before doing so, we make some modifications of the data set. First, we rescale the contrast of the target, TargCntr, by a factor of 1,000 and store the rescaled values in a new column of the data frame, TrgCnt1000. This will put the estimated coefficients on similar scale to that of the intercept. Second, we recode the flanker contrast as a factor in a variable FCtxCnt. While it is a "numeric" variable, there are only four levels. An alternative would be to treat it as an ordered factor, which would permit fitting orthogonal polynomials, but here we stay with this simpler parameterization. Then, we fit a GLM for each observer using lmList. The linear predictor includes the target contrast as a covariate, the flanker contrast as a factor and the interaction of these two terms. The estimates of each term with 95% confidence intervals are graphed by plotting the output of the confint method. Specifying the argument order to the plot method reorders the factor levels for the plots for each term by the size of the Intercept. This can be useful for revealing patterns in the estimates.

Examining Fig. 9.5 indicates the need for several random effects. The confidence intervals for both the Intercept and the slope of TrgCnt1000 fail to overlap suggesting the need for random intercepts to model these terms. In addition, similar behavior is shown for each of the levels of the factor FCtxCnt. In fact, the pattern is quite similar across levels and similar to that for the slope of TrgCnt1000 and inversely for the intercept, suggesting that these variables may all be correlated. The evidence for a random effect associated with the interaction term is less clear, suggested most perhaps by observer A. Note also that the magnitude of the intercept seems to covary with the size of the confidence interval.

```
> Context <- within(Context,  {
+    TrgCnt1000 <- 1000 * TargCntr
+    FCtxCnt <- factor(ContCntr)
+    resp <- cbind(NumYes, NumNo)
+    } )
> Context.lst <- lmList(resp ~ TrgCnt1000 * FCtxCnt | Obs,
+    Context, binomial(probit))
> ord <- order(coef(Context.lst)[[1]])
> print(
+    plot(confint(Context.lst), order = ord) )
```

We begin with as complex a model as we can specify to define the linear predictor. Next, we evaluate how to simplify the random effects; before turning our attention to the fixed effects. This is not necessarily the best strategy to adopt with mixed-effects models as the number of parameters to be estimated for sufficiently complex models can easily outstrip what the data can support. Often, it is better to start with simpler models of the random effects and to work up to more complex ones. Here, the model is not too complex and we can explore a range of random structures to find one that describes the data best.

$$\mathsf{E}[P(R = \text{'Yes'})] = (\beta_0 + b_i) + (\alpha_j + a_{ij}) + (\beta_1 + b_i)Tc + (\beta_{12} + b_{ij})Tc : Fc_j \quad (9.7)$$

The fixed effects define an ANCOVA model with the target contrasts, Tc, as a covariate and the flanking contrast, F_c, as a factor. We allow for the possibility of differences in slope of the covariate by including an interaction term between the target and flanking contrast terms. We treat the random effect of flanker contrast as nested within observer, just as we did for the spatial frequency variable in the analysis of the ModelFest data set in Chap. 2. Every term has a fixed effect indicated by a Greek symbol and a random effect indicated by a Roman symbol. Specifically, the fixed effect intercept, β_0 is complemented by a random intercept, b_i, that depends on observer, i. The main effect of flanker contrast, α_j depends on flanker condition, j and is modulated by a random effect of flanker within Observer, a_{ij}. Fixed and random effects, β_1 and b_i, respectively, are coded for the covariate, Tc by observer and for the interaction with flanker term by flanker condition within observer. It is assumed that each of the random effects is normally distributed with mean zero and a variance to be estimated. In the full model, we allow for correlations between the random effects. This model is fit using glmer with the following call.

```
> Context.glmm1 <- glmer(cbind(NumYes, NumNo) ~
+   FCtxCnt * TrgCnt1000 + (TrgCnt1000 | Obs/FCtxCnt),
+   data = Context, family = binomial("probit"))
```

The left-hand side of the random effects term codes for a random slope and intercept term, the intercept being, as usual, implicit. The right-hand side specifies that both Obs and Obs:FCtxCnt should be treated as grouping factors for these random effects. Thus, there will be four variances to estimate. The fact that we include this specification within one term means that correlations will be estimated as well. Because we are concentrating on just the random effects at this point, we display only these terms. They can be extracted directly with the VarCorr method, but we display them in tabular format as in the summary method.

```
      Groups        Name  Variance Std.Dev   Corr
FCtxCnt:Obs (Intercept)  0.64926  0.8058
            TrgCnt1000   0.00914  0.0956  -0.935
       Obs (Intercept)  0.04846  0.2201
            TrgCnt1000   0.00436  0.0661   0.427
```

There are four random variances and two correlations, giving six estimated random parameters. The correlation between the intercept and slope for the flanker contrast within observer random effect is very high, suggesting considerable redundancy in the information provided by these two terms. To test whether either of these can be eliminated from the model, we refit the model with each correlation eliminated, in turn. We do this by separating the random effects into independent terms for the intercept and slope components. These models are nested with respect to the full model so we can perform a likelihood ratio test.

```
> Context.glmm2a <- glmer(cbind(NumYes, NumNo) ~
+   FCtxCnt * TrgCnt1000 + (TrgCnt1000 | Obs) +
+   (1 | Obs:FCtxCnt) + (TrgCnt1000 + 0 | Obs:FCtxCnt),
+   data = Context, family = binomial("probit"))
> Context.glmm2b <- glmer(cbind(NumYes, NumNo) ~
+   FCtxCnt * TrgCnt1000 + (1 | Obs) +
+   (TrgCnt1000 + 0 | Obs) +
+   (TrgCnt1000 | Obs:FCtxCnt),
+   data = Context, family = binomial("probit"))
> anova(Context.glmm1, Context.glmm2a)

Data: Context
Models:
Context.glmm2a: cbind(NumYes, NumNo) ~ FCtxCnt * TrgCnt1000 +
Context.glmm2a:     (TrgCnt1000 | Obs) + (1 | Obs:FCtxCnt)
Context.glmm2a:       + (TrgCnt1000 + 0 | Obs:FCtxCnt)
Context.glmm1: cbind(NumYes, NumNo) ~ FCtxCnt * TrgCnt1000 +
Context.glmm1:     (TrgCnt1000 | Obs/FCtxCnt)
              Df AIC BIC logLik Chisq Chi Df Pr(>Chisq)
```

```
Context.glmm2a 13 262 298    -118
Context.glmm1  14 247 286    -110   16.5      1     4.8e-05

Context.glmm2a
Context.glmm1  ***
---
> anova(Context.glmm1, Context.glmm2b)
Data: Context
Models:
Context.glmm2b: cbind(NumYes, NumNo) ~ FCtxCnt * TrgCnt1000 +
Context.glmm2b:    (1 | Obs) + (TrgCnt1000 + 0 | Obs) +
Context.glmm2b:    (TrgCnt1000 | Obs:FCtxCnt)
Context.glmm1: cbind(NumYes, NumNo) ~ FCtxCnt * TrgCnt1000 +
Context.glmm1:    (TrgCnt1000 | Obs/FCtxCnt)
                Df AIC BIC logLik Chisq Chi Df Pr(>Chisq)
Context.glmm2b 13 245 281    -110
Context.glmm1  14 247 286    -110  0.08      1        0.78
```

The results indicate that the high negative correlation between the intercept and slope random effects of flanker contrast within observer should be retained while the lower correlation between the intercept and slope for observer is not significant. As discussed previously, zeroing a variance component forces the fit to be at the boundary of the parameter space which renders the p-value conservative, possibly too large by a factor as great as two [11, 141]. In this case, the conclusions would be unaffected by reducing the p-values by a factor of 2.

We now test whether the random variances specific to Obs are necessary. As before, we do this by refitting the model without each component and term and performing a likelihood ratio test on the nested models.

```
> Context.glmm3a <- glmer(cbind(NumYes, NumNo) ~
+   FCtxCnt * TrgCnt1000 +
+   (1 | Obs) + (TrgCnt1000 | Obs:FCtxCnt),
+   data = Context, family = binomial("probit"))
> Context.glmm3b <- glmer(cbind(NumYes, NumNo) ~
+   FCtxCnt * TrgCnt1000 +
+   (TrgCnt1000 + 0 | Obs) + (TrgCnt1000 | Obs:FCtxCnt),
+   data = Context, family = binomial("probit"))
> anova(Context.glmm2b, Context.glmm3a)
> anova(Context.glmm2b, Context.glmm3b)

Data: Context
Models:
Context.glmm3a: cbind(NumYes, NumNo) ~ FCtxCnt * TrgCnt1000 +
Context.glmm3a:    (1 | Obs) + (TrgCnt1000 | Obs:FCtxCnt)
Context.glmm2b: cbind(NumYes, NumNo) ~ FCtxCnt * TrgCnt1000 +
```

```
Context.glmm2b:        (1 |Obs) + (TrgCnt1000 + 0 | Obs) +
Context.glmm2b:        (TrgCnt1000 | Obs:FCtxCnt)
               Df AIC BIC logLik Chisq Chi Df Pr(>Chisq)
Context.glmm3a 12 248 281    -112
Context.glmm2b 13 245 281    -110  4.76         1         0.029 *
---
Signif. codes:  0
> anova(Context.glmm2b, Context.glmm3b)
Data: Context
Models:
Context.glmm3b: cbind(NumYes, NumNo) ~ FCtxCnt * TrgCnt1000 +
Context.glmm3b:        (TrgCnt1000 + 0 | Obs) +
Context.glmm3b:        (TrgCnt1000 | Obs:FCtxCnt)
Context.glmm2b: cbind(NumYes, NumNo) ~ FCtxCnt * TrgCnt1000 +
Context.glmm2b:        (1 | Obs) + (TrgCnt1000 + 0 | Obs) +
Context.glmm2b:        (TrgCnt1000 | Obs:FCtxCnt)
               Df AIC BIC logLik Chisq Chi Df Pr(>Chisq)
Context.glmm3b 12 244 277    -110
Context.glmm2b 13 245 281    -110  0.53         1         0.47
```

The results indicate that we can eliminate the random effect of intercept for Obs but not of the slope. To summarize at this stage, we have a model that includes a random intercept for the grouping factor Obs:FCtxCnt and random slopes for both Obs and Obs:FCtxCnt. We also model the correlation between the intercept and the slope of the Obs:FCtxCnt grouping factor.

Having arrived at a model of the random effects, we turn to the fixed effects with Context.glmm3b as our base model. We begin by testing the interaction FCtxCnt:TrgCnt1000.

```
> Context.glmm4 <- glmer(cbind(NumYes, NumNo) ~
+    FCtxCnt + TrgCnt1000 +
+    (TrgCnt1000 + 0 | Obs) + (TrgCnt1000 | Obs:FCtxCnt),
+    data = Context, family = binomial("probit"))
> anova(Context.glmm3b, Context.glmm4)

> Context.glmm4 <- glmer(cbind(NumYes, NumNo) ~
+    FCtxCnt + TrgCnt1000 +
+    (TrgCnt1000 + 0 | Obs) + (TrgCnt1000 | Obs:FCtxCnt),
+    data = Context, family = binomial("probit"))
> anova(Context.glmm3b, Context.glmm4)
Data: Context
Models:
Context.glmm4: cbind(NumYes, NumNo) ~ FCtxCnt + TrgCnt1000 +
Context.glmm4:        (TrgCnt1000 + 0 | Obs) +
Context.glmm4:        (TrgCnt1000 | Obs:FCtxCnt)
```

```
Context.glmm3b: cbind(NumYes, NumNo) ~ FCtxCnt * TrgCnt1000 +
Context.glmm3b:      (TrgCnt1000 + 0 | Obs) +
Context.glmm3b:      (TrgCnt1000 | Obs:FCtxCnt)
                Df AIC BIC logLik Chisq Chi Df Pr(>Chisq)
Context.glmm4   9 243 268   -113
Context.glmm3b 12 244 277   -110  5.66      3       0.13
```

While the likelihood ratio tests of fixed effects in mixed models tend to be anti-conservative [141], i.e., give *p*-values that are too low, here we can feel reasonably assured about the above results. Further support for dropping the interaction term, however, is given by the AIC and by examining the marginal statistics for the terms in the coefficient table from the summary method (left as an exercise). Further tests showed that each of the remaining fixed effects was necessary in the model (also left as an exercise).

At this point, one should examine various diagnostic plots associated with the fitted model as demonstrated previously. As of this writing, there is no plot method defined for objects of class "mer," but fitted values and residuals are easily extracted from the object and useful plots constructed. We leave this as an exercise for the reader. Instead, we demonstrate how to plot the fitted model against the experimental data.

We first examine the fixed effects and compare the fitted curves with the average response rate across the four flanker conditions. The fixed effects coefficients are extracted from the model object with the fixef method. As with other modeling functions, "treatment contrasts" are the default for factors. Thus, to obtain the intercept term for the linear predictor of each flanker condition, we must add the first fixed effect coefficient to all succeeding ones, except for the last one that gives the slope of the covariate TrgCnt1000. To obtain a smooth looking curve, we generate a series of 200 points over the range of target contrasts and use this in the linear predictor with the lines function when drawing the graph. The predicted values on the linear predictor scale are transformed to probabilities with the inverse link given here by the pnorm function. In addition, we readjusted the scale to be in the original units. The comparison of the fitted curves with the average data is shown in Fig. 9.6.

```
> xx <- seq(0, 0.01, len = 200)
> gm4.fe <- fixef(Context.glmm4)
> intc <- c(gm4.fe[1], gm4.fe[1] + gm4.fe[-c(1, 5)])
> plot(c(0, 0.011), c(0, 1), type = "n",
+    xlab = expression(paste("Target Contrast ", C[t])),
+    ylab = expression(paste("Yes rate P(yes | ", C[t], ")")))
+    )
> for (ix in 1:4) {
+    points(unique(Context$TargCntr), Pc.mean[, ix],
+        ylim = c(0, 0.01), cex = 1.4,
+        pch = ix)
+    lines(xx, pnorm(intc[ix] + gm4.fe[5] * xx * 1000),
+        lty = 5 - ix, lwd = 3)
```

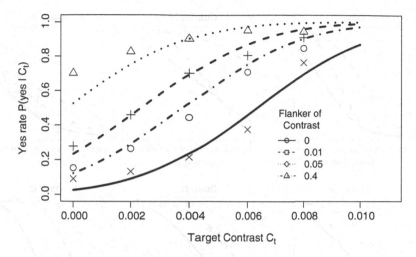

Fig. 9.6 Comparison of prediction of fixed effects component of model with average proportion of "Yes" responses for the four conditions of flanker contrast

```
+  }
> legend(0.007, 0.45, c(0, 0.01, 0.05, 0.4), pch = 21:24,
+    lty = 1:4, bty = "n",
+    title = "Flanker of n Contrast ")
```

To display the predictions for the individual curves, the random effects must be incorporated in the calculations. The random effects are returned in a list of data frames by the method `ranef`. The first component contains the intercept and slope random effects for the `Obs:FCtxCnt` grouping variable. There are 24 rows (6 observers × 4 flanker contrasts). The second component contains the six random slope components for `Obs` as a grouping factor. The linear predictor is calculated by including the appropriate random effects in the calculation and then transformed with the `pnorm` function. The individual predictions are shown in Fig. 9.7.

In the final model, there is a fixed effect of the flanker contrast on the intercept but no dependence of the slope on flanker contrast. This accords with the conclusions of the authors. To account for individual differences, however, we include random effects of the intercept by flanker contrast within observer and a random variation of slope for observer and flanker contrast within observer. The variability of the psychometric functions requires this last effect even though there is no fixed effect interaction of the flanker contrast with the slope of the psychometric function given by the `TrgCnt1000` term.

```
> gm4.re <- ranef(Context.glmm4)
> opar <- par(mfrow = c(2, 3))
> for (obs in levels(Context$Obs)[c(4:6, 1:3)]){
+    NObs <- which(LETTERS %in% obs) - 1
+    plot(Pc ~ TargCntr, Context,
```

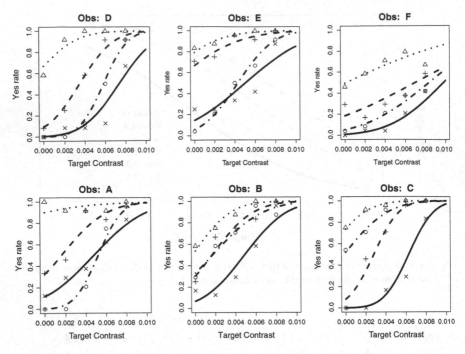

Fig. 9.7 Predictions for individual observers for each of the four conditions of flanker contrast

```
+            pch = rep(1:4, each = 5),
+            xlim = c(0, 0.01), ylim = c(0, 1),
+            subset = Obs == obs, cex = 1.4,
+            xlab = "Target Contrast",
+            ylab = "Yes rate",
+            main = paste("Obs: ",obs))
+        for (Cond in 1:4) {
+        nn <- 4 * NObs + Cond
+        lines(xx, pnorm(intc[Cond] + gm4.re[[1]][nn, 1] +
+            (gm4.fe[5] + gm4.re[[1]][nn, 2] +
+                gm4.re[[2]][obs, 1]) * xx * 1000),
+            lty = 5 - Cond, lwd = 3) } }
> par(opar)
```

9.3.2 Multiple-Alternative Forced-Choice Data

One evident question is, how can these analyses be extended to multiple-alternative forced-choice experiments? Currently, this is not as simple as writing a new link function in R, as we did with glm. This is because the link functions used in the **lme4.0** package are written into the C++ code underlying the functions it provides. Of course, since R is open source, the option is always available to modify the code and reinstall the package, but this avenue, we suspect, is not likely for the average user. A second option is to specify the lower asymptote as an offset in the model as described in Sect. 5.1.1, if this approach is appropriate for the data set. The glmer function accepts the argument offset for that purpose.

It is possible to exploit the specialized link functions of the **psyphy** package using some other functions for fitting GLMMs. One example would be the glmmPQL function in the **MASS** package that fits a GLMM using penalized quasi-likelihood. While one sacrifices the rigor of having a likelihood function for comparing models, this approach can be quite useful under some circumstances. For an interesting example, see the article by Williams et al. [194].

These link functions also work with the development version of the **lme4** package (available from the Rforge repository at http://r-forge.r-project.org/projects/lme4/). This capability will eventually be integrated into the main version of the package on CRAN.

9.4 MLDS

In this section, we explore three strategies for analyzing MLDS experiments as mixed-effects models using glmer. We will apply these only to the Method of Quadruples, but in principle they are easily extended to the Method of Triads.

The situations for which mixed-effects models are appropriate are the same as those evoked above, for example, when data are collected from several observers, or when data are collected from the same observer over several sessions. In each of these cases, there are likely to be randomly varying components of the responses that we would like to quantify along with the fixed effects components. Observers may vary in their sensitivity to stimulus differences, in the perceptual strategies that enter their decision or simply in the attentional resources devoted to the task. Similar factors may influence the responses of an individual observer across sessions.

9.4.1 Regression on the Decision Variable: Fixed Functional Form

Recall that in Chap. 7, we examined the three data sets of an MLDS experiment repeated on the same observer, kk1, kk2, and kk3, in which he judged differences in correlation between pairs of pairs of scatterplots. We begin by comparing the perceptual scales obtained from the individual runs. We store the three data sets in a list, making it simple to fit each one in turn with mlds using lapply and to store the

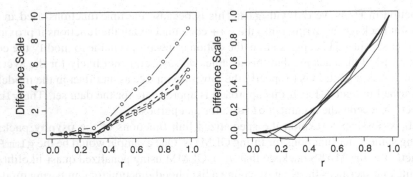

Fig. 9.8 Perceptual scale of perceived correlation in scatterplots obtained by MLDS. *Left*. The *thin lines* with *points* correspond to three repeated runs on the same observer. The *thick solid line* is the mean of the three perceptual scales. The *dashed curve* is the scale obtained by combining the three data sets and estimating one scale. *Right*. The *thin lines* indicate the individual scales replotted on the standard scale. The *thick line* is a parabola passing through 0 and 1

resulting model objects in a second list. We then plot each unconstrained scale with points connected by thin lines in the left panel of Fig. 9.8. Two of the scales are quite similar but the third is different, with maximum value nearly twice as high as the others. Recall that the differences on the unconstrained scale are proportional to d'. Thus, the observer shows greater sensitivity on that run than on the other two. This is unlikely to be due to learning, since it is the second run, not the third that shows this effect. The solid line shows the mean of the three runs. Note that this is different than the scale obtained by simply combining the three runs into a single data set, shown as the dashed line. The individual scales appear to be simply multiples of each other. This is supported by plotting the three runs as standard scales, i.e., normalized to 0 and 1 at the endpoints, displayed as the thin lines in the right panel of Fig. 9.8.

```
> library(MLDS)
> kk.lst <- list(R1 = kk1, R2 = kk2, R3 = kk3)
> kk.mlds.lst <- lapply(kk.lst, mlds)
> opar <- par(mfrow = c(1, 2))
> plot(0:1, c(0, 10), type = "n",
+    xlab = "r",
+    ylab = "Difference Scale")
> for(ix in 1:3) lines(kk.mlds.lst[[ix]], type = "b")
> lines(kk.mlds.lst[[1]]$stimulus,
+    rowMeans(sapply(kk.mlds.lst, "[[", "pscale")),
+    lwd = 3)
> lines(mlds(do.call(rbind, kk.lst)), lty = 2, lwd = 2)
> plot(0:1, c(0, 1), type = "n",
+    xlab = "r",
+    ylab = "Difference Scale")
> for(ix in 1:3) lines(kk.mlds.lst[[ix]], standard.scale = TRUE)
```

```
> x <- seq(0, 0.98, len = 100)
> lines(x, (x/0.98)^2, lwd = 3)
> par(opar)
```

We would like to estimate an average scale as a fixed effect and model the run-to-run variations as a random effect. Recall from (7.11) that the GLM model for MLDS is a sum of covariate terms as follows

$$\Phi^{-1}(\mathsf{E}[Y]) = \beta_2 X_2 + \beta_3 X_3 + \cdots + \beta_p X_p, \tag{9.8}$$

where each covariate is associated with a physical level of the stimulus presented in the experiment. On any given trial, only four of the stimulus levels appear, so that all but these values are set to zero in the corresponding row of the model matrix. The four stimulus levels of a given trial are coded by the weights of their contributions to the hypothesized decision rule for this task.

$$\delta(a,b;c,d) = \psi_d - \psi_c - \psi_b + \psi_a \tag{9.9}$$

The model has no intercept and the covariate corresponding to the first stimulus level is dropped from the design matrix to render the model identifiable but also to fix the first level at 0. Thus, a random intercept does not make sense for this model. What would be ideal is a random slope that is common across all of the $p-1$ covariates. We have not been able to devise a way to do this using glmer in terms of this parameterization of the model.

We can obtain an alternate parameterization, however, by taking advantage of the fact that the perceptual scale for this particular stimulus dimension appears to be well described by a parabola, i.e., the square of the stimulus dimension. This is shown in the right panel of Fig. 9.8 as the thick solid curve. Suppose, then, that for each trial, we calculate the decision variable, (9.9), on the standard scale of r^2.

$$DV = r_d^2 - r_c^2 - r_b^2 + r_a^2. \tag{9.10}$$

Note that the decision variable is linear in r^2. Thus, we can specify a model using this decision variable as a covariate with a random slope term.

$$\Phi^{-1}(\mathsf{E}[Y]) = (\beta + b_i)DV, \tag{9.11}$$

where $b_i \sim \mathcal{N}(0, \sigma^2)$. Now, the random effect applies across the scale. The following code fragment illustrates how we fit this model with glmer.

```
> Stim <- attr(kk1, "stimulus")
> kk <- do.call(rbind, kk.lst)
> kk$Run <- factor(rep(names(kk.lst), each = nrow(kk1)))
> kk$dv <- matrix(Stim[unlist(kk[, -c(1, 6)])]^2,
+    ncol = 4) %*% c(1, -1, -1, 1)
> kk.glmm <- glmer(resp ~ dv + (dv + 0 | Run) - 1, kk,
```

```
+    binomial("probit"))
> kk.glmm
Generalized linear mixed model fit by the Laplace approximation
Formula: resp ~ dv + (dv + 0 | Run) - 1
   Data: kk
 AIC BIC logLik deviance
 651 661   -323      647
Random effects:
 Groups Name Variance Std.Dev.
 Run     dv   2.28     1.51
Number of obs: 990, groups: Run, 3

Fixed effects:
   Estimate Std. Error z value Pr(>|z|)
dv    6.604     0.962     6.86  6.7e-12 ***
---
Signif. codes:  0 '***' 0.001 '**' 0.01 '*' 0.05 `.' 0.1 ' ' 1
```

The fixed effect is the coefficient of the curve of r^2 so it should provide an estimate of the unconstrained maximum. As such, it provides a closer estimate to the maximum of the mean of the individual scales, 6.43, than to the scale obtained from combining all runs, 5.57. Averaging over the variances at each scale value and taking the square root gives a value that is close to the standard deviation of the random effect, 1.31. However, the basis function that we have chosen to model the data, r^2, is only an approximation and is not likely to predict the observer's responses as well as the scale values obtained from the individual MLDS fits.

9.4.2   Regression on the Decision Variable: Rescaling to a Common Scale

In this and the next section, we make use of a data set from Fleming et al. [63] in which observers judged synthetic images of a transparent pebble-like object that was rendered to vary along a physical scale of index of refraction of the material. The two most extreme examples of the stimuli tested are shown in Fig. 9.9. MLDS was used with the Method of Quadruples, and data were collected from six observers, thus providing a data set with a more reasonable sample of levels for estimating random effects. The data set can be found in a data frame named Transparency in the **MLDS** package.

The data set contains six components. The first five consist of the standard results from a difference scaling experiment using the Method of Quadruples: the response and the indices of the four stimuli on each trial. The last component is a factor identifying the observer. There were 10 values along the physical scale tested and each quadruple was tested twice, so that there are 420 trials for each observer.

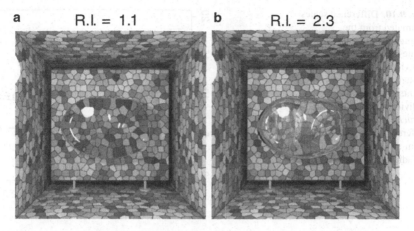

Fig. 9.9 Two examples of stimuli used in the difference scaling experiments of Fleming et al. [63]. The difference between the two images is the refractive index of the rendered pebble-like object

```
> data(Transparency)
> str(Transparency)
'data.frame':   2520 obs. of  6 variables:
 $ resp: num  0 0 1 1 1 1 1 0 1 0 ...
 $ S1  : int  1 1 1 1 1 1 1 1 1 1 ...
 $ S2  : int  2 2 2 2 2 2 2 2 2 2 ...
 $ S3  : int  3 3 3 3 3 3 3 4 4 4 ...
 $ S4  : int  4 5 6 7 8 9 10 5 6 7 ...
 $ Obs : Factor w/ 6 levels "01","02","03",..: 1 1 1 1 1 1..
 - attr(*, "stimulus")= num  1.1 1.23 1.37 1.5 1.63 1.77 1..
```

The estimated individual and average perceptual scales are displayed in Fig. 9.10. The observers' scales fall into three groups, two with low sensitivity, one with high sensitivity, and three in the middle. The thick dashed line is the scale obtained from fitting the whole data set with a single scale. As with the scatterplot example, this scale underestimates the average over individuals.

```
> Stim <- attr(Transparency, "stimulus")
> Tr.mlds.lst <- lapply(levels(Transparency$Obs),
    function(obs) {
+   tmp.df <- subset(Transparency, Obs == obs, select = 1:5)
+   mlds(as.mlds.df(tmp.df, st = Stim))
+   })
> Tr.mlds.cb <- mlds(as.mlds.df(Transparency[, -6], st = Stim))
> plot(Stim, rowMeans(sapply(Tr.mlds.lst, "[[", "pscale")),
+   type = "l", lwd = 3,  ylim = c(0, 20),
+   xlab = "Index of Refraction", ylab = "Difference Scale")
```

Fig. 9.10 Difference scales
estimated from the
`Transparency` data set.
Scales estimated for
individual observers are
plotted as *thin lines* with
points. The average of these
individual curves is plotted as
a *thick solid line*. The
difference scale estimated
from the aggregated data set
is indicated as a *dashed line*

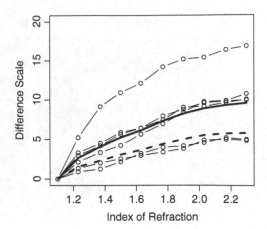

```
> for(md in Tr.mlds.lst) lines(md, type = "b")
> lines(Tr.mlds.cb, lty = 2, lwd = 3)
```

It is not always possible to find a simple function to describe the perceptual scale
that can be used as a basis term for the decision variable. The `Transparency` data
set provides an example. The average scale, as well as several of the individual
ones, display a kink in the middle that renders it difficult to describe them by a
simple basis function like a power function.

To motivate an alternate approach that does not assume that the scale follows a
particular functional form, we begin by noting that the actual spacing of the physical
values along the continuum being tested plays no role in the MLDS analysis.[1] In the
scatterplot experiment, the physical scale values tested were equally spaced along
the correlation axis but by re-spacing them as r^2, we render the perceptual scale
approximately linear (see Fig. 7.3). Thus, we ask whether there exists a transform
that would linearize the estimated perceptual scales.

One possibility is to calculate the decision variable in terms of the average
standard scale and let `glmer` find the correct scaling for each observer of this
average function as a random effect. Thus, we would use the average scale as we
used the quadratic function in the last example. We find that this approach tends to
underestimate the random effect for some observers, as in the preceding example,
because it will never predict his responses as well as his own scale does.

We overcome this difficulty if, instead, we use each observer's own standard scale
to calculate the decision variable. This procedure maps each observer's scale onto
the line segment from $(0,0)$ to $(1,1)$ although with individualized spacing. Then, we
fit a GLMM to decision variables obtained from these normalized scales. In effect,
we factor out shape differences between individual scales in order to observe more
clearly the amplitude differences between them. This entails a two-step approach

---

[1]With the exception to this in the use of the formula method.

in which first the perceptual scale is estimated individually for each observer using mlds. Then, each individual's scale is used to calculate his decision variables on each trial. The concatenated individualized decision variables enter as the covariate in a GLMM to predict the binary choices. The following code fragment performs these steps and fits the GLMM using glmer.

```
> stanScale <- sapply(Tr.mlds.lst, "[[", "pscale")
> stanScale <- apply(stanScale, 2, function(x) x/x[length(x)])
> colnames(stanScale) <- levels(Transparency$Obs)
> Tr.DV.lst <- lapply(levels(Transparency$Obs), function(obs) {
+       tmp.df <- subset(Transparency, Obs == obs, select = 1:5)
+       RX <- make.ix.mat(tmp.df)
+       MX <- RX[, -1]
+       MX <- cbind(stim.1 = ifelse(rowSums(MX), 1, 0), MX)
+       as.matrix(MX) %*% stanScale[, obs]
+       })
> Transparency$DV <- do.call(rbind, Tr.DV.lst)
> Tr.glmm <- glmer(resp ~ DV + (DV + 0 | Obs) - 1,
+       Transparency, binomial("probit"))
> Tr.glmm
Generalized linear mixed model fit by the Laplace approximation
Formula: resp ~ DV + (DV + 0 | Obs) - 1
   Data: Transparency
  AIC  BIC logLik deviance
 1485 1497  -741     1481
Random effects:
 Groups Name Variance Std.Dev.
 Obs    DV   13.7     3.69
Number of obs: 2520, groups: Obs, 6

Fixed effects:
   Estimate Std. Error z value Pr(>|z|)
DV    9.39       1.57       6  2e-09 ***
- - -
Signif. codes:  0 '***' 0.001 '**' 0.01 '*' 0.05 '.' 0.1 ' ' 1
```

The fixed effect coefficient for the term DV is a close estimate to that of the maximum value of the mean perceptual scale, 9.59, from Fig. 9.10. The random effect standard deviation should provide an estimate of the variability of the scale amplitudes across observers. It is similar to the square root of the average variance of the scale values, 3.59. We compare the individual scales with the predicted functions by rescaling each observer's standard perceptual scale by the sum of the fixed and random effects. We store these results in a data frame so that we can access them more easily when producing the plots with xyplot.

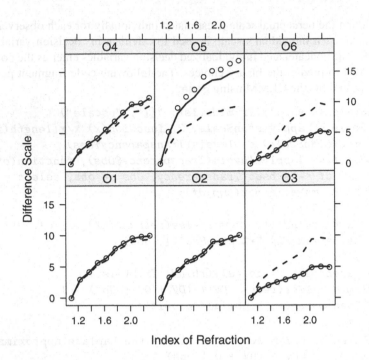

Fig. 9.11 Perceptual scales estimated by `mlds` for transparency perception (*points*) with random effect predictions for model `Tr.glmm` (*solid lines*) and fixed effect predictions (*dashed lines*)

```
> re <- ranef(Tr.glmm)$Obs$DV
> fe <- fixef(Tr.glmm)
> pref.df <- data.frame(Stimulus = rep(Stim, 6),
+    FEpsc = fe * c(stanScale),
+    REpsc = (rep(re, each = 10) + fe) * c(stanScale),
+    pscale = as.vector(sapply(Tr.mlds.lst, function(x)
+        x$pscale)),
+    Obs = rep(paste("O", 1:6, sep = ""), each = 10)
+    )
```

The individual graphs are shown in Fig. 9.11 as the solid lines. The points correspond to the individual perceptual scales. The dashed lines show the observer's scale adjusted by only the fixed effect coefficient. All of the curves provide good estimates of the amplitudes of the individual curves except for the underestimation of observer O5.

```
> print(
+ xyplot(pscale ~ Stimulus | Obs, pref.df,
+    xlab = "Index of Refraction", ylab = "Difference Scale",
+    subscripts = TRUE, ID = pref.df$Obs,
+    panel = function(x, y, subscripts, groups, ...){
+        panel.xyplot(x, y, col = "black")
```

```
+        panel.lines(x, pref.df$REpsc[subscripts],
+            col = "black", lwd = 2)
+        panel.lines(x, pref.df$FEpsc[subscripts],
+            lty = 2, lwd = 2, col = "black")
+        } )
+ )
```

This approach seems to provide a reasonable estimate of individual differences in observer sensitivity but, as already noted, at the expense of information concerning the shape of the scale.

9.4.3 Regression on the Estimated Coefficients

Finally, we demonstrate a third approach that consists of simply fitting a linear mixed-effects model to the individual perceptual scales. This is also a two-step approach, as first the perceptual scales are estimated individually with mlds and subsequently modeled using a linear mixed-effects model, here, using the lmer function from the **lme4.0** package. The stimulus dimension is used as a covariate. It may be treated as a factor with the stimulus levels as factor levels, but, if possible, we can also model the scales as a function of the physical stimulus values, thereby attempting to describe the data more parsimoniously. If the perceptual scale is approximately linear, then the stimulus values enter as a simple covariate for which we estimate the slope. There will be no intercept in the model as all the scales are constrained to pass through zero at the lowest stimulus level, by construction. If the perceptual scale is not linear, as here, then we may attempt to model it as a specific function of the stimulus values, if we know one, or using polynomial or spline terms. If the function is nonlinear in its parameters, then the nlmer or nlme functions can be used from packages **lme4.0** and **nlme**, respectively. The spline approach can be attractive within an additive model framework to control the degree of smoothness of the fitted function. Tools for fitting (generalized) additive mixed-effects models can be found within the **mgcv** [198] and **gamm4** [197] packages. The former depends on lme in the **nlme** package and the latter uses lmer in the **lme4** package. Here, we will demonstrate the use of orthogonal polynomials. The common formula interface of most modeling functions makes it relatively straightforward to explore the analyses using these other approaches.

To begin, we create a data frame with the coefficients from the perceptual scales fit to each individual by mlds.

```
> cStim <- mean(Stim[-1])
> Tr.df <- data.frame(
+        Stim = Stim[-1] - cStim,
+        Obs = rep(factor(paste("O", 1:6, sep = "")), each = 9),
+        logDS = c(log10(sapply(Tr.mlds.lst, coef)))
+        )
```

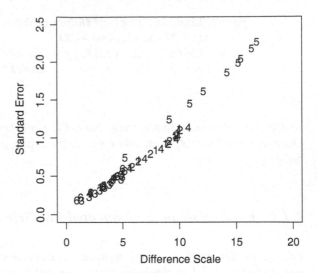

Fig. 9.12 Standard errors as a function of the individual estimated scale values. Each observer's data are plotted as a number. Note that observer 5 stands out on this plot

We include the stimulus values as a covariate but drop the first level at which all the scales equal zero and center the scale with respect to its mean. Then, we specify an identifier for each observer in the factor Obs. The function sapply is used to extract the individually estimated perceptual scale values from the list in which we previously stored the results from each individual, Tr.mlds.lst, and transformed logarithmically. Our justification for this transformation is two-fold. First, this conveniently transforms estimation of a multiplicative coefficient of the perceptual scales to that of an additive intercept. Second, this transformation may reduce heterogeneity in the variance of the estimated perceptual scale values. Recall that the function mlds uses glm as the default method to fit the MLDS model and stores the model object as a component, obj. We extract the perceptual scale estimates and the estimated standard errors with the coef and summary methods and note that the linear relation between the two suggests that the logarithm would be an appropriate variance stabilizing transform (Fig. 9.12).

```
> plot(c(0, 20), c(0, 2.5), type = "n",
+    xlab = "Difference Scale",
+    ylab = "Standard Error")
> for(ix in 1:6)
+    points(coef(summary(Tr.mlds.lst[[ix]]$obj))[, 1:2],
+        pch =as.character(ix))
```

We begin by testing the random effects. We specify the fixed effects in the model with orthogonal polynomials using the poly function. Its first argument is the vector over which the polynomial coefficients for each term will be calculated, here the vector Stim. The second argument indicates the degree of the highest term, here 6, which we expect to be higher than required to describe the variation on 10 points. Below, when we test the fixed effects, we will see how much this term can be reduced. The function `poly` returns a matrix of coefficients for each of the terms of the polynomial, excluding the intercept. We could specify the random terms using `poly`, as well, but this complicates testing the independence of the random effects, so we use the covariate `Stim` from the data frame which is the previously defined normalized stimulus dimension.

```
> P3 <- lmer(logDS ~ poly(Stim, degree = 6) +
+    (Stim + I(Stim^2) + I(Stim^3) | Obs), Tr.df,
+    REML = FALSE)
> P2 <- lmer(logDS ~ poly(Stim, degree = 6) +
+    (Stim + I(Stim^2) | Obs), Tr.df, REML = FALSE)
> P1 <- lmer(logDS ~ poly(Stim, degree = 6) +
+    (Stim | Obs), Tr.df, REML = FALSE)
> P0 <- lmer(logDS ~ poly(Stim, degree = 6) +
+    (1 | Obs), Tr.df, REML = FALSE)
> anova(P0, P1, P2, P3)
Data: Tr.df
Models:
P0: logDS ~ poly(Stim, degree = 6) + (1 | Obs)
P1: logDS ~ poly(Stim, degree = 6) + (Stim | Obs)
P2: logDS ~ poly(Stim, degree = 6) + (Stim + I(Stim^2) |
P2:      Obs)
P3: logDS ~ poly(Stim, degree = 6) + (Stim + I(Stim^2) +
P3:      I(Stim^3) | Obs)
   Df  AIC  BIC logLik Chisq Chi Df Pr(>Chisq)
P0  9 -126 -108   71.9
P1 11 -157 -135   89.7 35.46      2     2e-08 ***
P2 14 -162 -135   95.3 11.20      3     0.011 *
P3 18 -158 -123   97.2  3.94      4     0.414
---
Signif. codes:  0 '***' 0.001 '**' 0.01 '*' 0.05 '.' 0.1 ' ' 1
```

We specify REML = FALSE to avoid potential complications in performing likelihood ratio tests between the models.[2] The likelihood ratio tests indicate that model P3 does not fit significantly better than model P2 so we retain a model with random coefficients of the intercept, linear, and quadratic terms. Examination of the summary output indicates some high correlations between these random effects. We test whether these correlations can be eliminated by specifying the random effects as independent terms and comparing with model P2.

```
> P2.a <- lmer(logDS ~ poly(Stim, degree = 6) +
+    (Stim | Obs) + (I(Stim^2) + 0 | Obs), Tr.df,
+    REML = FALSE)
```

---

[2]The REML criterion leads to less biased variance estimates. However, some models are not nested under this criterion. See [141] for details.

```
> P2.b <-  lmer(logDS ~ poly(Stim, degree = 6) +
+   (1 | Obs) +
+   (Stim + 0 | Obs) + (I(Stim^2) + 0 | Obs), Tr.df,
+   REML = FALSE)
> anova(P2, P2.a, P2.b)


> P2.a <- lmer(logDS ~ poly(Stim, degree = 6) +
+   (Stim | Obs) + (I(Stim^2) + 0 | Obs), Tr.df,
+   REML = FALSE)
> P2.b <-  lmer(logDS ~ poly(Stim, degree = 6) +
+   (1 | Obs) +
+   (Stim + 0 | Obs) + (I(Stim^2) + 0 | Obs), Tr.df,
+   REML = FALSE)
> anova(P2, P2.a, P2.b)
Data: Tr.df
Models:
P2.b: logDS ~ poly(Stim, degree = 6) + (1 | Obs) + (Stim +
P2.b:    0 | Obs) + (I(Stim^2) + 0 | Obs)
P2.a: logDS ~ poly(Stim, degree = 6) + (Stim | Obs) +
P2.a:    (I(Stim^2) +  0 | Obs)
P2: logDS ~ poly(Stim, degree = 6) + (Stim + I(Stim^2) |
P2:    Obs)
      Df  AIC  BIC logLik Chisq Chi Df Pr(>Chisq)
P2.b 11 -160 -138   91.1
P2.a 12 -162 -138   92.9  3.55      1      0.060 .
P2   14 -162 -135   95.3  4.70      2      0.095 .
---
Signif. codes:  0
```

The results indicate borderline nonsignificance of the correlations among the random effects. However, as noted earlier, the parameters of interest are forced to the border of the parameter space and the p-values can be too high by a factor of two. Thus, we choose to retain the correlations in the model specification.

Since we transformed the scale values by a logarithm prior to analyses, these random effects are multiplicative on the original scale. The linear term on this scale, then, becomes exponential and the quadratic, Gaussian. These random effects might correspond to different strategies of cue integration by observers in making their judgments. The negative correlation between the linear and quadratic components suggest that observers base their judgments on some combination of cues with some giving more weight to a cue that varies exponentially along the stimulus dimension and others one that varies as a Gaussian. Confirmation of this hypothesis would require a much larger sample of observers.

The summary output of model P2 suggests that fixed effect terms including the third degree might be significant. We test this with a series of nested models of decreasing degree of the highest term and with the random effects decided upon above.

```
> P2.2 <- lmer(logDS ~ poly(Stim, degree = 2) +
+    (Stim + I(Stim^2) | Obs), Tr.df)
> P2.3 <- lmer(logDS ~ poly(Stim, degree = 3) +
+    (Stim + I(Stim^2) | Obs), Tr.df)
> P2.4 <- lmer(logDS ~ poly(Stim, degree = 4) +
+    (Stim + I(Stim^2) | Obs), Tr.df)
> P2.5 <- lmer(logDS ~ poly(Stim, degree = 5) +
+    (Stim + I(Stim^2) | Obs), Tr.df)
> anova(P2, P2.5, P2.4, P2.3, P2.2)
Data: Tr.df
Models:
P2.2: logDS ~ poly(Stim, degree = 2) + (Stim + I(Stim^2) |
P2.2:    Obs)
P2.3: logDS ~ poly(Stim, degree = 3) + (Stim + I(Stim^2) |
P2.3:    Obs)
P2.4: logDS ~ poly(Stim, degree = 4) + (Stim + I(Stim^2) |
P2.4:    Obs)
P2.5: logDS ~ poly(Stim, degree = 5) + (Stim + I(Stim^2) |
P2.5:    Obs)
P2: logDS ~ poly(Stim, degree = 6) + (Stim + I(Stim^2) |
P2:    Obs)
     Df  AIC  BIC logLik Chisq Chi Df Pr(>Chisq)
P2.2 10 -163 -143   91.5
P2.3 11 -167 -145   94.5  6.03      1     0.014 *
P2.4 12 -166 -142   95.1  1.24      1     0.266
P2.5 13 -164 -138   95.1  0.04      1     0.838
P2   14 -162 -135   95.3  0.25      1     0.619
---
Signif. codes:  0 '***' 0.001 '**' 0.01 '*' 0.05 '.' 0.1 ' ' 1
```

The likelihood ratio tests confirm the need for a cubic term and none higher.

We should check various model diagnostics at this point (and leave this as an exercise for the reader). Instead, we proceed to plot the results for the individual curves predicted from the model. The fixed effect prediction is obtained by first extracting the fixed effect coefficients from the model using the fixef method. These are coefficients for the polynomial basis terms that are obtained as a matrix with the poly function defined on the Stim covariate and with degree = 3. The fixed effects include an intercept term so an intercept column is added to the basis matrix.

```
> feP23 <- fixef(P2.3)
> fePred23 <- cbind(1, poly(Tr.df$Stim, degree = 3)) %*% feP23
> Tr.df$fit23 <- fitted(P2.3)
```

As a shortcut, we directly use the fitted values at the stimulus values tested rather than combining fixed and random effects to generate the individual curves. We simply extract the fitted values from the model object with the method `fitted`.

```
> print(
+ xyplot(10^logDS ~ I(Stim + cStim) | Obs, Tr.df,
+    ylim = c(0, 20), xlab = "Index of Refraction",
+    ylab = "Difference Scale", subscripts = TRUE,
+       ID = Tr.df$Obs,
+    panel = function(x, y, subscripts, groups, ...){
+     panel.xyplot(x, y, col = "black")
+     panel.lines(x, 10^(Tr.df$fit23[subscripts]), col = "black",
+       lwd = 2)
+     panel.lines(x, 10^fePred23[subscripts], lty = 2,
+       lwd = 2, col = "black")
+     }), more = TRUE, split = c(1, 1, 2, 1) )
> print(
+ xyplot(logDS ~ I(Stim + cStim) | Obs, Tr.df,
+    xlab = "Index of Refraction", ylab = "Log Difference Scale",
+    subscripts = TRUE, ID = Tr.df$Obs,
+    panel = function(x, y, subscripts, groups, ...){
+     panel.xyplot(x, y, col = "black")
+     panel.lines(x, Tr.df$fit23[subscripts], col = "black",
+       lwd = 2)
+     panel.lines(x, fePred23[subscripts], lty = 2, lwd = 2,
+       col = "black")
+     }), more = FALSE, split = c(2, 1, 2, 1))
```

The individual scales (points) and the model predictions for both the random effects (solid lines) and the fixed effect (dashed lines) are shown in Fig. 9.13. The results have been plotted in two ways. On the left, the scale values were transformed back to their original units in which equal differences correspond to equal perceptual steps. On the right, the scales are in the logarithmic units, which correspond to the scales on which the above analyses were performed. The model appears to provide a good description of the scale shape as well as the individual variations across the sample both in sensitivity and shape.

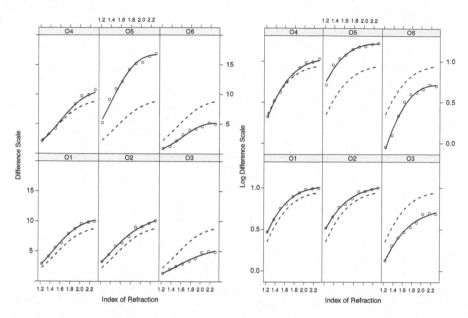

Fig. 9.13 Perceptual scales estimated by `mlds` for transparency perception (*points*) and predictions for the random effects (*solid lines*) and the fixed effects (*dashed lines*) obtained by fitting a linear mixed-effects model directly to the perceptual scales. On the *left*, the data are plotted in the original scale units. On the *right*, the log of the scales is plotted

9.5   Whither from Here

The examples presented in this somewhat lengthy chapter provide only a brief overview of what can be obtained by using mixed-effects models with psychophysical data. A full treatment would certainly require a book, such as the excellent one devoted to mixed-effects models applied to data from the field of ecology by Zuur et al. [207].

Some extensions of the above approaches are relatively straightforward. For example, the underlying structure of the decision rules for MLDS and MLCM are sufficiently similar that the approaches for the former should be able to be adapted easily to the latter. Specifying mixed-effects models for classification images should be simple as well, though one is likely to confront fitting difficulties from memory limitations. In any case, few experiments are performed in this domain with a sufficient number of observers for this modeling approach to be sensible.

Our hope is that we have whetted the reader's appetite for using R sufficiently that (s)he will be motivated not just to follow the examples of the book but to explore, as well, other methods that we have not covered for lack of space, all with the goal of extracting the maximum amount of information from their data to clarify the scientific questions at hand.

Appendix A
Some Basics of R

A.1 Installing R

The first order of business is to download and install R on your computer (if you haven't already). Make sure you are connected to the Internet, and go to the R web site, http://www.r-project.org/. This site is worth exploring. You will find considerable useful information here concerning R including a list of books treating elementary and advanced topics in statistics with extensive examples in R. But right now our goal is to download and install the most recent version of R that is appropriate for your type of computer.

To access the site for downloading, click on the link labeled CRAN on the list at the left of the page. CRAN stands for Comprehensive R Archive Network. Then, choose a mirror site that is geographically close to you.

R may be compiled from source but it is strongly recommended to begin by installing a pre-compiled, binary version. Choose your platform by clicking on one of Linux, Mac OS X or Windows and follow any links to the binaries for the base distribution. For example, in the case of Linux, you will need to proceed further to choose your distribution. Download the binary and follow normal procedures for installation. In the case of Windows and Mac OS X, this means just double clicking on the icon for the binary installation and following the instructions. Linux users will follow the installation procedures that are specific to their distribution, for example using rpm for Fedora distributions. After a few minutes, depending on the speed of your Internet connection, you will be ready to open an R session.

Running R will depend on your platform. For Windows and Mac OS X, there is a GUI interface that can be opened by simply double clicking on the R icon that has been installed by the above procedures. Under unix (e.g., linux and in a terminal window under Mac OS X), R can be run from a terminal shell command line by entering the command "R." A popular option under unix is to run R from within an editor that has macro and syntax coloring capabilities, such as emacs using the ESS (Emacs Speaks Statistics) macros, providing a more powerful interface than the command line on its own. Documentation can be found by searching for "ESS"

K. Knoblauch and L.T. Maloney, *Modeling Psychophysical Data in R*, Use R! 32,
DOI 10.1007/978-1-4614-4475-6,
© Springer Science+Business Media New York 2012

on http://www.r-project.org/ under "Related Projects" or more directly at the ESS homepage http://ess.r-project.org/. Rstudio (http://rstudio.org/), a relatively new project at the time of this writing, provides a platform independent GUI with a uniform interface and a number of additional features to help the user interact with R.

You only need to download and install an executable copy of R by following the instructions above. This copy includes extensive documentation, about 30 packages that are ready to be loaded and used in an R session, including a large collection of sample data sets used in illustrating commands.

A.2   Basic Data Types

Everything in R is an object, even the functions that operate on them. Here, we introduce some of the basic object types in R and how to manipulate them. A more detailed and definitive explanation will be found in the documentation that comes with R.

A.2.1   Vectors

The simplest data objects in R are vectors. These are indexed sets of elements of all the same class. The simplest vector types are called atomic and the most frequently encountered of these have types "numeric," "integer," "character," "logical." There are also atomic vectors of type "complex" and "raw."

Vectors are created in numerous ways. The simplest uses the function c

```
> ( x <- c(1.2, 4, 5.1, 7.2, pi) )

[1] 1.20 4.00 5.10 7.20 3.14
```

which here assigns a 5-element vector of class "numeric" to the variable x. Note that the variable pi is predefined in R. There are no scalar variables, but only vectors of length 1.

```
> ( y <- 5 )
[1] 5
```

In a similar fashion, we can create vectors of other classes.

```
> a <- c("Over", "the", "rainbow")
> d <- x < 4
> ls.str()
a :    chr [1:3] "Over" "the" "rainbow"
d :    logi [1:5] TRUE FALSE FALSE FALSE TRUE
x :    num [1:5] 1.2 4 5.1 7.2 3.14
y :    num 5
```

We created two vectors, respectively, of class "character" and "logical" that we will use in subsequent examples in this section. We, also, used the function ls.str to show the structures of the four vectors that we have added to the workspace. The vector a has three elements, character strings, each of class "character."

Note how we also created a logical vector d from a conditional operation on a vector of another class. In mathematical operations, the logical values of FALSE and TRUE take on the numeric values of 0 and 1, respectively.

```
> sum(d)
[1] 2
```

R contains standard comparison operators for creating logical vectors, such as >, <, >=, etc. We note that the negation operator in R is ! and the "not equal" operator !=. To test equality, we use the double equal sign, ==, as in many other computer languages. However, equality comparisons between "numeric" variables should be performed with circumspection because of digital limitations in the representation of floating point numbers.

```
> sqrt(3)^2 == 3
[1] FALSE
```

For this, R has the all.equal function to compare objects of near equality.

```
> all.equal(sqrt(3)^2, 3)
[1] TRUE
```

R contains two operators for logical AND. The first is & which performs comparisons element-by-element between a pair of vectors of numeric, logical or complex type. When used on numeric or complex values, a nonzero entry counts as TRUE, a zero entry as FALSE.

```
> c(1,0,2,0) & c(1,2,3,4)
[1]  TRUE FALSE  TRUE FALSE
```

The operator && is used for a comparison between a single pair of elements numeric, logical, or complex variables. If && is applied to a pair of vectors, only the first pair of elements from each will be compared.

```
> c(1,0,2,0) && c(1,2,3,4)
[1] TRUE
```

which would probably not be the intended purpose. A similar pair of operators exists for OR operations: | for pairwise comparison of vector elements and || for a single comparison.

There are numerous other functions that are useful for generating vectors with specific structure. To generate sequences,

```
> seq(1, 10, 2)
[1] 1 3 5 7 9
```

```
> seq(0, 1, length.out = 5)
[1] 0.00 0.25 0.50 0.75 1.00
```

or sequences that include repeated values

```
> rep(a, 3)
[1] "Over"    "the"     "rainbow" "Over"    "the"
[6] "rainbow" "Over"    "the"     "rainbow"
> rep(a, each = 3)
[1] "Over"    "Over"    "Over"    "the"     "the"
[6] "the"     "rainbow" "rainbow" "rainbow"
> rep(a, 1:3)
[1] "Over"    "the"     "the"     "rainbow" "rainbow"
[6] "rainbow"
```

using the character vector a previously defined on p. 305. The : operator is a shortcut for seq(1, 3) for which the third argument takes the default value of 1. The various forms of the second argument of rep allow one to generate different patterns of repetition. We will introduce and use these and similar functions throughout the book for coding explanatory variables that will be useful in simulation and modeling.

All vector types can include missing data indicated by the value NA (for not available).

```
> ( z <- c(3.1, 1.8, NA, 0.24) )
[1] 3.10 1.80   NA 0.24
```

Thus, the elements of logical vectors can be in any of three states {TRUE, FALSE, NA}. Some care must be taken with missing data, however, as all operations with input including an NA yield NA. Many functions include an argument na.rm that, when set to TRUE, conveniently causes the missing values to be ignored.

```
> mean(z)
[1] NA
> mean(z, na.rm = TRUE)
[1] 1.71
```

The elements of vectors are accessed using square brackets, as in

```
> x[4]
[1] 7.2
> a[c(2, 3)]
[1] "the"     "rainbow"
> x[ x > 4]
[1] 5.1 7.2
```

In the second and third examples above, the indices are a vector, permitting selection of several elements at once. In the third example, this vector was generated by a condition, which results in the selection of the elements for which the condition is TRUE. A particularly useful feature is the use of negative indices to specify which elements to exclude.

```
> v <- 1:10
> v[-seq(1, 5, 2)]
[1]  2  4  6  7  8  9 10
```

Positive and negative indices cannot be mixed, however.

There are several special functions for accessing the structure of the elements of a character string. For example,

```
> nchar(a[3])
[1] 7
> substring(a[1], 2, 4)
[1] "ver"
```

the effects of which should be obvious from the examples.

Ordinary arithmetic operations work element by element on vectors of the same length.

```
> x + 1:5
[1]  2.20  6.00  8.10 11.20  8.14
> x * rep(c(2, 3), c(3, 2))
[1]  2.40  8.00 10.20 21.60  9.42
```

With vectors of different lengths, the shorter one is automatically recycled, but a warning will be generated if the shorter length is not a multiple of the longer one.

```
> x + y
[1]  6.20  9.00 10.10 12.20  8.14
> x + 1:2
[1] 2.20 6.00 6.10 9.20 4.14

Warning message: In x + 1:2 :
 longer object length is not a multiple of shorter object length
```

A.2.1.1   Attributes and Indexing

A powerful feature of R is that objects may have *attributes* attached to them. These can influence how the object is treated by functions. For example, a vector can have a names attribute that allows the elements to be indexed by these names.

```
> w <- c(a = 5, b = 16, c = 4.1)
> w
```

```
    a    b    c
  5.0 16.0  4.1
> w["b"]
  b
 16
> w[letters[2:3]]
    b    c
 16.0  4.1
```

The names attribute affects how vectors are printed (i.e., by how they are displayed by their print method, implicitly called when we give the variable name). We, also, see the use of a predefined vector, letters, that contains the set of lowercase letters, to index the named elements of w. A similar vector LETTERS is predefined for the uppercase letters.

In R, matrices are just vectors that have a dim attribute.

```
> u <- 1:8
> dim(u) <- c(2, 4)
> u
     [,1] [,2] [,3] [,4]
[1,]    1    3    5    7
[2,]    2    4    6    8
```

The dim attribute has even changed the class of the vector, which you can check with the function class. The matrix is filled column by column.

Arrays are vectors with more than 2 dimensions.

```
> dim(u) <- rep(2, 3)
> class(u)
[1] "array"
> str(u)
 int [1:2, 1:2, 1:2] 1 2 3 4 5 6 7 8
```

Although assigning a dim attribute suffices for transforming a vector to a matrix (array), normally one would create a matrix (array) using the matrix (array) function(s).

```
> m <- matrix(1:9, nrow = 3, ncol = 3)
> arr <- array(1:16, dim = c(2, 2, 4))
```

Matrix elements are indexed or subset as with vectors using brackets but with an index for each dimension.

```
> m[2, 3]
[1] 8
> m[3, ]
[1] 3 6 9
```

```
> m[, -2]
     [,1] [,2]
[1,]   1    7
[2,]   2    8
[3,]   3    9
```

Leaving a dimension blank is a shortcut for including all of its values, and negative indexing can be used as well.

Matrix multiplication is defined by the operator %*% as in

```
> m %*% m[, -1]
     [,1] [,2]
[1,]   66  102
[2,]   81  126
[3,]   96  150
```

but, of course, only between matrices with conforming dimensions.

A.2.2   Factors

Factors are a special class of vector that are used for representing categorical variables. The factor can take on values from a finite set of categories called levels. The levels are attached to the variable as an attribute, but it is not necessary that all of the levels occur in the vector. For example, suppose that we wanted to categorize observers with respect to their type of color vision. The levels might be

```
> ColVisLevs <- c("Normal", "Protan", "Deutan", "Tritan")
```

and we define a factor

```
> ColVisType <- factor(rep(ColVisLevs, c(5, 1, 2, 0)),
+        levels = ColVisLevs)
> ColVisType
[1] Normal Normal Normal Normal Normal Protan Deutan Deutan
Levels: Normal Protan Deutan Tritan
```

We use the function factor to coerce a character vector to class "factor." The level "Tritan" does not occur in our vector but using the levels = ColVisLevs argument, we ensure that our factor has four levels.

Ordinarily, factors are printed in terms of the labels given to their levels, but, in fact, underlying the labels, the levels are coded as integers, here 1–4. This is revealed with the unclass function.

```
> unclass(ColVisType)
[1] 1 1 1 1 1 2 3 3
attr(,"levels")
[1] "Normal" "Protan"  "Deutan" "Tritan"
```

The argument levels = ColVisLevs, specified in the factor function above, ensured that the levels were assigned to the codes in the order given, rather than the default alphabetical order. The difference between the underlying codes of the factors and the labels of their levels is critical to keep in mind when manipulating factors, or error will surely ensue. The great value of factors is in modeling data. They are used in specifying categorical variables in a specialized language for specifying models in a simplified formula (e.g., see Chap. 2) and exploited internally by modeling functions to construct the design matrix for estimating the model parameters.

In some cases, categorical variables can be ordered. For example, a finer categorization of color deficient observers might indicate them as "simple anomalous," "extreme anomalous," and "dichromatic." There is a natural order among these categories in terms of capacity to make fine discriminations in certain parts of color space, which parts depending upon which type of color deficiency. Such information is included in a factor using the function ordered.

```
> DiscriminationLevs <- c("simple", "extreme", "dichromat")
> ColDiscType <- ordered(DiscriminationLevs[c(1,
+       1, 3, 2, 1, 2, 2, 3)], levels = DiscriminationLevs)
> ColDiscType
[1] simple    simple    dichromat extreme   simple
[6] extreme extreme   dichromat
Levels: simple < extreme < dichromat
```

Examining the class of such objects indicates that they have both classes "ordered" and "factor," so that they inherit the properties of ordinary factors. They are of value primarily when modeling. For example, we will use them in Chap. 3 for the representation of observer ratings.

A.2.3 Lists

Vectors, matrices, and arrays are limited to containing elements of all the same type. Lists, however, may contain objects of different types, even other lists.

```
> a.lst <- list(A = LETTERS[1:5], B = (1:5)^2,
+   state = c(TRUE, FALSE, TRUE),
+   f = factor(c("Male", "Female", "Male", "Female", "Female")))
> a.lst
$A [1] "A" "B" "C" "D" "E"

$B [1]  1  4  9 16 25

$state
[1]  TRUE FALSE  TRUE
```

```
$f [1] Male    Female Male    Female Female
Levels: Female Male
```

a.1st contains four components, each one a vector of a different class and length. There are several ways to access the components. Here, the components are named, but if they are not, then individual components can be accessed using double brackets.

```
> a.1st[[2]]
[1]  1  4  9 16 25
```

which in this case returns a vector of "numeric." Unlike for vectors and matrices, only one index, not a range, is permissible. Some care is required because single brackets may be used, but they return a list whose first component is the object, not the object itself.

```
> a.1st[2]
$B [1]  1  4  9 16 25
```

which returns a list with component B, a vector of "numeric."

When the components are named, they may be accessed using the $ operator.

```
> a.1st$state
[1]  TRUE FALSE  TRUE
```

The name may also be used with the double brackets if quoted.

```
> a.1st[["state"]]
[1]  TRUE FALSE  TRUE
```

Lists are quite flexible and many of the exotic objects one runs into in R are simply lists that have been assigned a specific class and for which specific method functions have been written.

A.2.4 Data Frames

The data frame is a class that is ubiquitous in R because it conveniently structures the data in diverse circumstances. In essence, it is a list with components that are vectors of all the same length. A data frame is usually created with the data.frame function. Here is a simple example,

```
> d.df <- data.frame(x = c(0.12, 0.15, 0.11, 0.13, 0.10, 0.25,
+                 0.22, 0.27), Sex = c("Female", "Male", "Female",
+                 "Female", "Male", "Male", "Male", "Male"),
+                 CVT = ColVisType
+                 )
> d.df
```

```
     x    Sex     CVT
1 0.12 Female Normal
2 0.15   Male Normal
3 0.11 Female Normal
4 0.13 Female Normal
5 0.10   Male Normal
6 0.25   Male Protan
7 0.22   Male Deutan
8 0.27   Male Deutan
```

This data frame has three components, a "numeric" vector x, and two factor vectors, Sex and CVT that we already defined above. Note that we did not specify that Sex would be a factor but it was coerced to one automatically. Because the components all have the same length, the data frame has a rectangular structure. It is displayed by its print method in tabular format. Typically, each row will correspond to an experimental observation and each column to a characteristic of the observation, such as the response and values of the observation on other explanatory variables, covariates, and factors.

Because it is a list, a data frame can be accessed by any of the methods used for lists.

```
> d.df$Sex
[1] Female Male Female Female Male Male Male Male Levels:
Female Male
> d.df[[1]]
[1] 0.12 0.15 0.11 0.13 0.10 0.25 0.22 0.27
> d.df[["CVT"]]
[1] Normal Normal Normal Normal Normal Protan Deutan Deutan
Levels: Normal Protan Deutan Tritan
```

Additionally, because of their tabular structure, methods have been defined so that data frames can also be accessed as if they were matrices.

```
> d.df[5, 3]
[1] Normal Levels: Normal Protan Deutan Tritan
> d.df[, 1]
[1] 0.12 0.15 0.11 0.13 0.10 0.25 0.22 0.27
> d.df[4, ]
     x    Sex     CVT
4 0.13 Female Normal
```

Note, however, that the first two examples above return vectors whereas the third returns a 1-row data frame. This makes sense since the items in a column are all of the same type, as each column *is* a vector. In contrast, different columns are likely to be of different types, preventing a row from being represented as a vector.

A data frame can be created manually as above or read-in from data files, as we will see in Sect. A.5.1.

A.2.5 Other Types of Objects

The descriptions above cover the built-in data types that one is most likely to encounter when beginning to use R. A few others are treated in the main text (e.g., formula objects and functions). Other more exotic types of built-in objects, such as environments, will not be treated in this book.[1] It can be useful to understand something about them, however, and the reader is encouraged to read the documentation that comes with R to learn more as necessary. There is a much larger set of derived objects defined as the output to modeling functions. A complication is that there are two object systems operational in R, designated as S3 and S4.

Many of the objects defined with respect to the S3 system are simply lists that have been assigned a special class. Specific methods for such a class will typically be defined to allow manipulating and extracting information from such objects, although in the event that a method does not exist, components can be accessed as if the object were an ordinary list.

The S4 system is more formalized. Objects defined using S4 methods have *slots* rather than components. Slots are accessed using the @ symbol rather than the $ symbol. Knowledge of S4 methods beyond this will not be necessary in this book.

A.3 Simple Programming Constructs

For programming, R includes standard conditional constructs, such as if and else for testing a single condition, but also ifelse for testing a vector.

```
> x <- 1:6
> ifelse(x > 3, "red", "green")
[1] "green" "green" "green" "red"   "red"   "red"
```

The switch function is used to choose an action based on multiple values of a variable.

Iterative looping can be programmed with for, while and repeat constructs. R also includes a family of apply functions, apply, tapply, sapply, etc., that can loop over matrices, factors, lists, etc. Examples of their use and explanation are scattered throughout the book.

A.4 Interacting with the Operating System

It is generally recommended to consecrate a separate directory for each project that you undertake in R. This permits keeping data sets and scripts related to the same project together. The function getwd returns the path to the current directory as

---

[1] The work space is an environment, as are each of the elements on the search path.

a character string, and the function setwd will change the working directory to the path name given by a character string as its first argument. The file names in the working directory are returned in a character vector by the function dir but it can also be used to return the names or the names with their path from the current directory or names containing a particular pattern of text. This is particularly useful in automating the read-in and analysis of sets of data files all at once.

Another useful function for interacting with the operating system is system which takes a system command as a text string and runs it in the operating system. It may return the result to R if so specified. See its help page. Also, the function pipe can be used to read data to and from a connection given as a character string (See Sect. A.5.7).

A.5   Getting Data into and Out of R

Nearly all the data sets in this book are retrieved either from packages, such as the package for this book, **MPDiR**, or generated by simulation. Most packages contain sample data sets for demonstration of the types of data to which their methods apply. In addition, R comes by default with over 100 data sets available in memory (try ls("package:datasets")) that provide a great variety of sample data with which to play, with which to model and from which to learn. Many packages use the LazyData option, so that the data sets are available by name on the search path as data frames as soon as the package has been loaded with the library function. If the package does not exploit the LazyData option, then data sets can still be read into the workspace using the data function, as demonstrated throughout the book. Alternatively, data sets can be read into the workspace without first loading the package by specifying the package name as an additional argument. For example, to load the HSP data set into the workspace without loading the **MPDiR** package, one executes

```
> data(HSP, package = "MPDiR")
```

where the package name is given as a character vector.

Most users will want to work on their own data, however, and will require understanding how to store on disk and retrieve them into memory. R is equipped with a diverse set of functions for accessing different formats and sources of data. In practice, just a few of these functions will probably serve nearly all of your needs. In the rest of this chapter, we will describe some of the facilities in R for accessing data. For a more thorough introduction, see the document "R Data Import/Export" which comes with any R installation. Nearly all books on R cover this material to some extent, but in particular see [132, 162] for a comprehensive exposition of the subject.

A.5.1 Writing and Reading Text Files

A.5.2 write.table

Text files are extremely versatile for storing data as they can be created and accessed by many different programs. They provide a simple medium for storing a data frame and restoring it to the work space. To write a data frame to a storage medium, use the write.table function. For example, the following code suffices to store HSP on disk, in a text file, supposing that it is still in the work space.

```
> write.table(HSP, file = "HSP.txt")
```

The name of the object and the file name to which it will be written must be specified. If no file by that name exists in the working directory, then one will be created. If one exists, it will be overwritten, unless the argument append = TRUE is included. This argument allows the user to append multiple data frames to one file. There are several other arguments that allow additional control of the format of the output file, for example, specifying the field separator, the end of line character, whether row and column names should be included, etc. As usual, see the help page for the details.

The output file is an ordinary text file that can be edited or displayed. If the argument row.names = FALSE is specified, then the file structure is rectangular with the same number of fields on each line, possibly with the column names as the first line. If row and column names are included in the output, then the first line of column names will include one less field (it has no row name). This presents no problem as the read.table command to which we will now turn will automatically interpret the first line as column names when it has one fewer field than the succeeding lines.

A.5.3 read.table

As indicated above, to read in a text file in a tabular format, possibly with a header line, the read.table function is extremely versatile. To read in the file that we just saved is as simple as

```
> HSP <- read.table("HSP.txt")
> str(HSP)
'data.frame':    30 obs. of  5 variables:
 $ Q  : num  46.9 73.1 113.8 177.4 276.1 ...
 $ p  : num  0 9.4 33.3 73.5 100 100 0 7.5 40 80 ...
 $ N  : int  35 35 35 35 35 35 40 40 40 40 ...
 $ Obs: Factor w/ 3 levels "MHP","SH","SS": 2 2 2 2 2 2 2 ..
 $ Run: Factor w/ 2 levels "R1","R2": 1 1 1 1 1 1 2 2 2 2 ..
```

The first argument is the file name specified as a character string. The succeeding function call shows that `read.table` returns a data frame. The character variables are read in as factors by default, but character variables may, in appropriate circumstances, be read-in as logical, complex, or even numeric. If "character" is the preferred class for the character columns in the file, then the argument `as.is = TRUE` should be included. The function distinguishes between columns that contain numbers with a decimal point, read in as numeric, and those without, read in as integers. If the text file does not have a column of row numbers, and the first line is a row of column names, then the argument `header = TRUE` should be included. The columns will be given default names `V1, V2, ...` when `header = FALSE` is specified. The help page for `read.table` describes 22 possible arguments that provide a great deal of flexibility in the format of files that it may read, including options that permit handling non-rectangular file formats.

The first argument is not limited to file names of text files, however. R deals with input and output devices through objects with class "connection." The first argument can refer to any connection that can be read-in as text. The function will automatically open the connection, read in the data, and close the connection. For example, the following code uses a URL to read a data set based on the Stiles and Burch color matching fundamentals from the Colour & Vision Database (http://www.cvrl.org/) over the web.

```
> SB2deg.df <- read.table(
+    "http://www.cvrl.org/database/data/cones/linss10e_5.csv",
+    sep = ",")
```

Note that the extension of the file indicates that the fields are comma separated so we include the `sep = ","` argument; alternatively, we could have used the `read.csv` function directly.

Beginning with line 47, the fourth column of the data set is blank, reflecting the dichromacy of human color matches in this portion of the spectrum. The data for these values are read in as `NA`. We can substitute zeros for the NA values by either

```
> SB2deg.df[is.na(SB2deg.df[, 4]), 4] <- 0
```

or since these are the only `NA` values in the data frame,

```
> SB2deg.df[!complete.cases(SB2deg.df), 4] <- 0
```

where the function `complete.cases` conveniently returns a logical vector indicating the rows that do not contain values of `NA`. We, therefore, negate this vector with the negation operator, "!," to select the rows that do.

Another very useful operation with `read.table` is to read data from the clipboard, the buffer in which data are temporarily stored during a copy operation from the keyboard. Under Windows the user can type

```
> read.table("clipboard")
```

and the contents of the last copy operation will be returned. An alternate syntax is also available under Mac OS X when not using X11.

> `read.table(pipe("pbpaste"))`

`pbpaste` is a command, only available under Mac OS X, that writes the clipboard to stdin. The function `pipe` returns the contents of stdin. A complementary function `pbwrite` will permit writing to the clipboard as well. These operations are described under `help(connections)`. Reading from the clipboard is probably the simplest way to transfer data from a spreadsheet program to R or from a browser window. However, in the latter case, one must beware of hidden character codes that can mess-up the transfer.

A.5.4   *scan*

Sometimes one wants to access data that are simpler than a table, for example, a sequence of numbers or characters in a file or a matrix. The function `scan` provides a simple means of accomplishing this task. It takes a file name or text connection as first argument. By default, it assumes that the input is numeric, but this can be set otherwise by the `what` argument. If successful, it returns a vector of the given type. If reading a matrix from a file, the input can be formatted directly in R but care must be taken because matrices are filled column-wise, by default. `scan` can be used to read from the clipboard using the same arguments as described above.

A.5.5   *save and load*

Table formatted files do not suffice to store all types of data used by R. Objects can have attributes, which are arbitrary lists attached to them. These are not saved in the typical text representation. Also, arbitrary lists and objects cannot be written and read with the above described functions.

The function `save` writes an object or a series of objects to a file, by default in a compressed binary format. The stored representation is an exact copy of the object(s), with, for example, attributes intact. Typically, such files are given an extension of ".Rdata" and are transferrable across different computing platforms. The data files used throughout this book were created using `save` but with the extension ".rda."

The data in a file created with `save` can be retrieved with the function `load`, which minimally takes the filename as its argument. On some operating systems (e.g., Mac OS X), a double mouse click on the file icon that had been saved in this fashion is sufficient to open the R GUI (if not already open and) to run the `load` function.

The functions `dput` and `dump` can be used to create ASCII representations of objects more exotic than a data frame. However, it is not guaranteed that certain characteristics of the object will be restored when sourced into memory.

The Posting Guide (at http://www.r-project.org/posting-guide.html) outlines suggested etiquette for interacting with the online R-Help mailing lists and suggests using dump to communicate complex data structures to the list.

A.5.6   Interacting with Matlab

One of the major platforms for analyzing data from experiments in vision (and possibly in other psychophysical domains) is Matlab and, thus, a great quantity of data is likely stored in ".mat" files.

The package **R.matlab** [12] provides readMat and writeMat functions for files of this format. As of Matlab v7, the data compression scheme changed and, in order to read these files, one will also need the **Rcompression** [106] package available from the Omegahat repository (http://www.omegahat.org/R).[2]

The **R.matlab** package also includes facilities for setting up a client/server relationship between R and Matlab. One can then transfer data back and forth between R and Matlab and R may send commands to Matlab, as well. This functionality is only available for Matlab v6 or higher.

A.5.7   Interacting with Excel Files

We indicated in Sect. A.5.3 how to transfer data from a spreadsheet to R through a keyboard copy operation. Another strategy is to save the file from Excel in a "comma separated value" format (with a ".csv" extension). Then, it can be read in with one of the functions read.csv or read.csv2 which are simply wrapper functions for read.table with the appropriate default arguments. The second version of the function is set up to read formats in which a comma is used as the decimal point.

Several packages cater to reading and writing directly from Excel spreadsheet files. The data must generally be in tabular format, however. The package **gdata** [181] includes a read.xls function, and the packages **WriteXLS** [160] and **dataframes2xls** [163] write files in a format that can be opened by Excel. The package **XLConnect** allows more sophisticated interactions between R and Excel [70].

A.5.8   Reading in Other Formats

Data files come in many formats, and it sometimes seems as if every statistical package has its own special data encryption, thereby complicating transfer of

---

[2]As of version 1.5.0 of this package, **RCompression** is only necessary for R versions older than 2.10.0.

data between platforms. The recommended package **foreign** [145] contains several functions for reading from and writing to formats for several popular statistical packages.

Databases with their associated query languages provide a powerful means of storing, accessing, and manipulating data in a manner that can be independent of the statistical package operating on the data. They are of some interest in R as providing a buffer for dealing with data sets that would exhaust or strain computer memory. There is a growing number of packages that cater to reading and writing to databases. The list, **DBI** [81], **RJDBC** [177], **RMySQL** [83], **RODBC** [149], **ROracle** [84], **RPostgreSQL** [39], **RSQLite** [82], is a sample obtained by a search on the term "database" at the CRAN package web site (http://cran.r-project.org/web/packages/).

Appendix B
Statistical Background

B.1 Introduction

In this appendix we summarize without proof some basic results of mathematical statistics that we use in the main chapters of the book. We do assume that the reader has some previous experience with elementary probability theory and statistics and is familiar with hypothesis testing and linear regression, etc. Any undergraduate statistics book would cover enough material to serve as the prerequisite.

There is no need to read this appendix from start to finish. There are numerous references to sections of this appendix in each chapter and the reader interested in a particular chapter can read it, referring to material in this appendix only as needed.

The models used in simulating psychophysical observers are typically parametric, i.e., they are completely specified except for a small number of "free" parameters. The experimenter who summarizes his (or her) data set by estimating the mean and variance of a data sample is effectively fitting a normal (Gaussian) distribution and can use the fitted distribution to estimate confidence intervals, quantiles, etc. Of course, there are other distributions besides the Gaussian, and we will review some of them below.

Any possible model of the observer results from selecting values of the free parameters and the first task in any psychophysical application is to select the parameter values so that the model best matches a subject's behavior in a particular experimental condition.

Figure B.1a, for example, is a histogram of 75 choice response times of a single observer from a perceptual task. On each trial, the observer viewed an array of dots in which pairs of the dots could be oriented either randomly or along virtual concentric circles, forming a Glass Pattern, in the latter case [69]. The observer's task was to decide as rapidly as possible whether the concentric pattern was present or not. Each response is the time in seconds between the start of a trial and the touch response.

The Gaussian distribution would be a poor choice as a model for these response times since they appear to be skewed (asymmetric), and we know from long

K. Knoblauch and L.T. Maloney, *Modeling Psychophysical Data in R*, Use R! 32,
DOI 10.1007/978-1-4614-4475-6,
© Springer Science+Business Media New York 2012

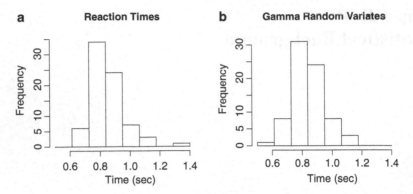

Fig. B.1 (a) Response times in a choice response task. Unpublished data provided by Elodie Mahler. (b) Histogram of values simulated from a Gamma distribution with best fit parameters to the data in part (a)

experience that response times are often skewed. We first select a parametric model of the observer's choice reaction times and in this case we use the family of Gamma distributions with two free parameters. In Fig. B.1b we show the results for a simulated observer based on the Gamma distribution in R, a parametric family of distributions. We selected the parameters so that the histogram of simulated data, generated by numerical methods with R (using the rgamma function; see below), resembles the actual histogram. The number of data points for the simulated and actual observers is the same. If we have succeeded in matching the model to the observer, then the simulated data "could have been" the actual data.

Of course, the model of the observer on which we base a simulation is, first of all, an hypothesis concerning the observer, that a Gamma distribution is an appropriate model, and, second of all, a claim that we have estimated the free parameters of the Gamma that "best" match the observer. Stating the model clearly and comparing it to actual data allows us to test and, if need be, revise the model. Perhaps there is no Gamma distribution that adequately captures human performance. We may also have a number of possible models of the observer in mind, and we are trying to decide which model offers the best description of the observer's behavior, a process referred to as *model selection* [24, 206]. This appendix summarizes basic results from mathematical statistics concerning *parametric model fitting* and *parametric model selection*.

The fitting problem is the heart of statistics. What is novel here is the emphasis we place on building actual simulations of observers based on explicit models of the observer. Of course, some such model is typically presupposed in any parametric estimation or model selection problem. Even if we chose to analyze data using a standard statistical package and a standard method such as multiple regression, there are models of how data were generated underlying and justifying the fitting method.

There are several advantages to emphasizing simulation in the way that we do. If, for example, the experimenter cannot specify such a simulation program, then his failure may mean that he has not thought through his assumptions about what the observer is doing in the task. Or, even though he can write the simulation program, he may find that he has serious misgivings about the model as representative of human performance, once it is written down and all of its assumptions are made clear. With many statistical packages, for example, it is easy to fit data to very sophisticated models without paying much attention to assumptions about the model implicit in the fitting procedure. It is currently too easy to fit data to indefensible models.

Last, if we accept that the simulation is a surrogate for the actual experiment, the experimenter can test out different experimental designs and fitting methods before collecting data. Like a good athlete, the experimenter can "practice his moves" before the actual competition. He may discover, for example, that his experimental design is not adequate to answer the questions that interest him and that an experimental design involving more trials is needed. These sorts of questions come under the heading of "power" in mathematical statistics [129] and the simulation-fitting loop is a convenient tool for analyzing the power of statistical tests. We will find that the simulation-fitting loop has other advantages. In particular it allows us to routinely use resampling methods and the bootstrap to estimate confidence intervals and biases for parameter estimates [55].

B.2 Random Variables and Monte Carlo Methods

Since Fechner [59], human responses in psychophysical tasks are often modeled by *random variables*. We are primarily concerned with two kinds, *finite discrete random variables* and *continuous random variables*. We concentrate on univariate random variables, where the random variable takes on only single real numbers (they are *univariate real*), although we also consider random variables that return a vector of real numbers. Most books on probability theory or statistics will define several types of random variables and their properties [34,153]. The handbooks by Johnson, Kotz, and colleagues [85–87] are encyclopedic treatments of commonly used (and not so commonly used) random variables and are very useful. We recommend them as bedtime reading. We will define only the minimum number of random variables needed in the remainder of a book.

B.2.1 Finite, Discrete Random Variables

A univariate finite discrete random variable X is defined by listing the set of values that it can take on $\{x_1, x_2, \ldots, x_n\}$ and the probability $p_i = p(x_i) \geq 0$ that each value can occur. The probabilities are constrained to sum to 1, $\sum_{i=1}^{n} p_i = 1$ ("something always happens"). The function $p(x_i)$ is referred to as the *probability mass function (pmf)* of the random variable.

B.2.2 Bernoulli Random Variables

For our purposes, the most important example of a finite discrete random variable is a *Bernoulli random variable B* with values $\{0, 1\}$ and corresponding probabilities defined by $P[B = 1] = p$ and $P[B = 0] = 1 - p$. The *Bernoulli random variable* is a useful model for binary decisions between two possibilities (such as signal present, signal absent) where the outcome effectively is just a choice between two alternatives. The importance of the Bernoulli random variable is just that experimenters like to design experiments where the observer's response is limited to Yes/No, Present/Absent, Right/Left, Salty/Sweet, etc. We will further discuss the Bernoulli random variable and a generalization of it, the Binomial, below.

Simple as it is, the Bernoulli is an example of a *parameterized family* of random variables that we could denote by $\mathscr{B}(p)$ which includes Bernoulli random variables for all choices of the parameter $0 \leq p \leq 1$. If we specify p in $\mathscr{B}(p)$, then we have specified a particular random variable in the family. Later on, when we consider parametric estimation problems, we will try to decide which random variable in a family is the best match to an observer given his or her performance in a psychophysical task.

B.2.3 Continuous Random Variables

A univariate continuous random variable X is specified by giving its *probability density function (pdf)* $f(x) \geq 0$ with

$$\int_{-\infty}^{\infty} f(x)\,dx = 1. \tag{B.1}$$

The pdf effectively determines the probability that the random variable will be in any interval (a, b) with $a < b$ (see Fig. B.2)

$$P[X \in (a,b)] = P[a < X < b] = \int_{a}^{b} f(x)\,dx. \tag{B.2}$$

Probability density should not be confused with probability. For a continuous random variable, the probability that it takes on a particular value a is 0, in symbols, $P[X = a] = 0$. The evident paradox is that the random variable must take on *some* value, but each particular value has probability 0. These sorts of anomalies make probability theory confusing at first and the best approach to the subject involves suppressing intuitions and paying close attention to definitions. For us then, a pdf is a way to specify the probability that the random variable is in any interval and no more.

We will usually denote random variables by uppercase letters, for example, X, and the values that a random variable can take on by the corresponding lower-case

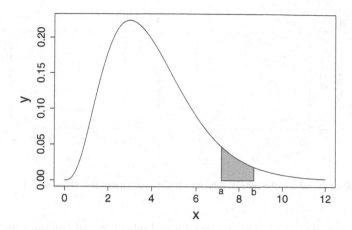

Fig. B.2 The probability that a random variable will be in the interval (a, b) is the area under the pdf over this interval

letter, e.g. x. The *cumulative distribution function* (cdf) of a continuous random variable is defined as

$$F_X(x) = \int_{-\infty}^{x} f(x)dx \qquad (B.3)$$

and since $f(x) = F'(x)$, either the pdf or cdf can be used to characterize a continuous random variable. The cdf of a discrete random variable is defined in the same way. The inverse of the cdf function of a random variable is referred to as the *quantile function*. If, for example, $F(x)$ is a distribution, then $F^{-1}(0.75)$ is the value of x that falls at the 0.75 quantile (0.75th percentile) of the distribution.

B.2.4   Discrete vs Continuous Random Variables

The distinction between discrete and continuous random variables is artificial. In more advanced probability texts the pmf of a discrete random variable would be written as a pdf in terms of Dirac (impulse) generalized functions $\delta(x)$ [19]. We can write the pdf of a finite discrete random variable with outcomes $x_1, \ldots, x_n$ and corresponding probabilities as

$$p(x) = \sum_{i=1}^{n} p_i \delta(x - x_i) \qquad (B.4)$$

The resulting pdf takes on the value p_i precisely at x_i and is 0 otherwise. The single great advantage of using Dirac notation is that we can refer to finite discrete random

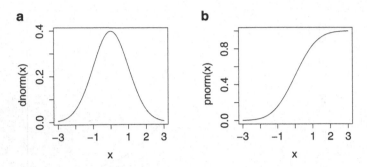

Fig. B.3 The pdf and cdf of a normal random variable with mean $\mu = 0$ and $\sigma = 1$

variables and continuous random variables and other types of random variables that are partly discrete and partly continuous using the same terminology and definitions.

We can define random variables (X, Y) that are bivariate and take values in the real plane. We define the pdf (pmf) by a function $f(x, y) \geq 0$ with

$$1 = \int_{-\infty}^{\infty} \int_{-\infty}^{\infty} f(x, y) \, dx \, dy \qquad \text{(B.5)}$$

and we can define cdfs, inverse cdfs, etc. for such random variables. The extension to multivariate random variables is easily made but we will have little use for multivariate random variables $(X_1, \ldots, X_n)$ except to occasionally use multivariate notation and write the pdf as $f(x_1, \ldots, x_n)$.

B.2.5 Normal (Gaussian) Random Variables

If the reader has previously encountered a random family of continuous distributions, it is almost certainly the Gaussian. In the **stats** package, a recommended package that is normally attached to the search path at startup by default, four functions are defined for manipulating and simulating from each of a selection of 18 random variables. These four functions conventionally are prefixed with the letters "d," "p," "q" and "r," for the density (pdf), probability (cdf) and quantile functions (inverse cdf) and for random variate generation, respectively. All the functions for any random variable follow this same convention and all four functions are documented in the same help file. For example, if we type

```
> help("rnorm")
```

we obtain documentation for the four functions, dnorm, pnorm, qnorm, and rnorm for working with normal (Gaussian) random variables. The function dnorm, for example, is the pdf of the normal distribution while pnorm is the cdf. These are both plotted in Fig. B.3 using the default values of their arguments. qnorm

is the inverse cdf or quantile function $F^{-1}(x)$. If $p = F(x)$, then $x = F^{-1}(p)$.
For example,

```
> qnorm(pnorm(1))
[1] 1
```

and

```
> pnorm(qnorm(0.8413447))
[1] 0.841
```

We can invert $F(x)$ only when it is a strictly increasing (or decreasing) function and the solution to the equation $p = F(x)$ is unique. However, if we are careful (and we always are) we can define an inverse for any cdf. We saw the use of the quantile function in estimating thresholds in Sect. 1.3.2.

B.2.6 Location-Scale Families

Recall that we are interested in parametric families of distributions. The normal family is a two-parameter family with parameters μ and $\sigma > 0$ that are referred to, respectively, as the "mean" and "standard deviation" of the normal random variable. In the call to pnorm above, $\mu = 0$ and $\sigma = 1$ are the default values. These defaults are overridden by adding additional arguments, for example, pnorm(x, mean = 2, sd = 0.5). These parameters are examples of location and scale parameters, respectively, and the names are self-explanatory, at least as soon as we plot the pdf for several choices of $\mu = 0, 1, 2$ with $\sigma = 1$ and for $\sigma = 1, 2, 3$ and $\mu = 0$. The pdf is "located" at μ and its width is scaled by choice of σ.

We can take any pdf, $f(x)$, and turn it into a location-scale family by defining $f(x; \mu, \sigma) = f\left(\frac{x-\mu}{\sigma}\right)$. If $f(x)$ is already in a parameterized family, we just add μ and σ to the list of parameters. In the case of a Bernoulli random variable $B(p)$ that takes on values $\{0, 1\}$, we could change it to a generalized Bernoulli $B(p, \mu, \sigma)$ that takes on values $\{a, b\}$ with probabilities $1 - p, p$, respectively. For example, with $\mu = -0.5$ and $\sigma = 0.5$ the generalized Bernoulli now takes on the value 1 with probability p and otherwise -1. Notice that, as the scale parameter is $1/2$, the Bernoulli has been "stretched" so that the spacing between the two possible outcomes is 2, not 1.

In any location-scale family, the location parameter determines the location of the cdf as well. In Fig. B.4a we show the cdf of the normal for different values of μ and $\sigma = 1$. The scale parameter also scales the cdf. In the normal case (Fig. B.5b), different values of σ affect the steepness of the rise of the cdf near $\mu = 0$. A little bit of calculus reveals that the slope of the tangent to the cdf of the normal at μ is $(\sqrt{2\pi}\sigma)^{-1}$. The cdfs of location-scale random variables will be useful for modeling performance in psychophysical tasks, such as in estimating *psychometric functions*.

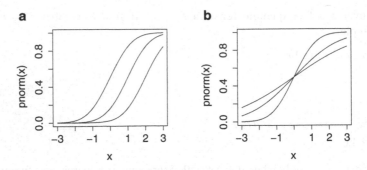

Fig. B.4 Normal distribution cdfs differing in (**a**) location and (**b**) scale

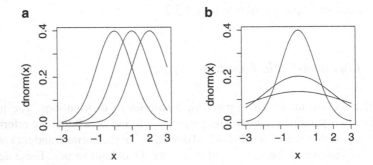

Fig. B.5 Normal distributions differing in (**a**) location and (**b**) scale

B.3   Independence and Samples

The lastfunction in the collection of R functions associated with the normal distribution is `rnorm`. It simulates a realization or "draw" of a normal random variable. We can think of a random variable as a box with a button on the side. Every time we push the button a real number pops out of the box.[1] We do not know in advance what number will pop out (that's why it is a *random* variable), but the pdf (or cdf) characterizes how likely it is that the realization is in any specified range. The first argument of `rnorm` "n" is the size of the *sample* desired, how many times we want to press the button. A sample is denoted by subscripted uppercase letters such as $X_1, \ldots, X_n$. This notation is ambiguous. It could refer to the future sample that we have not taken yet or the sample already collected which is a list of specific numbers. We will live with this ambiguity and we will call attention to it when it might cause confusion.

---

[1] Actually, only those real numbers that can be represented as floating point numbers in the computer.

A sample is, by definition, a series of *independent* draws from the same random variable. A definition of independence can be found in almost any probability book; intuitively, the elements in the sample are independent in that any one outcome is unaffected by the outcomes of the other draws. Alternatively, knowledge of some values of a sample tells us nothing about the remaining values that we did not know. We can also say that a sample is *independent, identically-distributed (iid)* to emphasize that all the sample values are drawn from the same distribution.

A sample can be considered as a multivariate random variable, $(X_1, \ldots, X_n)$, and independence is equivalent to the claim that its pdf is just $f_n(x_1, x_2, \ldots, x_n) = f(x_1)f(x_2), \ldots, f(x_n)$: that is, the multivariate pdf of the sample is just the product of n copies of the pdf $f(x)$ of the univariate random variable X. If X is part of a parameterized family with parameter θ then $f_n(x_1, x_2, \ldots, x_n; \theta) = f(x_1; \theta)f(x_2; \theta), \ldots, f(x_n; \theta)$.

As an example of a sample, we can type

```
> (X <- rnorm(10, mean = 0, sd = 1))
 [1]  0.0911  1.2501 -0.3081 -0.1473  0.7015  0.6408 -0.1959
 [8]  0.5594 -0.5278 -1.3072
```

and we obtain 10 values drawn from the normal distribution with parameters 0 and 1. A second draw,

```
> (Y <- rnorm(10, mean = 10, sd = 1))
 [1] 8.47 11.33  9.25  8.61  9.16  9.07  8.86  9.00 11.69  8.21
```

produces very different values, "roughly" 10 units greater than the first sample.

Independence is one of the deepest concepts in statistics and probability. Paradoxically, it is very unlikely that any sample of data from a psychophysical observer is really independent or identically distributed since the brain is, in effect, a large bowl of neural soup where everything affects everything and each trial permanently changes the brain and the outcome of the next. There are standard methods such as randomization of the order of trials to minimize the consequences of sequential effects. See Kingdom and Prins [91]. In a recent article, Fründ et al. [66] analyze the consequences of sequential dependence in psychophysical data.

B.3.1 Expectation and Moments

Given a random variable X with pdf $f(x)$, and any function $g(x)$, we compute the *expected value* of $g(X)$ as

$$E[g(X)] = \int_{-\infty}^{\infty} g(x)f(x)\,dx \tag{B.6}$$

If the random variable is discrete, with outcomes x_i and probabilities p_i for $i = 1, \ldots, n$, the equation immediately above becomes

$$E[g(X)] = \sum_{i=1}^{n} g(x_i) p_i. \tag{B.7}$$

If $g(x) = x$, then the resulting expected value is the expected value of X, sometimes referred to as the *population mean* and often denoted μ_X. If $g(x) = (x - \mu_X)^2$, then the resulting expected value is the *variance* of X. The expected value $E[(X - \mu_X)^n]$ is the nth central moment of X. The variance of X is also its second central moment and is often denoted σ_X^2.

When a random variable is part of a parametric family, expectations may depend on the parameters of the family. For example, it is reassuring that, for a normal random variable X from the family $\mathcal{N}(\mu, \sigma^2)$, $E(X) = \mu$ and $\mathrm{Var}(X) = \sigma^2$. We must be doing something right.

We can approximate the integral in (B.6) numerically [143], or we can approximate it by Monte Carlo methods as follows. If, for example, we wanted to know the fourth central moment of a normal random variable $\mathcal{N}(0, 1)$, then we can simply take a very large sample using `rnorm(n, mu = 0, sigma = 1)`, compute $g(X) = (X - \mu)^4$, and average these values. A sample of size $n = 1,000,000$ gives us a Monte Carlo estimate of 3.004, not too far from the true value of 3. We will use Monte Carlo methods extensively to explore and approximate complex formulas involving random variables. After a while, we will think of calculus as the refuge of those with small computers.

In using computational methods in this way we are making use of a second key idea in probability theory, *convergence*. Large samples, in a very precise sense, characterize the distribution they are drawn from. We return to this point below in discussing *resampling methods*.

If the integral in (B.6) does not converge, then $E[g(X)]$ is not defined. In particular, not every random variable has a mean or variance. A good example of such a distribution is the Cauchy. The reader is invited to a estimate its nonexistent mean by Monte Carlo methods with progressively larger samples, using the function `rcauchy`.

We define expectation for bivariate and multivariate random variables similarly [34] and we can define bivariate and multivariate moments as well. For example, given the bivariate random variable (X, Y) with pdf $f(x, y)$, and any function $g(x, y)$, we can compute the *expected value* of $g(X, Y)$ as

$$E[g(X, Y)] = \int_{-\infty}^{\infty} \int_{-\infty}^{\infty} g(x, y) f(x, y) \, dx \, dy. \tag{B.8}$$

If $g(X, Y) = X$, the resulting expected value is the *marginal expected value* of X denoted μ_X and μ_Y is defined similarly. If $g(X, Y) = (X - \mu_X)(Y - \mu_Y)$, the resulting expected value is the *covariance* of X and Y denoted $\mathrm{Cov}(X, Y)$. See, for example, [153] or any introductory statistics text for details.

We can define multivariate expectation and moments by analogy to bivariate. Some functions for working with multivariate samples are available in R, for example, for multivariate normal distributions in the packages **MASS** [178] and **mvtnorm** [67] and for multivariate t-distributions in the latter.

B.3.2 Bernoulli and Binomial Random Variables

The second example of a discrete random variable that we consider is the *Binomial random variable*. We can type ?Binomial or ?rbinom to read about its "four functions." A Binomial random variable is part of a parameterized distribution family $Bi(s,p)$, the Binomial family. It has two parameters "size" s, often denoted "n", and "prob," often denoted "p." In the special case where "size"$=1$, the Binomial family is just the Bernoulli family with parameter p. One way to generate a single Binomial random variable X of size s with probability p is to generate a sample of s Bernoulli random variables with probability p: $B_1, B_2, \ldots, B_s$. Then set $X = \sum_{i=1}^{s} B_i$. Each of the Bernoulli variables is either 0 or 1 and the sum must be an integer between 0 and s. The resulting Binomial is a discrete random variable with outcomes $\{0, 1, 2, \ldots, s\}$ and probabilities

$$P[X = k] = \binom{s}{k} p^k (1-p)^{s-k}. \tag{B.9}$$

The first term on the left hand is the Binomial coefficient. In R, this result can be obtained more directly using the function rbinom, taking arguments, "n" for the number of random values to generate, "size" for the parameter s and "prob" for the probability of a success. Note: for large samples, rbinom is considerably faster than adding up many Bernoulli random variables.

B.3.3 Uniform Random Variables

The uniform distribution on $(0,1)$ is part of a location scale family $\mathcal{U}(a,b)$ that comprises distributions uniform on intervals (a,b) with pdf[2]

$$f(x; a, b) = \frac{1}{b-a}, \qquad a < x < b \tag{B.10}$$

---

[2]It is tedious to write "0 otherwise" as we would have to do for many of the distributions starting with the uniform above. When we define distributions from now on it is understood that the pdfs are 0 at points not specified in the definition and the cdfs are 0 or 1 as appropriate.

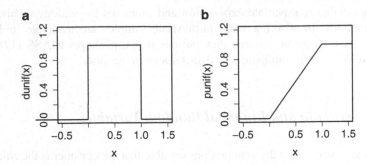

Fig. B.6 The probability density function (pdf) and cumulative distribution function (cdf) of a uniform random variable $\mathcal{U}(0,1)$

and cdf

$$F(x;a,b) = \frac{x-a}{b-a}, \qquad a < x < b. \tag{B.11}$$

The pdf and cdf are plotted in Fig. B.6.

The family of functions dunif, punif, qunif, and runif i n R allow the user to work with the uniform distribution. The logical expression runif(size) < p generates a logical vector of length size. Each entry is TRUE with probability p and otherwise FALSE. The function sum treats TRUE as 1 and FALSE as 0 (a Bernoulli variable $\mathcal{B}(p)$ with 1 replaced by TRUE, 0 replaced by FALSE) and so B is a number from 0 to size distributed as a Binomial random variable. Of course, it is more efficient to use rbinom to generate such random values.

Reparameterization

The family $\mathcal{U}(a,b)$ is actually a location-scale family but a,b are not its "natural" location and scale parameters. To put the family in location and scale form we would define new parameters such as its "center" $m = (a+b)/2$ and its "width" $s = (b-a)$.

The $\mathcal{U}(m,s)$ family with this change of parameters represents all the same distributions but now with an obvious location parameter and an obvious scale parameter. This change is an example of reparameterization. We have simply replaced one set of parameters by another using a one-to-one transformation from (a,b) to (m,s). We demonstrate that the transformation is one-to-one by writing down the inverse transformation: $b = m + s/2$ and $a = m - s/2$. The original parameters (a,b) are so useful and easy to work with that you will rarely see this location-scale family parameterized as a location scale family. When we discuss estimation of parameters, the issue of reparameterization will arise again in Sect. B.4.3 below.

B.3.4 Exponential and Gamma Random Variables

The exponential distribution is part of a family known as the Gamma family with parameters n and λ. When $n = 1$, the Gamma random variable is also called the *exponential random variable*. An exponential random variable has pdf

$$f(x;\lambda) = \lambda e^{-\lambda x}, \qquad x > 0 \tag{B.12}$$

and cdf

$$F(x;\lambda) = 1 - e^{-\lambda x}, \qquad x > 0. \tag{B.13}$$

The exponential is widely used to model random delays in time and the equation above often appears with t (denoting time) instead of x as the argument. A very common reparameterization of the exponential is to replace λ by $\tau = \lambda^{-1}$. The exponential is often used to model the completion time of decision processes in cognition and is closely connected to the study of reaction times (response times). If we think of the occurrence of the exponential random variable as an "event" in time, we can assign units of "mean events per unit time" to λ and "mean time to event" to τ.

With reparameterization, the exponential distribution is a scale family with scale parameter τ and it could be made into a location-scale family by adding a location parameter.

The Gamma family $\Gamma(n, \tau)$ contains random variables that are the sums of n independent, identically distributed exponential random variables with rate parameter τ. Its pdf is

$$\gamma(x;n, \tau) = C_n \tau^n x^{n-1} e^{-x/\tau}, \qquad x > 0. \tag{B.14}$$

The constant C_n is chosen so that the area under the pdf is 1 ($C_1 = 1$). When $n = 1$ the pdf of the Gamma is the pdf of the exponential with parameterization in terms of τ. Its cdf does not have a closed form except when $n = 1$.

Some examples of the pdf for different values of τ are given in Fig. B.7 The parameter n is an example of a *shape parameter*. When n increases the shape of the Gamma pdf changes and the change is not just a simple change in location or scale. The parameter n of the binomial $\mathscr{B}(x;n, p)$ is also a shape parameter.

A sample from the Gamma distribution is shown in Fig. B.1b above to simulate human response times. The Gamma function has also been used to model the duration of the states of bi-stable figures [124].

B.3.5 The Exponential Family

The *exponential family* [109, 127, Sect. 1.5] is a broad class of distributional families all of whose pdfs can be written in the form

$$f(x|\theta) = h(x)g(\theta)e^{\eta(\theta)T(x)}, \tag{B.15}$$

Fig. B.7 The probability
density function (pdf),
$\gamma(x; n, \tau)$, of a Gamma
random variable, $\Gamma(n, \tau)$, for
various values of n and $\tau = 1$

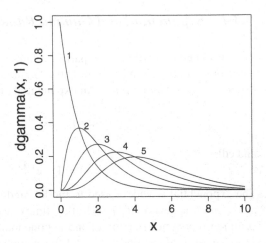

where y is a random variable, and the θ are unknown parameters to be estimated.
For the Gaussian distribution, for example $\theta = (\mu, \sigma^2)$. and h, g, η, and T are
known functions [53]. The choice of functions h, g, η, and T specifies a particular
parametric family of distributions with parameters θ such as for the Gaussian. Any
distributional family whose pdf can be written in the form of the equation above
is part of the exponential family. Many commonly used distributional families
including the Gaussian, exponential, Gamma, χ^2, Beta, Bernoulli, binomial, and
Poisson are part of the exponential family. The t-distribution is an example of
a distribution that is *not* in the exponential family. The exponential family is
sometimes referred to as the *exponential class* but the term "exponential family"
is far more common. It is doubly unfortunate since its members are themselves
distributional families and the exponential distributional family (Sect. B.3.4) plays
no special role. It is just another exponential family distribution.

The equation above is often presented in alternative forms. If, for example, we
set $A(\theta) = \log(g(\theta))$ and $B(x) = \log(h(x))$, then the equation above becomes

$$f(x|\theta) = e^{\eta(\theta)T(x) + A(\theta) + B(x)}. \tag{B.16}$$

The change in form is purely cosmetic. It is sometime written including an
additional dispersion parameter, ϕ [127, 178], but this will be assumed to be known
in nearly every case that we consider here.

Exponential family distributions have desirable statistical properties that go
beyond the scope of this book. See [109, Sect. 1.5].

B.3.6   Monte Carlo Methods

The values generated by R functions such as `rbinom` and `rnorm` are not random. They are the output of algorithms that generate deterministic sequences of numbers that mimic many properties of random variables including their unpredictability. The original versions of these algorithms were inspired by the needs of scientists working on the Manhattan Project during the Second World War and were nicknamed "Monte Carlo" methods after the eponymous country and its famous casino. The key insight is that certain computations produce sequences of numbers that have many of the properties of random sequences while being completely deterministic. The sequence $0, 0, 1, 0, 0, 0, 1, 1, 0, 0, 0, 1, \ldots$ would pass most statistical tests for a Bernoulli random variable with $p = 0.5$ even though it is just the digits of π to base 10, classified as even (1) or odd (0). Designing such "pseudo-random number generators (pRNGs)" is difficult and considerable work has gone into developing pRNGs that are useful for simulation and data analysis [20, 52]. Knuth describes his amusing attempts to develop a "super random" pRNG and how it failed [100, Chap. 3].

Current computer languages (including, of course, R) have Monte Carlo procedures that have been carefully tested. Although they are all deterministic algorithms, they can be treated as random in modeling physical (including biological) stochastic processes. To learn about R's facilities for random number generation and how to modify them, see `?RNG`.

B.4   Estimation

Suppose that we are given a sample $X_1, \ldots, X_n$ from a random variable in a parametric family with parameters $\theta_1, \ldots, \theta_m$. For example, we have data from the random variable in the Gamma family but we do not know the value of its shape parameter n and its scale parameter τ. We want to estimate these "unknown parameters" from the sample of data. For simplicity, let's focus on the case where the distributional family has only one parameter θ, i.e., $m = 1$. An *estimator* of θ is any function of the data

$$\hat{\theta} = T_{\theta_i}(X_1, \ldots, X_n). \tag{B.17}$$

The first time you encounter this definition it may seem odd. We emphasize that any function of the data can be termed an estimator (also known as a *statistic* or *statistical estimator*). Even apparently pathological estimators such as $\hat{\theta} = T_{\theta_i}(X_1, \ldots, X_n) = 7$. This is the estimator that throws away the data and always estimates that θ is 7. This quirk in statistical terminology is intentional. It is easier to discuss the advantages and disadvantages of estimators if we first cast a broad net.

In effect, from among all of the estimators possible, we would like to find estimators that are good. And before doing so, we need to define what we consider to be good.

B.4.1   What Makes an Estimate "Good"?

There are multiple competing criteria for goodness in estimation, and consideration of different criteria and how to design statistics that achieve them would make up a rigorous course in mathematical statistics. Here we consider two of the most venerable, unbiasedness and minimum variance. An estimator is just a function of the random variables in the sample and is itself a random variable. We can consider its expected value for any specific value of θ:

$$E_\theta[T_\theta] = \int \int \cdots \int T_\theta(x_1, \ldots, x_n) f(x_1; \theta) f(x_2; \theta) \ldots f(x_n; \theta) \, d\theta. \qquad (B.18)$$

Note that this expected value is not a random variable. We integrated out all of the randomness in taking the expected values.

An estimator is *unbiased* if and only if $\theta = E_\theta[T_\theta]$. In words, a statistic is unbiased if and only if it is "on average" equal to what it is supposed to be estimating.

A second criterion uses the variance of an estimator.

$$Var_\theta[T_\theta] = \int \int \cdots \int (T_\theta(X_1, \ldots, X_n) - E_\theta[T_\theta])^2 f(x_1; \theta) f(x_2; \theta) \ldots f(x_n; \theta) \, d\theta. \qquad (B.19)$$

We can now cut down the universe of all possible estimators by focusing on estimators that are unbiased and that have very low variance for all possible estimates of θ. The last clause "...all possible estimates of θ" is important. Otherwise, the "silly estimator" $T_\theta(X_1, \ldots, X_n) = 7$ looks really good when $\theta = 7$ (zero variance!) but less so for other values of θ.

One particular class of estimators in common use are *uniform minimum variance unbiased estimators (UMVUE)*. These are estimators that for every possible value of θ are unbiased and, again for every possible value of θ have a variance no greater than that of any other unbiased estimator. For a particular problem, there need be no UMVUE and for many problems there is no easy way to prove that a particular estimator is or is not the UMVUE. An example of a UMVUE: suppose that we have a sample $X_1, \ldots, X_n$ from a normal distribution $\mathcal{N}(\mu, \sigma^2)$, and we want to estimate the location parameter μ. Then the mean of the sample

$$\hat{\mu} = \bar{X} = (X_1 + \cdots + X_n)/n \qquad (B.20)$$

is an UMVUE. Proving this is beyond the scope of this appendix. You can consult any book on mathematical statistics (e.g., [129]).

B.4.2 Maximum Likelihood Estimation

We will use a particular estimation method to derive many estimators in the applications presented in this book. There are other possible choices of methods for devising estimators and we will discuss one, *Bayesian estimation*, very briefly in a later section of this Appendix. Maximum likelihood estimators are easy to derive, and they can usually be computed numerically using numerical optimization methods, such as optim in R. Also, for many well-known problems, the resulting estimators are standard, i.e., the ones already in common use.

Suppose we have a sample of data $X_1, \ldots, X_n$ from a parametric distribution with pdf $f(x; \theta_1, \ldots, \theta_m)$. We compute the *likelihood* of the sample as,

$$\mathscr{L}(\theta_1, \ldots, \theta_m; X_1, \ldots, X_n) = \prod_{i=1}^{n} f(X_i; \theta_1, \ldots, \theta_m). \tag{B.21}$$

Two observations: First, the *variables* in the likelihood function are the unknown parameters $\theta_1, \ldots, \theta_m$. The sample, after we have collected the data, is just a collection of numbers that are fixed, the data we just collected. For that reason, we write the likelihood function as $\mathscr{L}(\theta_1, \ldots, \theta_m; X_1, \ldots, X_n)$, emphasizing that the focus of attention is now on the unknown values of the parameters $\theta_1, \ldots, \theta_m$. Second, if the random variable X is discrete, then the pdf $f(x; \theta_1, \ldots, \theta_m)$ is actually a pmf (see Sect. B.2.4) and the likelihood function is simply the probability that the observed sample $X_1, \ldots, X_n$ would occur if the parameters had the specific values $\theta_1, \ldots, \theta_m$. If, however, the random variable is continuous, the likelihood function cannot be interpreted as a probability. Indeed, it is not even a pdf in its own right since if we integrated over the parameters $\theta_1, \ldots, \theta_m$ the result is very unlikely to be 1. The likelihood function is the likelihood function: we are interested in "mining" it for information about the parameters.

There is evidently some information about the parameters in the likelihood function. If, for example, we are considering the uniform family $\mathscr{U}(a,b)$ with unknown parameters a and b and one of our sample values $X_1 = 2.31$, then it is clear that $a < 2.31 < b$. Otherwise, the sample value could not have occurred.

To turn the likelihood function (B.21) into an estimation method, we simply decree that our estimators of the unknown parameters $\theta_1, \ldots, \theta_m$ are the values $\hat{\theta}_1, \ldots, \hat{\theta}_m$ that maximize the likelihood function. The maximum is usually unique, but if it is not then we have multiple choices of estimators. This almost never occurs in practice. We will typically reserve the notation $\hat{\theta}$ for the maximum likelihood estimator of θ.

Example

Suppose that we take a sample $X_1, \ldots, X_n$ from a normal distribution with pdf

$$f(x; \mu, \sigma) = \frac{1}{\sqrt{2\pi}\sigma} e^{-\frac{(x-\mu)^2}{2\sigma^2}} \tag{B.22}$$

and compute the likelihood function

$$\mathscr{L}(\mu,\sigma;X_1,\ldots,X_n) = \prod_{i=1}^{n} \frac{1}{\sqrt{2\pi}\sigma} e^{-\frac{(X_i-\mu)^2}{2\sigma^2}} \qquad (B.23)$$

It is scarcely obvious that we can maximize this likelihood equation by choice of μ and σ using calculus and get closed form solutions,

$$\hat{\mu} = \bar{X} = \frac{\sum_{i=1}^{n} X_i}{n} \qquad (B.24)$$

and

$$\hat{\sigma} = \sqrt{\frac{\sum_{i=1}^{n}(X_i-\bar{X})^2}{n}} \qquad (B.25)$$

The formula for $\hat{\mu}$ is likely familiar while that for $\hat{\sigma}$ may look a bit odd with n in the denominator instead of $n-1$. We will return to this point later.

Instead of maximizing likelihood we can instead maximize the logarithm to any base of likelihood by choice of $\hat{\theta}_1,\ldots,\hat{\theta}_m$. If the value $\hat{\theta}$ maximizes $\mathscr{L}(\theta_1,\ldots,\theta_m;X_1,\ldots,X_n)$, then it also maximizes $\log\mathscr{L}(\theta_1,\ldots,\theta_m;X_1,\ldots,X_n)$. We can take the logarithm since the likelihood function is nonnegative (we allow for $\log(0) = -\infty$ and treat it as any other number). The log likelihood function

$$\log\mathscr{L}(\theta_1,\ldots,\theta_m;X_1,\ldots,X_n) = \sum_{i=1}^{n} \log f(X_i;\theta_1,\ldots,\theta_m) \qquad (B.26)$$

is often much simpler to write down and sometimes we can maximize it by simple calculus. To return to the normal example above,

$$\log\mathscr{L}(\mu,\sigma;X_1,\ldots,X_n) = -\sum_{i=1}^{n} \frac{(X_i-\mu)^2}{2\sigma^2} - n\log\sigma + C \qquad (B.27)$$

where the constant term C contains the contributions of $\sqrt{2\pi}$. We are only planning to maximize likelihood or its logarithm and any constant terms do not affect maximization. They can be neglected and often we will leave them out altogether. The resulting log likelihood function for the normal sample is much simpler than the likelihood function and we can see clearly that it is written in terms of a sum of squared differences. Maximizing log likelihood for the normal will prove to be equivalent to minimizing the sum of squared differences and the well-known connection between normal estimation and so-called "least squares" emerges from the log likelihood equation. Moreover, the resulting ML estimate $\hat{\mu}$ of μ is UMVUE and the "MV" part is a bit more plausible given the connection to minimizing least squares.

Example

A second example of an ML estimate is that for a Bernoulli sample $B_1, \ldots, B_n$ from a $\mathscr{B}(p)$ distribution. Here there is only one parameter $m = 1$. The likelihood function is

$$\mathscr{L}(p; B_1, \ldots, B_n) = \prod_{i=1}^{n} n p^{B_i} (1-p)^{1-B_i} \tag{B.28}$$

and the log likelihood function is

$$\log \mathscr{L}(p; B_1, \ldots, B_n) = \sum_{i=1}^{n} n(B_i \log(p) + (1-B_i)\log(1-p)). \tag{B.29}$$

We can solve for the maximum of either by differentiating with respect to p. For the log likelihood function the derivative is

$$\frac{\mathrm{d}}{\mathrm{d}p} \log \mathscr{L}(p; B_1, \ldots, B_n) = \sum_{i=1}^{n} n \left(\frac{B_i}{p} - \frac{1-B_i}{1-p} \right) \tag{B.30}$$

which we set to 0 and solve for the maximum $\hat{p} = (\sum_{i=1}^{n} B_i / n)$, the proportion of Bernoulli outcomes that are 1. This outcome is scarcely surprising but certainly comforting. The intuitively obvious estimate is the MLE.

In log likelihood form we can see that each value in the sample contributes a "vote" to the log likelihood function of the form $\log f(X_i; \theta_1, \ldots, \theta_m)$ and that the overall log likelihood function is the sum of these votes. We can denote these separate contributions as L_i, the summands of the log likelihood function. In the normal case, each value in the sample contributes an inverted parabola

$$\log \mathscr{L}_i(\mu, \sigma : X_i) = -\frac{(X_i - \mu)^2}{2\sigma^2} - \log(\sigma) \tag{B.31}$$

and the value that maximizes $\log \mathscr{L}_i$ is obviously X_i. We show the parabolas for a sample of size 3 in Fig. B.8. The overall log likelihood function for $\sigma = 1$ is also an inverted parabola which is the sum of the $\log L_i$ as shown in bold in Fig. B.8.

Of course, the log likelihood function of the normal depends on both μ and σ. If we plot these functions for other values of σ the height of the parabolas and their width will change. But the basic picture remains. Each X_i is voting for values near X_i and the overall estimate (the thicker black curve) is a combination of these "votes." Notice that it is narrower. The accumulated support from the three sample values is converging on an estimate. The log likelihood function is sometimes referred to as the *support function* and it is clear from the figure how the data "support" different possible estimates.

If the likelihood function is a probability (the random variable X is discrete as in the Bernoulli example), then we are selecting the values of the parameters that are most likely to have given rise to the observed sample. Stated in this form, it is not obvious why one would employ such a criterion. Unlike criteria such as

Fig. B.8 The support
functions for each entry in a
sample of size 3 of a
Gaussian random variable
with $\sigma = 1$ and the sum of
these support functions into
an overall support function

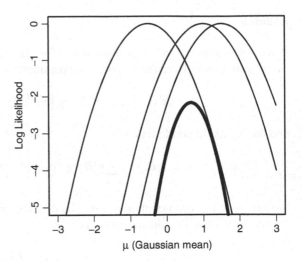

"minimum variance," it is not obvious that a maximum likelihood estimate has
any desirable property per se. In the next section we will consider the properties
that result from using maximum likelihood estimation. In any case, if the random
variable X is continuous, then the likelihood function is not a probability or even
a pdf, as mentioned above, and it is even less obvious why one would choose to
maximize likelihood.

B.4.3   Why Likelihood? Some Properties

Asymptotically UMVUE

ML estimates are, in general, *not* unbiased and *not* minimum variance. However, as
the sample size increases, they converge to estimates that are unbiased and minimum
variance. We say that ML estimators are asymptotically UMVUE. Of course, for
any specific application, the ML estimates may still exhibit appreciable bias and it
is little consolation to know that, in a similar application with a much larger data
set, the bias would be reduced. In practice then it is always appropriate to verify that
estimator bias is small enough to be neglected.[3]

---

[3]One reaction to such promises of asymptotic good performance is that of J. M. Keynes, "In the
long run we are all dead." Still, for particular examples such as estimating the parameter μ of the
normal, the ML estimator may be not only unbiased but also an UMVUE. Life is sometimes good.

Invariance Under Reparameterization

An important advantage of ML estimators is that they are *invariant under reparameterization*. The exact choice of parameters for any parametric family is arbitrary. A change in parameters is called *reparameterization* (see Sect. B.3.3). In considering ML estimation of parameters of the normal, we chose to estimate the parameter σ. Of course, we could have chosen to estimate the parameter $V = \sigma^2$. The former is the standard deviation of the normal distribution and is a scale parameter; the latter is its variance. Both parameterizations can be argued for and it is interesting that the normal is often denoted $\mathcal{N}(\mu, \sigma^2)$, effectively preserving both parameterizations! At first glance the choice of parameterization would seem to be mainly a matter of taste: "you say tomayto, I say tomahto." However, suppose that you have chosen σ as parameterization and estimated it as $\hat{\sigma}$ using maximum likelihood. We have taken the same sample and estimated the variance $\hat{\sigma^2}$. If your estimated squared $(\hat{\sigma})^2$ is not equal to our estimate, then we have a conflict. Our estimates should be consistent with one another but they disagree. The inconsistency is serious, as serious as getting different estimates of the location of a planet if we calculated in miles or in kilometers.[4] If, however, we both used ML estimates, then there can be no such disagreement: $(\hat{\sigma})^2 = \hat{\sigma^2}$ always. This is invariance under reparameterization and it always holds if we use ML estimators. Invariance under reparameterization can fail if we use Bayesian estimation methods (See below, Sect. B.4.4).

In an elementary introduction to estimation, we would normally not discuss invariance under reparameterization. However, in considering the psychophysical methods in several of the chapters, it will prove to be a very useful tool. If we can, for example, change parameterization and make a difficult estimation problem easy, then we will do so, knowing that within the framework of ML estimation, we are entitled to do so.

Example

We noted that the formula for the MLE of σ given a sample from a normal distribution was

$$\hat{\sigma} = \sqrt{\frac{\sum_{i=1}^{n}(X_i - \bar{X})^2}{n}} \tag{B.32}$$

and we can therefore write down the MLE of the variance parameter σ^2 as

$$\hat{\sigma^2} = \frac{\sum_{i=1}^{n}(X_i - \bar{X})^2}{n}. \tag{B.33}$$

---

[4]Which NASA once did, leading to the loss of the Mars Climate Orbiter spacecraft in 1999.

The reader is likely familiar with the alternative estimator

$$S^2 = \frac{\sum_{i=1}^{n}(X_i - \bar{X})^2}{n-1}$$ (B.34)

available on most hand calculators. The reason that the alternative is preferred is that it is unbiased $E[S^2] = \sigma^2$ and therefore the ML estimator is biased $E[\hat{\sigma}^2] = \frac{n-1}{n}\sigma^2$. We can see that as the sample size increases, $n - 1/n \to 1$ and the bias vanishes asymptotically.

While the ML estimator may be biased we can often correct for the bias either in closed form as in the example just given or by means of simulation using resampling methods to estimate bias. See [55].

B.4.4 Bayesian Estimation

An alternative to MLE is Bayesian estimation. The estimation problem is unchanged. We have a sample of data $X_1, \ldots, X_n$ from a parametric distribution with pdf $f(x; \theta_1, \ldots, \theta_m)$. However, now we assume that the true values of the parameters $\theta_1, \ldots, \theta_m$ are the realization of a random variable $\Theta = (\theta_1, \ldots, \theta_m)$ with its own pdf $\pi(\theta_1, \ldots, \theta_m)$ referred to as the *prior distribution*. We would like to use the prior together with the sample $X_1, \ldots, X_n$ to improve our estimates of $\theta_1, \ldots, \theta_m$. Of course, to do so, we need to know what the prior is.

Suppose, for example, that we are trying to estimate the peak sensitivity, λ_L, of the long-wavelength (L) photoreceptor of a specific individual. We collect data in a psychophysical task and we could use an ML method to estimate λ_L. In doing so, we use only the data and the assumptions made in modeling the psychophysical task. But suppose that we know that the individual was drawn "at random" from a particular population and that we have estimates of λ_L for many other individuals drawn from the same population as shown in Fig. B.9. These estimates are not a prior (they are just a histogram) but we might be willing to use them to estimate a distribution (shown as a solid curve) that we will treat as a prior. Because of the prior, we know something about λ_L even before we collect the data.

The first step in computing a Bayesian estimate is the same as in computing the ML estimate. We compute the likelihood of the sample,

$$\mathscr{L}(\theta_1, \ldots, \theta_m; X_1, \ldots, X_n) = \prod_{i=1}^{n} f(X_i; \theta_1, \ldots, \theta_m)$$ (B.35)

and then we multiply it by the prior $\pi(\theta_1, \ldots, \theta_m)$. The result need not be a probability density function (it may not integrate to 1), but we can normalize it to get the *posterior distribution*

$$p(\theta_1, \ldots, \theta_m; X_1, \ldots, X_n) = C^{-1} L(\theta_1, \ldots, \theta_m; X_1, \ldots, X_n) \pi(\theta_1, \ldots, \theta_m).$$ (B.36)

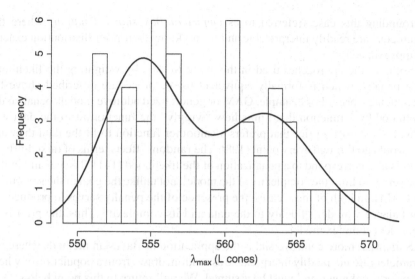

Fig. B.9 Histogram of peak wavelengths (λ_{max}) for the human L-cone photopigment as measured using microspectrophotometry of cone photoreceptors [45]. The *solid curve* is a best-fitting sum of Gaussians

The normalizing constant is just

$$C = \int_{-\infty}^{\infty} \cdots \int_{-\infty}^{\infty} p(\theta_1, \ldots, \theta_m; X_1, \ldots, X_n) \, d\theta_1, \ldots, \theta_m. \tag{B.37}$$

Equation (B.36) is a form of Bayes Theorem applied to probability density functions. We interpret the equation as a method for taking the prior (which contains all the information we have about the distribution before the data are collected), and combining it with the data, summarized by the likelihood function to get a new version of the prior, the posterior. We interpret the posterior as our best estimate of the pdf of the parameters, and, thereafter, we can derive estimates of the parameters from it. For example, by analogy to maximizing likelihood, the values of the parameters that maximize the posterior are the *maximum a posteriori* (MAP) *estimate* of the parameters. The MAP is just the mode of the posterior distribution. However, since we interpret the posterior $p(\theta_1, \ldots, \theta_m; X_1, \ldots, X_n)$ as a probability density function we can also consider other measures such as the mean of the posterior or its median as alternatives.

The key differences between the ML approach and Bayesian methods are typically not very great in practice. The key conceptual difference is the assumption that the parameters we seek to estimate are themselves random variables and that an intermediate goal of estimation is to estimate the distribution of the parameters given all available knowledge including the prior. As just presented, it would be difficult to criticize the Bayesian approach and, in fact, there is little controversy surrounding this case (referred to as *empirical Bayesian estimation*) where the parameters are readily interpretable and we do know their prior distribution exactly or approximately.

Some of the approaches used in this book fit data by weighting the likelihood by a penalty, which is formally equivalent to the procedure described above for introducing a prior. For example, GAM or generalized additive models penalize the likelihood by a function that limits how "wiggly" the function fit to the data will be and, thus, uses a prior that prefers a smoother function to fit the data than one that would go through every point [198]. The random effects terms of mixed-effects models also correspond to a penalization of the likelihood [141]. The penalty in this case serves to limit the complexity of the model, not unlike the goal of the smoothing for GAMs. In both of these cases, the presence of the penalty serves to produce fits that balance a model's fidelity to the data and its complexity. Thus, the prior is to prefer less complex models.

Somewhat more controversial is the application of Bayesian methods where the parameters are not readily interpretable as random draws from a population or where the prior is unknown and must be assumed. We will return to this point below.

Example

We wish to estimate the probability that a particular coin will land "heads" when tossed. As in Sect. B.4.2, we will model the coin as a Bernoulli random variable $\mathscr{B}(p)$. Suppose that we toss the coin ten times and obtain "heads" three times and "tails" seven times. We saw above in Sect. B.4.2 that the MLE of p was $\hat{p} = 0.3$, the proportion of "heads." But now suppose that the coin is drawn from a large population of coins whose probabilities of landing "heads" are unknown but may vary. We denote the probability that our particular coin lands "heads" by P, following our usual convention that random variables are denoted by uppercase letters. P is a random variable since in drawing the coin from a population of coins we have effectively drawn P from all populations of possible probabilities. To apply Bayesian estimation we need a prior distribution for P. Let's begin by assuming that we are completely ignorant of how coins behave and to represent our ignorance we set the pdf of P to be $\mathscr{U}(0,1)$, the uniform distribution. Suppose that we toss the coin n times, modeling the outcomes as Bernoulli random variables, $B_1, B_2, \ldots, B_n$ and apply (B.36).

We find that the posterior

$$p(P; B_1, \ldots, B_n) = C \prod_{i=1}^{n} n p^{B_i} (1 - p)^{1 - B_i} \tag{B.38}$$

is just the likelihood function multiplied by a positive constant (the constant needed to insure that the posterior is a pdf). The resulting distribution is an example of a *Beta distribution*.

Fig. B.10 The posterior distribution of P, the parameter of a Bernoulli $\mathscr{B}(P)$ random variable following a sample with three 1s and seven 0s. The prior distribution was $\mathscr{U}(0,1)$. The resulting distribution is a Beta distribution

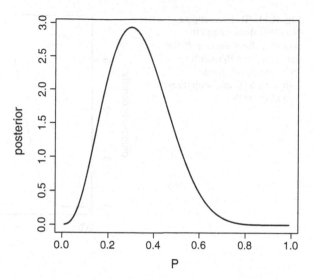

The Beta family has two parameters α, β The pdf of a Beta distribution is 0 outside of the interval $(0,1)$ and is of the form

$$f(x; \alpha, \beta) = p^{\alpha-1}(1-p)^{\beta-1} \qquad p \in (0,1). \tag{B.39}$$

In our example, α is one more than the number of heads, and β is one more than the number of tails. With three "heads" and seven "tails," the resulting posterior is shown in Fig. B.10. With a uniform prior, the MLE and MAP are identical (the positive constant cannot alter the location of the maximum), and they are both the proportion of "heads," 0.3. However, now we can interpret the likelihood/posterior as a pdf and use it to estimate the probability that $P < 0.5$, among other things.

Example

Suppose that we are uncomfortable with the assumption that the prior of P is uniform. After all, most coins we have encountered have probabilities p close to 0.5. Then we might assume a different prior. One clever way to do that is to pick a prior that when multiplied by the likelihood results in a posterior distribution that is easy to work with. We can, for example, pick the prior to be a Beta distribution with parameters $\alpha = 3$ and $\beta = 3$ or $\alpha = 5$ and $\beta = 5$ or $\alpha = 10$ and $\beta = 10$. The dotted and dashed curves in Fig. B.11 are the resulting priors, the broader dotted curve that for $\alpha = 3$ and $\beta = 3$ and the narrower dotted curve being that for $\alpha = 10$ and $\beta = 10$. All three are symmetric around 0.5. These are intended to capture our prior beliefs concerning the coin and we pick the middle one $\alpha = 5$ and $\beta = 5$ (dashed). We then toss the coin ten times and get three "heads" and seven "tails." We compute the posterior now incorporating our new prior and the resulting posterior is shown

Fig. B.11 Three candidate
prior distributions and the
posterior distribution of P, the
parameter of a Bernoulli
$\mathscr{B}(P)$ random variable
following a sample with three
1s and seven 0s

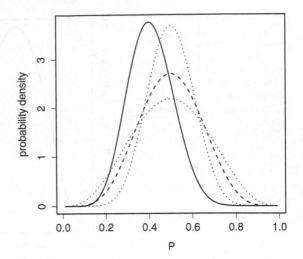

as a solid curve in Fig. B.11. The posterior is very different from that obtained with
a uniform prior and it is a Beta distribution with parameters $\alpha = 3 + 5 + 1$. and
$\beta = 7 + 5 + 1$ and the MAP is $0.4 = 8/20$. The new prior has led to an estimate that
is closer to 0.5 than the estimate obtained with the uniform prior. One controversial
aspect of the Bayesian approach is illustrated in these last two examples. The prior is
intended to capture the aspects of our knowledge concerning the parameters outside
of the sample. When, however, our knowledge is imprecise, it is not obvious that
introducing it as a prior produces a better estimate. If, however, the prior distribution
had been based on extensive analyses of the probabilities of actual coins drawn
from an actual population, then the use of the prior to improve the estimate would
generate little controversy.

A more subtle difficulty arises when we engage in Bayesian estimation using the
uniform distribution to represent ignorance. If the parameter P of the Bernoulli is
drawn from a uniform distribution $\mathscr{U}(0,1)$, then the reparameterization $R = 1/P$ is
not. Indeed, the cdf of R

$$P[R < x] = P[1/P < x] = P[P > 1/x] = 1 - 1/x, \qquad 1 \le x \le \infty \qquad (B.40)$$

and the pdf is x^{-2} over the interval $(1, \infty)$ and otherwise 0. The resulting distribution
is not uniform and there is, of course, no uniform distribution over the interval $1 \le
x \le \infty$. We seem to be simultaneously ignorant about P and quite knowledgeable
about its reciprocal.

A similar paradox arises when we consider Bayesian estimation and reparameterization. Suppose we collect a sample from the Bernoulli $\mathscr{B}(P)$ and form a maximum
likelihood estimate $\hat{P}$. As a consequence of invariance of reparameterization, the
MLE of R is just $1/\hat{p}$. However, the Bayesian estimate with *any* prior on P will not
be invariant under reparameterization. An example and discussion can be found in
[54, pp. 20–22].

B.5 Nested Hypothesis Testing

Hypothesis testing is a major division of mathematical statistics comparable to estimation and in the form of null hypothesis testing it is mired in controversy. We will avoid controversy by concentrating on the applications of hypothesis testing that are most common in psychophysics. These typically concern a comparison of two parametric models, one of which is a special case of the other. We will characterize two such models by their likelihood functions $\mathscr{L}_1(\theta_1,\ldots,\theta_m,\theta_{m+1},\ldots,\theta_{m+p};X_1,\ldots,X_n)$ and $\mathscr{L}_0(\theta_1,\ldots,\theta_m,\theta_{m+1} = C_1,\ldots,\theta_{m+p} = C_p;X_1,\ldots,X_n)$. The first model has a likelihood function with $m+p$ parameters and the second model is identical to the first except that p of the parameters have been set to specific constant values, $C_1,\ldots,C_p$. The notation above is cumbersome and an example will help.

Example

Suppose that the first model corresponds to the normal family $\mathcal{N}(\mu,\sigma^2)$ with two parameters, μ and σ. The second model is a subset of the distributions in the first family, namely those for which $\mu = 0$. The two likelihood functions are $\mathscr{L}_1(\mu,\sigma;X_1,\ldots,X_n)$, given above in (B.23) and $\mathscr{L}_0(\mu = 0,\sigma;X_1,\ldots,X_n)$ which is the same equation but with every occurrence of μ set to 0.

The maximum likelihood estimates $\hat{\mu}$ and $\hat{\sigma}$ are the values that maximize L_1 and therefore $\mathscr{L}_1(\hat{\mu},\hat{\sigma};X_1,\ldots,X_n)$ is the maximum value of likelihood achievable. Now, we set $\mu = 0$ in $\mathscr{L}_0$ and maximize its likelihood by varying the one remaining parameter σ. The resulting maximum likelihood estimate $\hat{\sigma}_0$ need not be the same as the previous ML estimate $\hat{\sigma}$. It almost certainly won't be the same since the latter is the maximum likelihood when both μ and σ are allowed to vary, and the former corresponds to the maximum when $\mu = 0$ by fiat.

The constrained ML cannot be bigger than the unconstrained:

$$\mathscr{L}_1(\hat{\mu},\hat{\sigma};X_1,\ldots,X_n) \geq \mathscr{L}_0(\mu - 0,\hat{\sigma}_0;X_1,\ldots,X_n) \tag{B.41}$$

If it were, then setting $\mu = 0$ and $\sigma = \hat{\sigma}_0$ would lead to a likelihood value for L_1 greater than that achieved by the maximum likelihood values, an impossibility.

The question we address is, should we reject the constrained model whose likelihood is L_0 in favor of the unconstrained L_1? We interpret the ratio of the likelihoods at their respective maxima as a measure of the improvement in accounting for the data,

$$\Lambda = \frac{\mathscr{L}_1(\hat{\mu},\hat{\sigma};X_1,\ldots,X_n)}{\mathscr{L}_0(\mu,0,\hat{\sigma}_0;X_1,\ldots,X_n)}. \tag{B.42}$$

If this value is large enough, we would be justified in rejecting the constrained model.

To develop a sense of how the resulting *likelihood ratio* statistic behaves, we can make use of a remarkable but asymptotic result known as Wilks' Theorem [193] that tells us that if the sample was in fact drawn from the constrained model/family (in our example, $\mu = 0$ in reality), then the statistic

$$W = 2\log\Lambda \tag{B.43}$$

will be distributed as a χ^2 random variable whose degrees of freedom (df) are the number of parameters constrained (the difference in number of parameters in the two models). In our example, only μ was constrained and the corresponding df is 1. (See [129, p. 441ff].) We would expect the W value to be distributed as a χ^2 variable with one degree of freedom χ_1^2. This result is remarkable since it holds true for any two nested models. For example, let's consider a nested hypothesis test based on the exponential distribution. The unconstrained hypothesis is that the data are drawn from an exponential distribution with parameter $\lambda > 0$. The constrained hypothesis is that $\lambda = 1$ We can verify Wilks' Theorem by drawing a sample of size 100 from an exponential distribution with $\lambda = 1$ and computing the maximum constrained and maximum unconstrained likelihoods. It is computationally convenient to compute the log likelihoods of the two models (to avoid exceeding the numerical precision) and taking their difference. The log likelihood of the sample is

$$\log\mathscr{L}(\lambda;X_1,\ldots,X_n) = -\lambda\sum_{i=1}^{n}X_i + n\log\lambda. \tag{B.44}$$

Given a sample $X_1,X_2,\ldots,X_n$ the unconstrained maximum likelihood estimate of $\lambda = 1/\bar{X}$, and this is the value we use to maximize the unconstrained likelihood. In computing the constrained likelihood we use the value $\lambda = 1$. We repeat this simulation 1,000 times and plot the histogram of the resulting W values with the χ_1^2 pdf superimposed in Fig. B.12. As one should begin to expect (we hope), we obtain the smooth pdf curve using the dchisq function. The match illustrates Wilks' Theorem and also provides assurance that the asymptotic distribution is a good model for the actual distribution of W for such a large sample.

Following the logic of hypothesis testing we would pick a size α for our test of the unconstrained model against the constrained and reject if the test statistic W falls above the $1 - \alpha$ quantile of the χ_1^2 distribution. In the general case, with p constrained parameters, we would replace χ_1^2 by χ_p^2.

In the discussion above, we used Monte Carlo methods to good effect. Given any asymptotic claim such as Wilks' Theorem, we typically have no guarantee that the sample size in any particular application is "large enough" to be "nearly asymptotic." However,we can simply simulate our experiment with the design we used or plan to use and check how a statistic such as W behaves.

Fig. B.12 Simulation of
distribution of Wilk's
likelihood ratio statistic. The
solid lines correspond to the
pdf of χ_1^2

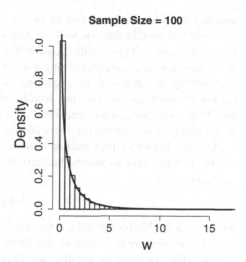

B.5.1   *Reparameterization: Nested Hypothesis Tests*

One very common form of the nested hypothesis test makes use of constraints
that do not involve setting parameters to specific values but instead considering
that in a list of parameters some of the parameters are identical. That is, in
the list $\theta_1,\ldots,\theta_p$ we wish to test whether $\theta_{p-1} = \theta_p$. We can easily recast this
problem as simply a reparameterization. We replace the parameters $\theta_1,\ldots,\theta_p$ by the
reparameterization $\theta_1,\ldots,\theta_{p-1},\Delta\theta$ where each θ_i in one parameterization is equal
to θ_i in the other, $i=1,\ldots,p-1$ and $\theta_p = \theta_{p-1}+\Delta\theta$. the reader can verify that we
can transform back and forth from one set of parameters to the other and that the
constraint $\theta_{p-1} = \theta_p$ in the first parameterization is just the constraint $\Delta\theta = 0$ in the
second. Nested hypothesis tests based on likelihood ratios are also invariant under
reparameterization and we can simply use the second parameterization in place of
the first.

B.5.2   *The Akaike Information Criterion*

In fitting a model to data, we hope to characterize as accurately as possible a
systematic, underlying component that depends on a set of explanatory variables. If
the model describes the data well, then the variation that remains after the fit can be
attributed to unsystematic factors that have not or cannot be controlled. The accuracy
with which the systematic component is characterized depends on the appropriate
choice of model for the data under consideration. The ML procedures attempt to
find parameter values for which the model is closest to the data, according to the ML
criterion. Adding additional parameters will lead to models with a greater likelihood

and that are a better description of the actual data. The danger in fitting the data too well, however, is that one will describe not only the systematic component but the noise as well. This is called *over-fitting* and leads to models that are poor at predicting new data. Likelihood ratio tests of nested models provide one method for approaching this problem, by excluding models for which the added parameters do not significantly increase the likelihood. They do not necessarily shield the user from over-fitting, however, since a parameter that increases the likelihood by serendipitously accounting for unsystematic variation in the data could be selected.

An (or Akaike's) Information Criterion (AIC) [6] is one of a number of information theoretic measures that have been proposed to address this issue. It is defined as

$$\text{AIC} = -2\log(\mathscr{L}) + 2k, \tag{B.45}$$

where $\mathscr{L}$ is the likelihood and k is the number of parameters estimated in fitting the model. The measure is defined so that better fit models correspond to lower values of AIC. The decrease in negative log likelihood obtained by adding parameters (adding complexity) is penalized by adding back in twice the number of parameters to counterbalance the tendency toward over-fitting. The lowest AIC corresponds to a balance between good fit of the model to the data and model parsimony as indicated by the number of free parameters. The formula for the AIC arises in a derivation of fitting the model to the expected value of the data rather than the data, themselves, and in this way suggests a model that is better at predicting future data than one based just on likelihood [198].

Model fitting functions in R based on likelihood typically provide an AIC value in the output or one that can be assessed through an AIC method. Some functions provide an additional measure, the Bayesian Information Criterion (BIC), that employs a harsher penalty for model complexity, based on a penalty of $k\log(n)$ where n is the number of observations.

References

1. Abbey, C.K., Eckstein, M.P.: Classification image analysis: estimation and statistical inference for two-alternative forced-choice experiments. J. Vis. **2**, 66–78 (2002)
2. Ahumada, A.J.: Perceptual classification images from vernier acuity masked by noise. Perception **25** ECVP Abstract Suppl. (1996)
3. Ahumada, A.J.: Classification image weights and internal noise level estimation. J. Vis. **2**(1), 121–131 (2002)
4. Ahumada, A.J., Lovell, J.: Stimulus features in signal detection. J. Acoust. Soc. Am. **49**, 1751–1756 (1971)
5. Ahumada, A.J., Marken, R., Sandusky, A.: Time and frequency analyses of auditory signal detection. J. Acoust. Soc. Am. **57**(2), 385–390 (1975)
6. Akaike, H.: Information theory and an extension of the maximum likelihood principle. In: Petrov, B.N., Csàki, F. (eds.) Second International Symposium on Inference Theory, pp. 267–281. Akadémia Kiadó, Budapest (1973)
7. Baayen, R.H., Davidson, D.J., Bates, D.M.: Mixed-effects modeling with crossed random effects for subjects and items. J. Mem. Lang. **59**, 390–412 (2008)
8. Bates, D.: Fitting linear mixed models. R News **5**, 27–30 (2005). http://www.r-project.org/doc/Rnews/Rnews_2005-1.pdf
9. Bates, D., Maechler, M., Bolker, B.: lme4: Linear mixed-effects models using S4 classes (2011). R package version 0.999375-42. http://CRAN.R-project.org/package=lme4. Accessed date on 2nd August 2012
10. Bates, D., Maechler, M., Bolker, B.: lme4.0: Linear mixed-effects models using S4 classes (2012). R package version 0.9999-1/r1692. http://R-Forge.R-project.org/projects/lme4/. Accessed date on 2nd August 2012
11. Bates, D.M.: lme4: Mixed-Effects Modeling with R. Springer, New York (in preparation). http://lme4.r-forge.r-project.org/book/
12. Bengtsson, H., Riedy, J.: R.matlab: Read and write of MAT files together with R-to-Matlab connectivity (2011). R package version 1.5.1. http://CRAN.R-project.org/package=R.matlab. Accessed date on 2nd August 2012
13. Bernstein, S.N.: Sur l'ordre de la meilleure approximation des fonctions continues par les polynômes de degré donné. Mémoires de l' Académie Royale de Belgique **4**, 1–104 (1912)
14. Bishop, Y.M.M., Fienberg, S.E., Holland, P.W.: Discrete Multivariate Analysis: Theory and Practice. MIT, Cambridge (1975)
15. Block, H.D., Marschak, J.: Random orderings and stochastic theories of responses. In: Olkin, I., Ghurye, S., Hoeffding, W., Madow, W., Mann, H. (eds.) Contributions to Probability and Statistics, pp. 38–45. Stanford University Press, Stanford (1960)

16. Boeck, P.D., Wilson, M. (eds.): Explanatory Item Response Models: A Generalized Linear and Nonlinear Approach. Springer, New York (2004)
17. Boring, E.G.: Sensation and Perception in the History of Experimental Psychology. Irvington Publishers, Inc., New York (1942)
18. Bouet, R., Knoblauch, K.: Perceptual classification of chromatic modulation. Vis. Neurosci. **21**, 283–289 (2004)
19. Bracewell, R.N.: The Fourier Transform and its Applications, 3rd edn. McGraw-Hill, New York (2000)
20. Bratley, P., Fox, B.L., Schrage, L.E.: A Guide to Simulation. Springer, New York (1983)
21. Brindley, G.S.: Two more visual theorems. Q. J. Exp. Psychol. **12**, 110–112 (1960)
22. Britten, K.H., Shadlen, M.N., Newsome, W.T., Movshon, J.A.: The analysis of visual motion: A comparison of neuronal and psychophysical performance. J. Neurosci. **12**(12), 4745–4765 (1992)
23. Broström, G., Holmberg, H.: glmmML: Generalized linear models with clustering (2011). R package version 0.82-1. http://CRAN.R-project.org/package=glmmML. Accessed date on 2nd August 2012
24. Burnham, K.P., Anderson, D.R.: Model Selection and Inference: A Practical Information-Theoretic Approach. Springer, New York (1998)
25. Canty, A., Ripley, B.D.: boot: Bootstrap R (S-Plus) Functions (2012). R package version 1.3-5
26. Carney, T., Tyler, C.W., Watson, A.B., Makous, W., Beutte, B., Chen, C.C., Norcia, A.M., Klein, S.A.: Modelfest: Year one results and plans for future years. In: Rogowitz, B.E., Pappas, T.N. (eds.) Proceedings of SPIE: Human vision and electronic imaging V, vol. 3959, pp. 140–151. SPIE, Bellingham (2000)
27. Carroll, L.: Through the Looking Glass and Ahat Alice Found There. Macmillan, Bassingstoke (1871)
28. Chambers, J.M., Hastie, T.J. (eds.): Statistical Models in S. Chapman and Hall/CRC, Boca Raton (1992)
29. Charrier, C., Maloney, L.T., Cherifi, H., Knoblauch, K.: Maximum likelihood difference scaling of image quality in compression-degraded images. J. Opt. Soc. Am. A **24**, 3418–3426 (2007)
30. Chauvin, A., Worsley, K., Schyns, P., Arguin, M., Gosselin, F.: Accurate statistical tests for smooth classification images. J. Vis. **5**, 659–667 (2005)
31. Christensen, R.H.B.: ordinal—regression models for ordinal data (2010). R package version 2011.09-14. http://www.cran.r-project.org/package=ordinal/. Accessed date on 2nd August 2012
32. Christensen, R.H.B., Brockhoff, P.B.: sensR—an R-package for sensory discrimination (2011). R package version 1.2-13. http://www.cran.r-project.org/package=sensR/. Accessed date on 2nd August 2012
33. Christensen, R.H.B., Hansen, M.K.: binomTools: Performing diagnostics on binomial regression models (2011). R package version 1.0-1. http://CRAN.R-project.org/package=binomTools. Accessed date on 2nd August 2012
34. Chung, K.L., Aitsahlia, F.: Elementary Probability Theory, 4th edn. Springer, New York (2006)
35. Clark, H.H.: The language-as-fixed-effect fallacy: A critique of language statistics in psychological research. J. Verbal Learn. Verbal Behav. **12**, 355–359 (1973)
36. Cleveland, W.S., Diaconis, P., McGill, R.: Variables on scatterplots look more highly correlated when the scales are increased. Science **216**, 1138–1141 (1982)
37. Cleveland, W.S., McGill, R.: Graphical perception: Theory, experimentation and application to the development of graphical methods. J. Am. Stat. Assoc. **79**, 531–554 (1984)
38. Cleveland, W.S., McGill, R.: The many faces of a scatterplot. J. Am. Stat. Assoc. **79**, 807–822 (1984)
39. Conway, J., Eddelbuettel, D., Nishiyama, T., Prayaga, S.K., Tiffin, N.: RPostgreSQL: R interface to the PostgreSQL database system (2010). R package version 0.1-7. http://www.postgresql.org. Accessed date on 2nd August 2012

40. Cook, R.D., Weisberg, S.: Residuals and Influence in Regression. Chapman and Hall, London (1982)
41. Cornsweet, T.: Visual Perception. Academic, New York (1970)
42. Crozier, W.J.: On the visibility of radiation at the human fovea. J. Gen. Physiol. **34**, 87–136 (1950)
43. Crozier, W.J., Wolf, E.: Theory and measurement of visual mechanisms. IV. On flicker with subdivided fields. J. Gen. Physiol. **27**, 401–432 (1944)
44. Dakin, S.C., Bex, P.J.: Natural image statistics mediate brightness filling in. Proc. Biol. Sci. **1531**, 2341–2348 (2003)
45. Dartnall, H.J.A., Bowmaker, J.K., Mollon, J.D.: Microspectrophotometry of human photoreceptors. In: Mollon, J.D., Sharpe, L.T. (eds.) Colour Vision: Physiology and Psychophysics, pp. 69–80. Academic, London (1983)
46. Davison, A.C., Hinkley, D.V.: Bootstrap Methods and their Applications. Cambridge University Press, Cambridge (1997). http://statwww.epfl.ch/davison/BMA/
47. Debreu, G.: Topological methods in cardinal utility theory. In: Arrow, K.J., Karlin, S., Suppes, P. (eds.) Mathematical Methods in the Social Sciences, pp. 16–26. Stanford University Press, Stanford (1960)
48. DeCarlo, L.T.: Signal detection theory and generalized linear models. Psychol. Meth. **3**, 186–205 (1998)
49. DeCarlo, L.T.: On the statistical and theoretical basis of signal detection theory and extensions: Unequal variance, random coefficient and mixture models. J. Math. Psychol. **54**, 304–313 (2010)
50. Delord, S., Devinck, F., Knoblauch, K.: Surface and edge in visual detection: Is filling-in necessary? J. Vis. **4**, 68a (2004). http://journalofvision.org/4/8/68/
51. Devinck, F.: Les traitements visuels chez l'homme : stratégies de classification de la forme. Ph.D. thesis, Université Lyon 2 (2003)
52. Devroye, L.: Non-uniform Random Variate Generation. Springer, New York (1986)
53. Dobson, A.J.: An Introduction to Generalized Linear Models. Chapman and Hall, London (1990)
54. Edwards, A.W.F.: Likelihood, expanded edn. Johns Hopkins University Press, Baltimore (1992)
55. Efron, B., Tibshirani, R.J.: An Introduction to the Bootstrap. Chapman Hall, New York (1993)
56. Emrith, K., Chantler, M.J., Green, P.R., Maloney, L.T., Clarke, A.D.F.: Measuring perceived differences in surface texture due to changes in higher order statistics. J. Opt. Soc. Am. A **27**(5), 1232–1244 (2010)
57. Falmagne, J.C.: Elements of Psychophysical Theory. Oxford University Press, Oxford (1985)
58. Faraway, J.J.: Extending Linear Models with R: Generalized Linear, Mixed Effects and Nonparametric Regression Models. Chapman and Hall/CRC, Boca Raton (2006)
59. Fechner, G.T.: Elemente der Psychophysik. Druck und Verlag von Breitkopfs, Leipzig (1860)
60. Finney, D.J.: Probit Analysis, 3rd edn. Cambridge University Press, Cambridge (1971)
61. Firth, D.: Bias reduction of maximum likelihood estimates. Biometrika **80**, 27–38 (1993)
62. Fisher, R.A.: The design of experiments. In: Bennett, J.H. (ed.) Statistical Methods, Experimental Design and Scientific Inference, 9th edn. Macmillan, New York (1971)
63. Fleming, R.W., Jäkel, F., Maloney, L.T.: Visual perception of thick transparent materials. Psychol. Sci. **22**, 812–820 (2011)
64. Foster, D.H., Bischof, W.F.: Bootstrap estimates of the statistical accuracy of thresholds obtained from psychometric functions. Spatial Vis. **11**(1), 135–139 (1997)
65. Foster, D.H., Żychaluk, K.: Nonparametric estimates of biological transducer functions. IEEE Signal Process. Mag. **24**, 49–58 (2007)
66. Fründ, I., Haenel, N.V., Wichmann, F.A.: Inference for psychometric functions in the presence of nonstationary behavior. J. Vis. **11**(6), 1–19 (2011)
67. Genz, A., Bretz, F., Miwa, T., Mi, X., Leisch, F., Scheipl, F., Bornkamp, B., Hothorn, T.: mvtnorm: Multivariate normal and t distributions (2011). R package version 0.9-9991. http://CRAN.R-project.org/package=mvtnorm. Accessed date on 2nd August 2012

68. Gescheider., G.A.: Psychophysical scaling. Annu. Rev. Psychol. **39**, 169–200 (1988)
69. Glass, L.: Moiré effect from random dots. Nature **223**, 578–580 (1969)
70. GmbH, M.S.: XLConnect: Excel Connector for R (2011). R package version 0.1-5. http://CRAN.R-project.org/package=XLConnect. Accessed date on 2nd August 2012
71. Gold, J.M., Murray, R.F., Bennett, P.J., Sekular, A.B.: Deriving behavioural receptive fields for visually completed contours. Curr. Biol. **10**, 663–666 (2000)
72. Green, D.M., Swets, J.A.: Signal Detection Theory and Psychophysics. Robert E. Krieger Publishing Company, Huntington (1966/1974)
73. Guilford, J.P.: Psychometric Methods, 2nd edn. McGraw-Hill, New York (1954)
74. Hadfield, J.D.: Mcmc methods for multi-response generalized linear mixed models: The MCMCglmm R package. J. Stat. Software **33**(2), 1–22 (2010). http://www.jstatsoft.org/v33/i02/
75. Hansen, T., Gegenfurtner, K.R.: Classification images for chromatic signal detection. J. Opt. Soc. Am. A **22**, 2081–2089 (2005)
76. Harrell, F.E.: rms: Regression modeling strategies (2011). R package version 3.3-1. http://CRAN.R-project.org/package=rms. Accessed date on 2nd August 2012
77. Hastie, T., Tibshirani, R.: Generalized Additive Models. Chapman and Hall, London (1990)
78. Hauck, W.W. Jr, Donner, A.: Wald's test as applied to hypotheses in logit analysis. J. Am. Stat. Assoc. **72**, 851–853 (1977)
79. Hecht, S., Shlaer, S., Pirenne, M.H.: Energy, quanta and vision. J. Gen. Physiol. **25**, 819–840 (1942)
80. Ho, Y.X., Landy, M.S., Maloney, L.T.: Conjoint measurement of gloss and surface texture. Psychol. Sci. **19**, 196–204 (2008)
81. James, D.A.: DBI: R Database Interface (2009). R package version 0.2.5. http://CRAN.R-project.org/package=DBI. Accessed date on 2nd August 2012
82. James, D.A.: RSQLite: SQLite interface for R (2011). R package version 0.10.0. http://CRAN.R-project.org/package=RSQLite. Accessed date on 2nd August 2012
83. James, D.A., DebRoy, S.: RMySQL: R interface to the MySQL database (2011). R package version 0.8-0. http://biostat.mc.vanderbilt.edu/RMySQL. Accessed date on 2nd August 2012
84. James, D.A., Luciani, J.: ROracle: Oracle database interface for R (2007). R package version 0.5-9. http://www.omegahat.org. Accessed date on 2nd August 2012
85. Johnson, N.L., Kemp, A.W., Kotz, S.: Univariate Discrete Distributions. Wiley, New York (2005)
86. Johnson, N.L., Kotz, S., Balakrishnan, N.: Continuous Univariate Distributions, vol. 1. Wiley, New York (1994)
87. Johnson, N.L., Kotz, S., Balakrishnan, N.: Continuous Univariate Distributions, vol. 2. Wiley, New York (1995)
88. Keppel, G.: Design & Analysis: A Researcher's Handbook, 2nd edn. Prentice-Hall, Englewood Cliffs (1982)
89. Kienzle, W., Franz, M.O., Scholkopf, B., Wichmann, F.A.: Center-surround patterns emerge as optimal predictors for human saccade targets. J. Vis. **9**, 1–15 (2009)
90. Kienzle, W., Wichmann, F.A., Schölkopf, B., Franz, M.O.: A nonparametric approach to bottom-up visual saliency. In: Saul, L.K., Weiss, Y., Bottou, L. (eds.) Advances in Neural Information Processing Systems 19: Proceedings of the 2006 Conference, pp. 689–696. MIT, Cambridge (2007)
91. Kingdom, F.A.A., Prins, N.: Psychophysics: A Practical Introduction. Academic, New York (2009)
92. Kleiber, C., Zeileis, A.: Applied Econometrics with R. Springer, New York (2008). http://CRAN.R-project.org/package=AER
93. Klein, S.A.: Measuring, estimating, and understanding the psychometric function: A commentary. Percept. Psychophys. **63**, 1421–1455 (2001)
94. Knoblauch, K.: psyphy: Functions for analyzing psychophysical data in R (2012). R package version 0.1-7. http://cran.r-project.org/web/packages/psyphy

95. Knoblauch, K., Maloney, L.T.: Estimating classification images with generalized linear and additive models. J. Vis. **8**, 1–19 (2008)
96. Knoblauch, K., Maloney, L.T.: MLDS: Maximum likelihood difference scaling in R. J. Stat. Software **25**, 1–26 (2008). http://www.jstatsoft.org/v25/i02
97. Knoblauch, K., Maloney, L.T.: MLCM: Maximum likelihood conjoint measurement (2011). R package version 0.0-8. http://CRAN.R-project.org/package=MLCM. Accessed date on 2nd August 2012
98. Knoblauch, K., Maloney, L.T.: MPDiR: Data sets and scripts for Modeling Psychophysical Data in R (2012). R package version 0.1-11. http://cran.r-project.org/web/packages/MPDiR
99. Knoblauch, K., Vital-Durand, F., Barbur, J.: Variation of chromatic sensitivity across the life span. Vis. Res. **41**, 23–36 (2001)
100. Knuth, D.E.: The Art of Computer Programming: Seminumerical Programming, 3rd edn. Addison-Wesley, New York (1997)
101. Komárek, A., Lesaffre, E.: Generalized linear mixed model with a penalized gaussian mixture as a random-effects distribution. Comput. Stat. Data Anal. **52**(7), 3441–3458 (2008)
102. Kontsevich, L.L., Tyler, C.W.: What makes Mona Lisa smile? Vis. Res. **44**, 1493–1498 (2004)
103. Kosmidis, I.: brglm: Bias reduction in binary-response GLMs (2007). R package version 0.5-6. http://www.ucl.ac.uk/~ucakiko/software.html. Accessed date on 2nd August 2012
104. Krantz, D.H., Luce, R.D., Suppes, P., Tversky, A.: Foundations of Measurement (Vol. 1): Additive and Polynomial Representations. Academic, New York (1971)
105. Kuss, M., Jäkel, F., Wichmann, F.A.: Bayesian inference for psychometric functions. J. Vis. **5**, 478–492 (2005). http://journalofvision.org/5/5/8/
106. Lang, D.T.: Rcompression: In-memory decompression for GNU zip and bzip2 formats. R package version 0.93-2. http://www.omegahat.org/Rcompression. Accessed date on 2nd August 2012
107. Lazar, N.A.: The Statistical Analysis of Functional MRI Data. Springer, New York (2008)
108. Legge, G.E., Gu, Y.C., Luebker, A.: Efficiency of graphical perception. Percept. Psychophys. **46**, 365–374 (1989)
109. Lehmann, E.L., Casella, G.: Theory of Point Estimation, 2nd edn. Springer, New York (1998)
110. Lemon, J: Plotrix: A package in the red light district of R. R-News **6**(4), 8–12 (2010)
111. Levi, D.M., Klein, S.A.: Classification images for detection and position discrimination in the fovea and parafovea. J. Vis. **2**(1), 46–65 (2002)
112. Li, Y., Baron, J.: Behavioral Research Data Analysis in R, 1st edn. Springer, New York (2011)
113. Lindsey, D., Brown, A.: Color naming and the phototoxic effects of sunlight on the eye. Psychol. Sci. **13**, 506–512 (2002)
114. Lindsey, D.T., Brown, A.M., Reijnen, E., Rich, A.N., Kuzmova, Y.I., Wolfe, J.M.: Color channels, not color appearance or color categories, guide visual search for desaturated color targets. Psychol. Sci. **21**, 1208–1214 (2010)
115. Link, S.W.: The Wave Theory of Difference and Similarity. Lawrence Erlbaum Associates, Hillsdale (1992)
116. Luce, R.D., Green, D.M.: Parallel psychometric functions from a set of independent detectors. Psychol. Rev. **82**, 483–486 (1975)
117. Luce, R.D., Tukey, J.W.: Simultaneous conjoint measurement: A new scale type of fundamental measurement. J. Math. Psychol. **32**, 466–473 (1964)
118. Macke, J.H., Wichmann, F.A.: Estimating predictive stimulus features from psychophysical data: The decision image technique applied to human faces. J. Vis. **10** (2010)
119. MacLeod, D.: Visual sensitivity. Annu. Rev. Psychol. **29**, 613–645 (1978)
120. Macmillan, N.A., Creelman, C.D.: Detection Theory: A User's Guide, 2nd edn. Lawrence Erlbaum Associates, New York (2005)
121. Maloney, L.T.: Confidence intervals for the parameters of psychometric functions. Percept. Psychophys. **47**(2), 127–134 (1990)
122. Maloney, L.T., Dal Martello, M.F.: Kin recognition and the perceived facial similarity of children. J. Vis. **6**, 1047–1056 (2006). http://journalofvision.org/6/10/4/

123. Maloney, L.T., Yang, J.N.: Maximum Likelihood difference scaling. J. Vis. **3**(8), 573–585 (2003). http://www.journalofvision.org/3/8/5

124. Mamassian, P., Goutcher, R.: Temporal dynamics in bistable perception. J. Vis. **5**, 361–375 (2005)

125. Mangini, M.C., Biederman, I.: Making the ineffable explicit: Estimating the information employed for face classifications. Cognit. Sci. **28**, 209–226 (2004)

126. Marin-Franch, I., Żychaluk, K., Foster, D.H.: modelfree: Model-free estimation of a psychometric function (2010). R package version 1.0. http://CRAN.R-project.org/package= modelfree. Accessed date on 2nd August 2012

127. McCullagh, P., Nelder, J.A.: Generalized Linear Models. Chapman and Hall, London (1989)

128. Mineault, P.J., Barthelme, S., Pack, C.C.: Improved classification images with sparse priors in a smooth basis. J. Vis. **9**, 1–24 (2009)

129. Mood, A., Graybill, F.A., Boes., D.C.: Introduction to the Theory of Statistics, 3rd edn. McGraw-Hill, New York (1974)

130. Murray, R.F.: Classification images: A review. J. Vis. **11**, 1–25 (2011)

131. Murray, R.F., Bennett, P.J., Sekuler, A.B.: Optimal methods for calculating classification images: Weighted sums. J. Vis. **2**(1), 79–104 (2002)

132. Murrel, P.: Introduction to Data Technologies. Chapman and Hall/CRC, Boca Raton (2009)

133. Nandy, A.S., Tjan, B.S.: The nature of letter crowding as revealed by first- and second-order classification images. J. Vis. **7**(2), 5.1–26 (2007)

134. Neri, P.: Estimation of nonlinear psychophysical kernels. J. Vis. **4**, 82–91 (2004). http:// journalofvision.org/4/2/2/

135. Neri, P.: How inherently noisy is human sensory processing? Psychonomic Bull. Rev. **17**, 802–808 (2010)

136. Neri, P., Heeger, D.J.: Spatiotemporal mechanisms for detecting and identifying image features in human vision. Nat. Neurosci. **5**, 812–816 (2002)

137. Neri, P., Parker, A.J., Blakemore, C.: Probing the human stereoscopic system with reverse correlation. Nature **401**, 695–698 (1999)

138. Newsome, W.T., Britten, K.H., Movshon, J.A.: Neuronal correlates of a perceptual decision. Nature **341**(6237), 52–54 (1989)

139. Obein, G., Knoblauch, K., Viénot, F.: Difference scaling of gloss: Nonlinearity, binocularity, and constancy. J. Vis. **4**(9), 711–720 (2004)

140. Peirce, J.W.: The potential importance of saturating and supersaturating contrast response functions in visual cortex. J. Vis. **7**, 13 (2007)

141. Pinheiro, J., Bates, D.: Mixed-Effects Models in S and S-PLUS. Springer, New York (2000)

142. Pinheiro, J., Bates, D., DebRoy, S., Sarkar, D., the R Development Core Team: nlme: Linear and nonlinear mixed effects models (2012). R package version 3.1-104

143. Press, W.H., Teukolsky, S.A., Vetterling, W.T., Flannery, B.P.: Numerical Recipes, 3rd ed.: The Art of Scientific Computing. Cambridge University Press, Cambridge (2007)

144. Quick, R.: A vector-magnitude model of contrast detection. Kybernetik **16**, 65–67 (1974)

145. R-core members, et al.: Foreign: Read data stored by Minitab, S, SAS, SPSS, Stata, Systat, dBase, … (2011). R package version 0.8-46. http://CRAN.R-project.org/package=foreign. Accessed date on 2nd August 2012

146. R Development Core Team: R: A language and environment for statistical computing. R Foundation for Statistical Computing, Vienna, Austria (2011). ISBN 3-900051-07-0. http:// www.R-project.org/. Accessed date on 2nd August 2012

147. Rensink, R.A., Baldridge, G.: The perception of correlation in scatterplots. Comput. Graph. Forum **29**, 1203–1210 (2010)

148. Rhodes, G., Maloney, L.T., Turner, J., Ewing, L.: Adaptive face coding and discrimination around the average face. Vis. Res. **47**, 974–989 (2007)

149. Ripley, B.: RODBC: ODBC Database Access (2011). R package version 1.3-3. http://CRAN. R-project.org/package=RODBC. Accessed date on 2nd August 2012

150. Ripley, B.D.: Pattern Recognition and Neural Networks. Cambridge University Press, Cambridge (1996)

151. Roberts, F.S.: Measurement Theory. Cambridge University Press, Cambridge (1985)
152. Ross, M.G., Cohen, A.L.: Using graphical models to infer multiple visual classification features. J. Vis. **9**(3), 23.1–24 (2009)
153. Ross, S.M.: A First Course in Probability Theory, 8th edn. Prentice-Hall, Englewood Cliffs (2009)
154. Rouder, J.N., Lu, J., Sun, D., Speckman, P., Morey, R., Naveh-Benjamin, M.: Signal detection models with random participant and item effects. Psychometrika **72**, 621–642 (2007)
155. Salsberg, D.: The Lady Tasting Tea: How Statistics Revolutionized Science in the Twentieth Century. W. H. Freeman, New York (2001)
156. Sarkar, D.: Lattice: Multivariate Data Visualization with R. Springer, New York (2008). http://lmdvr.r-forge.r-project.org
157. Schneider, B.: Individual loudness functions determined from direct comparisons of loudness intervals. Percept. Psychophys. **28**, 493–503 (1980)
158. Schneider, B.: A technique for the nonmetric analysis of paired comparisons of psychological intervals. Psychometrika **45**, 357–372 (1980)
159. Schneider, B., Parker, S., Stein, D.: The measurement of loudness using direct comparisons of sensory intervals. J. Math. Psychol. **11**, 259–273 (1974)
160. Schwartz, M.: WriteXLS: Cross-platform Perl based R function to create Excel 2003 (XLS) files (2010). R package version 2.1.0. http://CRAN.R-project.org/package=WriteXLS. Accessed date on 2nd August 2012
161. Solomon, J.A.: Noise reveals visual mechanisms of detection and discrimination. J. Vis. **2**(1), 105–120 (2002)
162. Spector, P.: Data Manipulation with R. Springer, New York (2008)
163. van Steen, G.: dataframes2xls: dataframes2xls writes data frames to xls files (2011). R package version 0.4.5. http://cran.r-project.org/web/packages/dataframes2xls. Accessed date on 2nd August 2012
164. Stevens, S.S.: On the theory of scales of measurement. Science **103**, 677–680 (1946)
165. Stevens, S.S.: On the psychophysical law. Psychol. Rev. **64**, 153–181 (1957)
166. Strasburger, H.: Converting between measures of slope of the psychometric function. Percept. Psychophys. **63**, 1348–1355 (2001)
167. Sun, H., Lee, B., Baraas, R.: Systematic misestimation in a vernier task arising from contrast mismatch. Vis. Neurosci. **25**, 365–370 (2008)
168. Suppes, P.: Finite equal-interval measurement structures. Theoria **38**, 45–63 (1972)
169. Tanner, W.P., Swets, J.A.: A decision-making theory of visual detection. Psychol. Rev. **61**, 401–409 (1954)
170. Teller, D.Y.: The forced-choice preferential looking procedure: A psychophysical technique for use with human infants. Infant Behav. Dev. **2**, 135–153 (1979)
171. Thibault, D., Brosseau-Lachaine, O., Faubert, J., Vital-Durand, F.: Maturation of the sensitivity for luminance and contrast modulated patterns during development of normal and pathological human children. Vis. Res. **47**, 1561–1569 (2007)
172. Thomas, J.P., Knoblauch, K.: Frequency and phase contributions to the detection of temporal luminance modulation. J. Opt. Soc. Am. A **22**(10), 2257–2261 (2005)
173. Thurstone, L.L.: A law of comparative judgement. Psychol. Rev. **34**, 273–286 (1927)
174. Tolhurst, D.J., Movshon, J.A., Dean, A.F.: The statistical reliability of signals in single neurons in cat and monkey visual cortex. Vis. Res. **23**(8), 775–785 (1983)
175. Treutwein, B., Strasburger, H.: Fitting the psychometric function. Percept. Psychophys. **61**(1), 87–106 (1999)
176. Urban, F.M.: The method of constant stimuli and its generalizations. Psychol. Rev. **17**, 229–259 (1910)
177. Urbanek, S.: RJDBC: Provides access to databases through the JDBC interface (2011). R package version 0.2-0. http://www.rforge.net/RJDBC/. Accessed date on 2nd August 2012
178. Venables, W.N., Ripley, B.D.: Modern Applied Statistics with S, 4th edn. Springer, New York (2002). http://www.stats.ox.ac.uk/pub/MASS4

179. Victor, J.: Analyzing receptive fields, classification images and functional images: Challenges with opportunities for synergy. Nat. Neurosci. **8**, 1651–1656 (2005)
180. Warnes, G.R.: gmodels: Various R programming tools for model fitting (2011). R package version 2.15.1. http://CRAN.R-project.org/package=gmodels. Accessed date on 2nd August 2012
181. Warnes, G.R., et al.: gdata: Various R programming tools for data manipulation (2010). R package version 2.8.1. http://CRAN.R-project.org/package=gdata. Accessed date on 2nd August 2012
182. Watson, A.: The spatial standard observer: A human vision model for display inspection. In: 353 SID Symposium Digest of Technical Papers, **37**, pp. 1312–1315 (2006)
183. Watson, A.B., Ahumada, A.J.: A standard model for foveal detection of spatial contrast. J. Vis. **5**, 717–740 (2005)
184. Watson, A.B., Pelli, D.G.: QUEST: A Bayesian adaptive psychometric method. Percept. Psychophys. **33**, 113–120 (1983)
185. Watson, G.A.: Approximation Theory and Numerical Methods. Wiley, New York (1980)
186. Westheimer, G.: The spatial grain of the perifoveal visual field. Vis. Res. **22**(1), 157–162 (1982)
187. Wichmann, F.A., Graf, A.B.A., Simoncelli, E.P., Bülthoff, H.H., Schölkopf, B.: Machine learning applied to perception: Decision-images for gender classification. In: Saul, L.K., Weiss, Y., Bottou, L. (eds.) Advances in Neural Information Processing Systems 17, pp. 1489–1496. MIT, Cambridge (2005)
188. Wichmann, F.A., Hill, N.J.: The psychometric function: I. fitting, sampling and goodness of fit. Percept. Psychophys. **63**, 1293–1313 (2001)
189. Wickens, T.D.: Elementary Signal Detection Theory. Oxford University Press, New York (2002)
190. Wickham, H.: ggplot2: Elegant Graphics for Data Analysis. Springer, New York (2009). http://had.co.nz/ggplot2/book
191. Wilkinson, G.N., Rogers, C.E.: Symbolic description of factorial models for analysis of variance. Appl. Stat. **22**, 392–399 (1973)
192. Wilkinson, L.: The Grammar of Graphics. Springer, New York (2005)
193. Wilks, S.S.: The large-sample distribution of the likelihood ratio for testing composite hypotheses. Ann. Math. Stat. **9**, 60–62 (1938)
194. Williams, J., Ramaswamy, D., Oulhaj, A.: 10 Hz flicker improves recognition memory in older people. BMC Neurosci. **7**, 21 (2006)
195. Winer, B.J.: Statistical Principles in Experimental Design, 2nd edn. McGraw-Hill, New York (1971)
196. Winship, C., Mare, R.D.: Regression models with ordinal variables. Am. Soc. Rev. **49**, 512–525 (1984)
197. Wood, S.: gamm4: Generalized additive mixed models using mgcv and lme4 (2011). R package version 0.1-3. http://CRAN.R-project.org/package=gamm4. Accessed date on 2nd August 2012
198. Wood, S.N.: Generalized Additive Models: An Introduction with R. Chapman and Hall/CRC, Boca Raton (2006)
199. Xie, Y., Griffin, L.D.: A 'portholes' experiment for probing perception of small patches of natural images. Perception **36**, 315 (2007)
200. Yang, J.N., Szeverenyi, N.M., Ts'o, D.: Neural resources associated with perceptual judgment across sensory modalities. Cerebr. Cortex **18**, 38–45 (2008)
201. Yee, T.W.: The vgam package for categorical data analysis. J. Stat. Software **32**(10), 1–34 (2010). http://www.jstatsoft.org/v32/i10
202. Yeshurun, Y., Carrasco, M., Maloney, L.T.: Bias and sensitivity in two-interval forced choice procedures: Tests of the difference model. Vis. Res. **48**, 1837–1851 (2008)
203. Yovel, Y., Franz, M.O., Stilz, P., Schnitzler, H.U.: Plant classification from bat-like echolocation signals. PLoS Comput. Biol. **4**, e1000032 (2008)

204. Yssaad-Fesselier, R., Knoblauch, K.: Modeling psychometric functions in R. Behav. Res. Meth. Instrum. Comp. **38**, 28–41 (2006)
205. Zhaoping, L., Jingling, L.: Filling-in and suppression of visual perception from context: A Bayesian account of perceptual biases by contextual influences. PLoS Comput. Biol. **4**, e14 (2008)
206. Zucchini, W.: An introduction to model selection. J. Math. Psychol. **44**, 41–61 (2000)
207. Zuur, A., Ieno, E.N., Walker, N., Saveiliev, A.A., Smith, G.M.: Mixed Effects Lodels and Extensions in Ecology with R. Springer, New York (2009)
208. Żychaluk, K., Foster, D.H.: Model-free estimation of the psychometric function. Attention Percept. Psychophys. **71**, 1414–1425 (2009)

Index